AF581642

Security of Cyber-Physical Systems

Editor
Arman Sargolzaei
Mechanical Engineering
Tennessee Technological
University
COOKEVILLE
United States

Editorial Office
MDPI
St. Alban-Anlage 66
4052 Basel, Switzerland

This is a reprint of articles from the Special Issue published online in the open access journal *Electronics* (ISSN 2079-9292) (available at: www.mdpi.com/journal/electronics/special_issues/security_cps_1).

For citation purposes, cite each article independently as indicated on the article page online and as indicated below:

LastName, A.A.; LastName, B.B.; LastName, C.C. Article Title. *Journal Name* **Year**, *Volume Number*, Page Range.

ISBN 978-3-0365-4146-4 (Hbk)
ISBN 978-3-0365-4145-7 (PDF)

Contents

About the Editor

Arman Sargolzaei

Dr. Arman Sargolzaei's expertise is in applying linear and nonlinear control methods, machine learning, and artificial intelligence to the field of cyber-physical systems. His mission is to enhance the quality of life for people by assuring safety, security, and privacy concerns through extensive collaboration among multi-disciplinary fields. Dr. Sargolzaei is the recipient of the NSF CAREER award for his research on testing and verifying the security of connected and autonomous vehicles. He was recognized with the honor of the "Faculty Research Excellence Award"for two consecutive years. He received his doctorate degrees in Mechanical Engineering and Electrical Engineering from the University of Florida, and Florida International University. He is currently an assistant professor of Mechanical Engineering at Tennessee Technological University. Before joining Tennessee Tech, he was director of the Advanced Mobility Institute (AMI) and an assistant professor of Electrical Engineering at Florida Polytechnic University. Dr. Sargolzaei has published more than 70 articles in high-impact factor journals, including IEEE Transactions on Automatic Control, Industrial Informatics, and Industrial Electronics. Dr. Sargolzaei has two active and one pending patent for his research on cyber-physical systems. He is largely involved in national and international fund-raising events to improve education and training on the security of control systems. He is a proud splash instructor who volunteers and actively participates in outreach events.

 electronics MDPI

Review

Blockchain Applications to Improve Operation and Security of Transportation Systems: A Survey †

Navid Khoshavi [1,2,*], Gabrielle Tristani [2] and Arman Sargolzaei [3]

1 Department of Computer Science, Florida Polytechnic University, Lakeland, FL 33805, USA
2 Department of Electrical and Computer Engineering, Florida Polytechnic University, Lakeland, FL 33805, USA; gtristani8152@floridapoly.edu
3 Mechanical Engineering Department, Tennessee Technological University, Cookeville, TN 38505, USA; asargolzaei@tntech.edu
* Correspondence: nkhoshavinajafabadi@floridapoly.edu

† Partial support of this research was provided by the National Science Foundation under Grant No. CNS-1919855. Any opinions, findings, and conclusions, or recommendations expressed in this material are those of the author(s) and do not necessarily reflect the views of the sponsoring agency.

Abstract: Blockchain technology continues to grow and extend into more areas with great success, which highlights the importance of studying the fields that have been, and have yet to be, fundamentally changed by its entrance. In particular, blockchain technology has been shown to be increasingly relevant in the field of transportation systems. More studies continue to be conducted relating to both fields of study and their integration. It is anticipated that their existing relationships will be greatly improved in the near future, as more research is conducted and applications are better understood. Because blockchain technology is still relatively new as compared to older, more well-used methods, many of its future capabilities are still very much unknown. However, before they can be discovered, we need to fully understand past and current developments, as well as expert observations, in applying blockchain technology to the autonomous vehicle field. From an understanding and discussion of the current and potential future capabilities of blockchain technology, as provided through this survey, advancements can be made to create solutions to problems that are inherent in autonomous vehicle systems today. The focus of this paper is mainly on the potential applications of blockchain in the future of transportation systems to be integrated with connected and autonomous vehicles (CAVs) to provide a broad overview on the current related literature and research studies in this field.

Keywords: blockchain; security; privacy; financial transactions; transportation systems; autonomous vehicles

Citation: Khoshavi, N.; Tristani, G.; Sargolzaei, A. Blockchain Applications to Improve Operation and Security of Transportation Systems: A Survey. *Electronics* **2021**, *10*, 629. https://doi.org/10.3390/electronics10050629

Academic Editor: Giuseppe Prencipe

Received: 3 January 2021
Accepted: 26 February 2021
Published: 9 March 2021

1. Introduction

Despite being widely anticipated and celebrated by many today, the field of connected and autonomous vehicles (CAVs) has also faced scathing criticism, disadvantage analysis, and distrust from wary organizations and people [1–7]. Security, safety, and privacy concerns have all been brought up and, as a result, there is a great deal of uncertainty surrounding just how beneficial CAVs can truly be to our society and overall health. A major force in this skepticism is that CAVs systems rely on online networks and, as connected devices, they may be susceptible to numerous software and hardware faults that can be exposed and exploited by attackers [8–11]. Attackers are known for using any means possible to compromise user data, overwhelm networks, and potentially cause dangerous situations for users. In the case of AVs, any form of lax security and safety measures can prove to be fatal due to the inherent hazards that are involved in driving and managing past and present location-sensitive information of users [12,13].

A centralized system is viewed by many to be incredibly ineffective, as well as dangerous, due to the constantly changing nature and wide-scale data management required in CAV operations [14]. Retrieving data in a timely manner is an absolute necessity in ensuring prompt response time, which may be difficult in the case of a single central system managing all user data. Another concern with this setup is the potential for attackers to take advantage of the single failure point for all CAVs by overwhelming the centralized system with requests or manipulating certain user data, at which point they would be able to potentially devastate the entire CAV network [15].

Over time, inherent flaws in totally centralized networks led to the search for alternative models, such as those that are depicted in Figure 1. The proposed model relies on a variety of interconnections between users and other entities instead of using a single, central node. Among the many possibilities, the most promising one was blockchain, proposed by a person or group of people using the name Satoshi Nakamoto in 2008 [16]. Blockchain has the potential to realize a large system, with its size being supported by a variety of peers, equipped with measures to validate, begin, and end transactions through its consensus and validation protocols. No longer restricted to the performance and security challenges of centralized network, blockchain provides all the same necessary features that are present in centralized networks, but, under a high throughput, scalable peer-to-peer based architecture. Despite its creation being fueled primarily by a desire to allow bitcoin technology to function, it has gained a lot of traction with researchers and the public in a wide variety of fields due to its benefits in providing secure, reliable transactions without the use of a single central entity in managing them.

As of now, the blockchain solution has already been incorporated across varying areas of study with great success, which has led many to wonder whether it can be applied similarly to other emerging fields [8]. Among these fields, the applications of blockchain in CAVs have been extensively studied [12–14]. Through these studies, a wide spectrum of methods have been proposed on how blockchain can be incorporated to add and expand on existing CAV functionality, coupled with a number of tests indicating the feasibility of implementations [17,18].

While not designed with a CAV system use in mind, blockchain has proven itself to have a number of benefits that could be successfully carried over to CAVs [19]. Its application could lend itself to use in enhancing the security of CAVs as well as improving the privacy of the users [20], increasing passengers' safety, and maintaining records on vehicle actions in the case of accidents to provide more accurate information for insurance and compensation purposes, as has been noted by a variety of researchers [21], allowing financial transactions to occur directly between a vehicle and a user or other device without the involvement of unnecessary parties [22,23], and providing CAVs with the ability to interact with other devices to offer additional relevant services to users. With its high level of use in various industries and noted security and operation-based benefits, its integration into CAVs could completely eliminate former concerns and provide a framework for entirely new capabilities.

Current blockchain faults in terms of energy and resource consumption, as well as financial cost and increased delay, when several users that are connected to a network have caused some researchers to view it as too inherently flawed to scale to such a large and time-sensitive system, but such concerns may soon prove to be unfounded. Blockchain is, when compared to a variety of other technologies today, relatively new, which means that there is still plenty of ability for improvement. In addition, there have been several proposals on how to resolve such issues, including the use of more lightweight approaches, coupled with the revision of the current Proof of Work algorithm that is responsible for many of the issues discussed [24,25]. With more time to study and improve on blockchain and the algorithms that it tends to use, it will likely prove to be fully capable to support the decentralized network discussed, with its production cost lowering in turn due to an increase in blockchain understanding between companies and develop.

The stride to achieve a safe and secure ecosystem where all-inclusive CAVs operate without a human input has been aligned with our motivation to demonstrate the potential benefits of leveraging blockchain to address the previous transportation-related threats.

Because the current literature is lacking a comprehensive study on the application of blockchain in operation and security of Transportation Systems, this paper surveys the most recent studies in this domain. In this survey, we briefly describe the structure of blockchain as well as highlighting the advantages of utilizing blockchain in CAVs. Furthermore, we thoroughly demonstrate how blockchain can be, and has already been, applied to CAV technology. More specifically, this study aims to discuss the following concerns relating to CAVs:

- The importance of future CAV applications in our day-to-day convenience, safety, security, and growth is illustrated.
- The current obstacles that hinder the public acceptance and development of CAVs, like safety, security, and speed concerns, are discussed.
- The common applications and attributes of blockchain technology and how they have been used for numerous projects in the past are included.
- The benefits for the implementation of blockchain technology in the transportation system's development are presented.

The remainder of this article is organized, as follows: Section 2 of the paper discusses the basic background information surrounding the history of blockchain and how it functions, Section 3 elaborates on how common blockchain applications and features could benefit CAVs if implemented effectively. Section 4 describes the application of Blockchain in collective decision. Section 5 presents the future research direction and challenges. Section 6 serves as the conclusion.

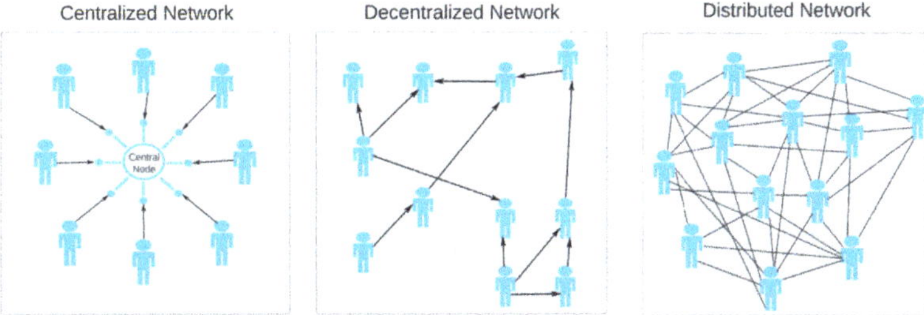

Figure 1. The different types of networks. Centralized networks have a central node that provides connection capabilities for the entire network. Decentralized networks have multiple connection paths between nodes, but some nodes can still lose connection with the network. Distributed networks have several communication paths between nodes, which drastically reduces the probability of a node disconnecting. Centralized is the most common, but decentralized and distributed are increasing in popularity.

2. Background

Before discussing specific blockchain applications, we need to briefly explain the terminologies, concepts, and algorithms that construct the blockchain methodology.

2.1. Ledger

The blockchain is best described as a distributed ledger, maintaining information regarding the transactions carried out and providing its services to all blockchain users. Under this system, every party maintains its own ledger copy, which allows them to check any past or present transaction record sent to them for security and validity. The network bundles transactions into blocks to facilitate the distribution of data across the nodes. Those blocks are then checked for authenticity and appended to the chain.

2.2. Block

A block is comprised of a number of different data-bearing segments, with each being necessary for upholding transactions. In the most general view, a block is composed of its header and its body. Its header contains important information regarding the specifics of the version, hash, and other relevant protocol details for the validation, while its body includes the actual transaction taking place, as well as its counters. One block can contain multiple transactions, with the total number of transactions being dependent on its size.

2.3. Proof-of-Work (PoW) and Hashes

Each block connects to the previous block by using hashes, as shown in Figure 2. Each hash is a set value, which is computed by assessing block contents and used to detect errors. As an extra mechanism to detect the alteration of previous blocks in the chain, the hash value of the prior block is included. Any entity that wants to send a block over the network needs first to compute an algorithm we will call a puzzle, Power-of-Work (PoW), and then send the solution to the network for approval. This requirement accomplishes two key goals: stopping attackers from generating and sending incorrect transaction data to the ledger and similarly limiting the number of concurrent transactions that a ledger can receive to prevent it from being overloaded.

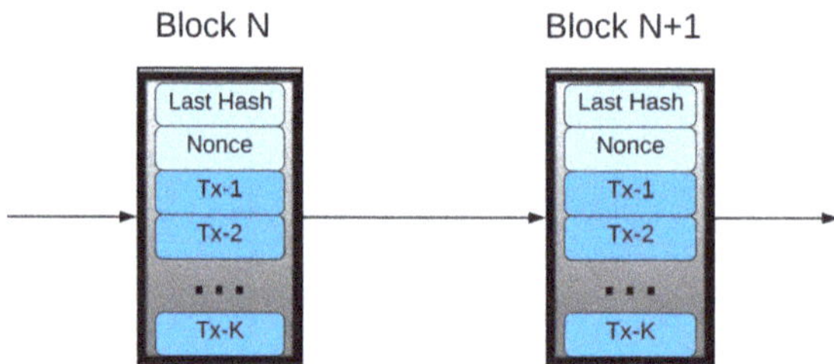

Figure 2. A diagram showing connections between blocks in blockchain technology, providing these connections through hashes.

2.4. Scalability

Scalability is one of the most highly desired features for any system that expects to expand over time or needs to be able to remain stable in the face of targeted attacks on its infrastructure as a whole. Centralized networks, while simple to set up and understand, are severely limited in this regard, since they rely on a single entity for all network operation, which provides a clear target for attackers, as well as allowing for the potential overload of the central node as more users join the network since these nodes do not come with their own resources to manage increased network traffic. This is demonstrated in Figure 3.

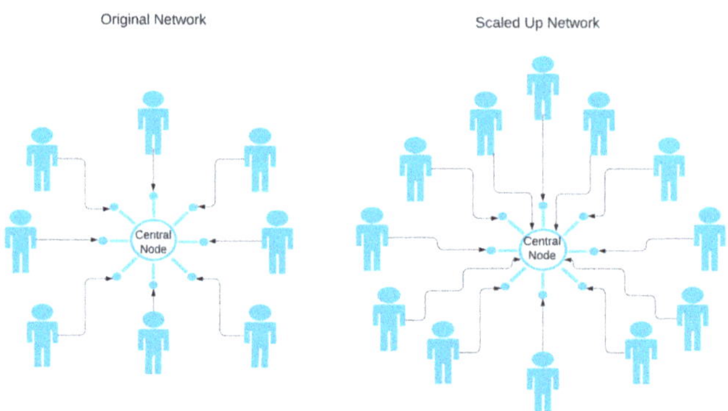

Figure 3. In a centralized network, a small percentage of the nodes handle a bulk of the work. To scale such a network, the primary nodes need to increase their ability to handle more traffic.

However, with blockchains, when a user is added to a blockchain, like in Figure 4, they come with their own set of resources. Besides the resource requirements, the complexity and run time of the consensus algorithms running on the network should also scale (either linearly or sublinearly) with the size of the network. This contribution of resources and the scalability of the consensus algorithm allows for enhanced network operation, since users will primarily take care of themselves and their own role in the chain, letting the blockchain scale easily when new users are added without straining a central server. The peer-to-peer-based architecture means that there is no single point in the network that is responsible for all normal operation of other nodes, meaning that the system lacks a single failure point, and is thus much more resilient. All of the nodes can rely on its interconnections between a variety of other nodes instead of choosing just one, providing a backup for all users over a network. This allows for the blockchain to retain overall connection and operation, even when several nodes are compromised.

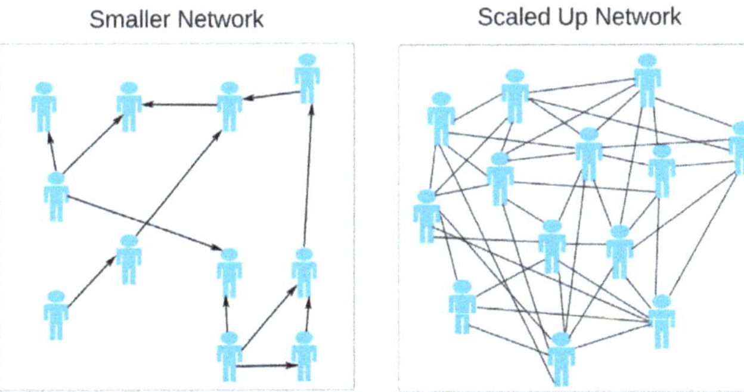

Figure 4. Scaling in a distributed network is simpler because new nodes contribute their own resources.

2.5. Privacy, Anonymity, and Keys

Instead of associating transactions with a fully-fledged identity, as is the case in fully centralized institutions, blockchain uses a pseudo-identifier. Commonly, this pseudo-identifier is just a cryptographic key. All of the transactions that would typically need a name, social security number, ID, and other associated information, now only require the key. This increases both privacy and risk proportionally, since, now, if an entity loses its key, it loses access to all of its information. Even worse, if a malicious party compromises an entity's key, it can then use that to either pretend to be the entity or access all the entities' records, credentials, and resources.

The cryptographic key is based on the famous public-private key architecture. In this architecture, a user generates two keys: the public key and private key. Private keys they keep to themselves, while public keys are broadcasted over the network whenever requested. In this system, private keys are used to encrypt the information, while public keys are utilized to decrypt encrypted messages. This allows the identity of the user to be easily verifiable. If the user encrypts a file and that file later comes into question, its authenticity can be verified, since the system trusts that a user will keep their private key secure.

2.6. Re-Purposing

Currently, the primary and most notable application of the blockchain protocol is Bitcoin. For processing transactions, users, referred to as miners, gain a digital monetary reward that is known as Bitcoin. As the number of miners on the network increase, the PoW increases proportionally. The appeal of a decentralized financial system that is unregulated by any government has led to its widespread popularity among the public.

Although blockchain was designed with application to bitcoin alone in mind, it is a very flexible system that can be applied to countless industries. Currently, blockchain technology is used in healthcare, online transaction security, and several other fields. Because blockchain is new, many researchers are eager to find new ways to apply this quickly growing and changing technology to other systems and areas.

Table 1 briefly demonstrates how blockchain can overcome the challenges of centralized systems.

Table 1. Blockchain solutions for centralized system flaws.

Centralized Problems	Blockchain Solutions
Requires Trusted Authority	Trustless System
Scalability Issues	System Scales with New Users
Information is Modifiable	Immutable Blocks Using Hashes
Identities are not Anonymous	Cryptographic Keys as Pseudo-Identifiers

3. Blockchain and Transportation

One of the main reasons CAV technology has not been fully embraced is an underlying safety concern. However, at the same time, many would agree that the most dangerous and unstable elements of transportation are the human drivers [26]. While not yet perfect, CAVs have the potential to compensate for the shortcomings of humans and fully prevent accidents. When discussing CAVs, there are several levels to consider: *level 0*, which has no automation, *level 1*, which has certain automation when needed for certain very specific and isolated functions, *level 2*, which has automation in the case of several different communicating functions, *level 3*, which has significantly limited, but still functional, self-driving capabilities that may require some user input, and *level 4*, which has the complete ability to operate and drive by itself. In the end, it is expected and desired that *level 4* CAVs will be developed, but, until then, the focus has been on enhancing the capabilities of previous levels, with the exception of level 0. Several companies, including Google, Uber, and Telsa, have recently made great strides in self-driving vehicles, and Tesla has recently announced that its shared autonomy fleet will go live within 2020.

Despite an overall reduction in accidents, there have still been many noted driving failures involving CAVs. Currently, CAVs are known to make incorrect decisions at times for two reasons: (1) because the technology is far from perfect, as it is challenging to meet the mandatory requirements, and (2) because the vehicles do not yet have enough information to process to avoid certain incidents.

Though many of the factors that cause such flaws and potential areas of improvement are well known, methods that outline how solutions can be implemented may not be as clear. Some examples of well studied problems in CAV technology are ensuring the abilities to both maintain and secure certain data, like physical and geographic location of vehicles [27–30], allow sufficient operation space for vehicles and control traffic flow [31–34], allowing and securing communication between vehicles and each other as well as other network-connected devices [35–39], providing collision warning and evasion techniques [40–44], providing security against attacks from malicious entities and faulty software or hardware [45–49], and offering safe and reliable availability of updates when needed [50]. The possibilities that are opened by CAV technology are too vast to be ignored, with broad applications to various fields to improve operation and user convenience. As noted, CAV technology lends itself to use in transport-based financial transactions, like public transportation systems [51].

CAVs are still relatively new, which leads them to suffer from a number of flaws that have lessened public support in the name of security and safety concerns for drivers and pedestrians. However, while CAVs have yet to gain full public trust, blockchain technology is viewed as one of the most secure methods used to maintain transactions and enable users to perform common operations as simply and safely as possible. In order to promote trust in CAVs and more fully demonstrate the full capabilities and possible extension of blockchain technology, the integration of blockchain technology and CAV systems is an idea that could prove advantageous to both fields. Figure 5 illustrates a summary of the leading research activities in each aspect of applications of blockchain. We explain the technical contribution of each work in the rest of the manuscript.

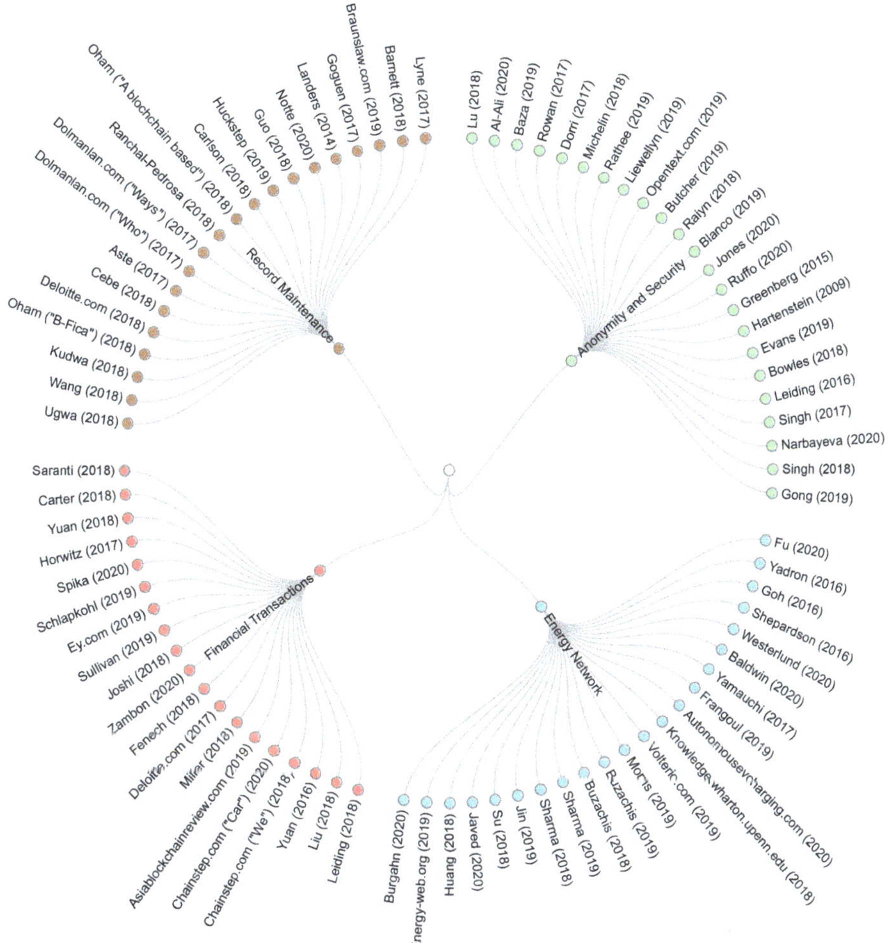

Figure 5. A circular dendrogram demonstrating the leading research activities in each domain.

3.1. Anonymity and Security

Figure 6 shows the attribute graph of the security concerns of CAVs that blockchain technology can address. In the case of CAVs, data and device security and anonymity are some of the largest places of failure, as well as the largest places that must be secured in vehicle operation, as brought up by a variety of researchers and wary consumers [52–56]. Because blockchain was built entirely to provide security in transactions, its role in providing security for CAV users is by far the most well studied and desired. Thus, the majority of this study will discuss this aspect, as well as the several approaches that utilize the blockchain to ensure security and user anonymity in CAV systems.

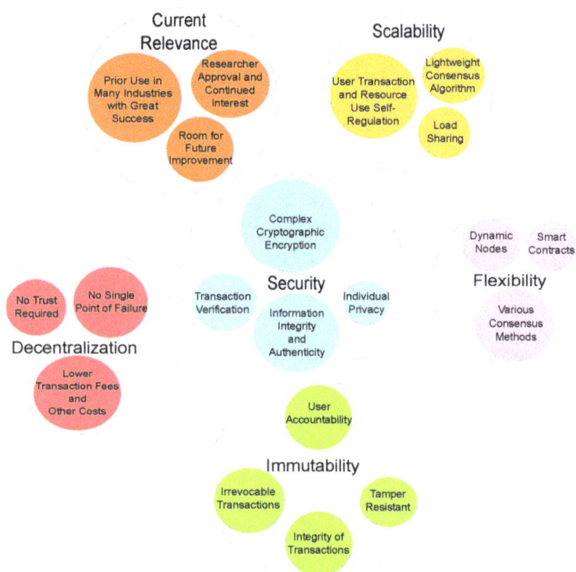

Figure 6. The attribute graph of the security concerns of connected and autonomous vehicles (CAVs) that blockchain technology can address.

Although data security and user anonymity are both highly prized features of any vehicle or device in use today, they tend to be ignored in favor of physical driver safety and more technically accurate operation. Recently, this has been extremely noticeable in CAV development, leading to user outrage from the fact that their personal data and vehicles' safety from attackers is not given necessary protection [57–61]. As an example, VANETs [62] seek to remedy the issue of numerous drivers, and the risk that is associated with them, by creating a network for cars to directly communicate with each other, sharing information on road safety issues, upcoming traffic stops, and general information to improve the efficiency of the whole vehicular system. However, while this sounds promising to many researchers and drivers, some users worry about the inherent risks involved, with concerns being based around the exposure of car, driver, and location information in these communications.

VANETs were not designed with the current developments of CAVs in mind, but could prove incredibly beneficial to an CAV network, possibly even more beneficial than it is to regular vehicle systems with the removal of human unpredictability. The use of consistent and constant communication between adjacent vehicles to help them prepare for future traffic events and situations would provide CAVs with all of the needed information to make decisions in a timely and accurate manner, which is why their incorporation is so essential. However, before they can be safely added, the previously mentioned risks to user privacy and security must be overcome.

Le et al. have outlined a system, called a Blockchain-based Anonymous Reputation System (BARS), which works using guidelines and defined operations to ensure trust throughout the network [10]. With this, Ref. [10] believe they have found a way to mitigate known security and privacy flaws in CAV systems. This BAR system works under its defined development in a number of connected phases, called steps, all of which were created to secure user privacy. The first step is to adjust blockchain features, so the preexisting, commonly used Public Key Infrastructure (PKI) can be extended to provide for a new authentication procedure that guarantees user data privacy. At this point, there is a distinct link from the communicating vehicle to its private key, which presents a huge security risk in allowing attackers to find the private key for a given user, and, in turn, gain access to their transaction information. This risk is accounted for and prevented with the addition of a Certificate Authority (CA), which serves to provide a new abstraction layer to protect the anonymity of all users, letting the whole network continue to function securely [10].

The next development step is in the researcher's design of a new algorithm to assess vehicle reputation and its level of trustworthiness based on its prior actions and broadcasts over the network, as well as nearby vehicles viewpoints [10]. Adversaries may attempt to spoof vehicles or manipulate vehicle data in such a way as to disturb the functionality of the network or the records of the victim vehicles. This can be incredibly dangerous to both the driver and the network relying on that vehicle's information. Table 2 lists the parameters that need to be considered in the design of a reliable consensus protocol in CAVs.

Currently, different trust models can be, and have been, applied to VANET systems, including entity-centric, data-centric, and combined trust models. The model used in [10] is entity-centric, meaning that it is concerned primarily with the vehicles themselves. Some potential assessment measures include a reputation-based system, as depicted in Figure 7, where every vehicle's weight is judged based on its past behavior in the network. The reputation of each vehicle transmitting data determines the validity of the transmission. There are three main message types in the network, including the following: beacon messages, which are sent periodically with simple driving status information, alert messages that are sent for emergencies and come in three levels, and disclosure messages, which are sent by witnessing vehicles and those with conflicting information [10]. By this model, underlying blockchain technology is the backbone of the system in regards to the security and stability of the system underlying vehicle operation in this model. Its inherent security, flexibility, and trust among users make it a natural choice, and one that could serve to solve many more common issues with current CAV technology.

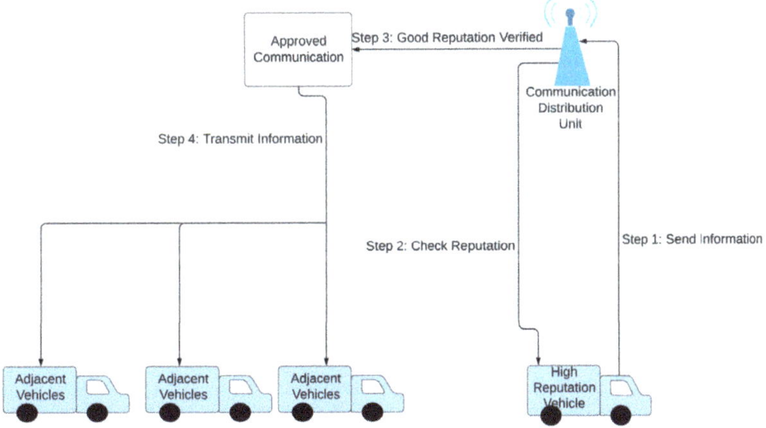

Figure 7. Reputation-based system for judging the validity of vehicles and their provided information on a network. This system is based on the past and present behavior of the vehicle and culminates in a score that is attributed to that vehicle.

Table 2. Considerations for consensus protocol involving CAVs.

Security against Known Exploits	Focused Validation Protocols
Low Communication Complexity	Vehicle Integrity Checks
Minimal Latency	Dynamic Node Tolerance
Resistant to DoS Attacks	Faulty Node Tolerance
Low Energy Cost	High Scalability
Consensus Finality	Fast Error Handling

The possibility of using blockchain in VANETs has also been noted by Leiding et al. [63], with them offering an approach based around peer-to-peer networks instead of the traditional centralized client-server architectural approach. Because blockchain inherently lends itself to decentralized approaches, it was noted as being a very promising potential choice. Decentralization is assured through the implementation of smart contracts, which function using an Ethereum blockchain implementation, with these smart contracts providing applications that make user vehicles perform operations that contribute to overall network decentralization by relying on several RSUs (road-side units) instead of a single entity [63]. Operations that are carried out may entail forcing vehicles to follow traffic rules or regulations, or presenting useful roadway condition information to drivers [63]. This heightened level of consistent control without reliance on a single party in operation ensures both the proper function of the system and connected vehicles and the stability and reliability of the underlying system, which makes this approach a very promising one to consider for future CAV use [63].

Any given VANET is heavily reliant on the reliable interaction between vehicles and, as a result, any malicious adversary interference could prove dangerous to the driver in question and other drivers in the system. To combat this risk, Singh and Kim [64] suggested implementing IV-TP into messages that are sent over the network, with their focus being to add needed security elements and data reliability. This element is represented as a singular, unpredictable number, which is chosen randomly and appended to any message sent in a particular communication. The researchers propose using a cloud storage solution that is based on blockchain to handle IV-TP communications. The authors note that the necessity for such a system derives from the fact that current vehicular ad-hoc networks use less secure forms of communication that can be accessed or manipulated by malicious adversaries, and that the proposed blockchain solution will provide a freely available and accessible ledger, secured via a Merkle tree and SHA-256 Hash with a consensus mechanism (PoW).

The new proposed system involves the use of blockchain technology, vehicular cloud computing (VCC), and a network-connected device (the vehicle). VCC operates as a form of hybrid technology, utilizing the resources that are owned by user-controlled vehicles, like their data maintenance and storage, computing power, and Internet-aided decision-making skills. In this case, the blockchain makes up much of the system and it has been divided into seven layers, similar to the popular OSI model that was used in the Internet.

An article written by Rathee et al. proposed a security method that made extensive use of blockchain technology in order to protect CAVs from exploitation [20]. Following this proposed method, blockchain technology would be used to protect user data security, as well as maintain a history of vehicle movement, decisions, and external conditions. By this implementation, blockchain technology is the main method of data protection, a job that has been noted to be extremely geared towards [20]. Similarly, Narbayeva et al. noted the applicability of blockchain technology to CAV systems [65]. The reasons for incorporating blockchain technology were supported in full with an analysis of past trends in technology to make predictions for currently developing CAV technology, as well as how many new technologies, like the Internet of Things and bitcoin, have made use of blockchain for data protection [65]. In addition to its user and producer trust, its ease of application and flexibility can be seen through its widespread use in a variety of fields to secure data and anonymity in transactions [65]. This flexibility, as discussed, makes it not only a safe and reliable, but a relatively easy and inexpensive to implement, technology [65].

In the past, blockchain technology has been applied with great success in security and it has been shown to be an effective means of facilitating the spread of information between several connected systems. Because of these traits, a study that was conducted by authors of [12] attempted to integrate blockchain technology with existing CAV traffic event validation systems to quickly and effectively secure vehicle information and eliminate misleading information exchanged by malicious vehicles [12]. To test the proposed system, the team tracked the number of attackers and the ability of the blockchain-based system to detect when users were generating malicious information, with the blockchain technology using a reputation-based system to track the trustworthiness of certain users based on their past actions [12]. The results gathered indicated that this system was very effective in distinguishing normal from attacker-generated data [12]. Its success heightens an area of particular importance for the use of blockchain technology in CAVs: attack mitigation and user safety. Based on these results, it can be said that blockchain technology has the potential to build greatly upon many previously concerning aspects of CAV operation and reliability.

Objects that are connected to a large-scale network are frequently subject to attacks, old and new, by malicious users, and CAVs have never been immune. Software vulnerabilities that are present in a given product are generally remedied through updates to fix known flaws, so the existence of a readily available system to provide such updates quickly and completely to all users is necessary. However, these updates must also be provided securely, with no chance of an attacker taking advantage of this system to infect CAVs. To meet security needs while ensuring that no vehicle is missed, Baza et al. [14] highlighted traits of blockchain technology that lend it to be used in such an application.

To explain the use of blockchain in finding a solution, the group outlined a firmware update scheme that uses blockchain technology to provide security in their releases and ensure that certain vehicles are not overlooked due to geographical location [14]. This system would use several distributors, vehicles that are rated highly based on their reputation in regards to its trustworthiness and driving history, to deliver new updates [14]. The reputation of vehicles would be recorded via the implementation of blockchain technology, ensuring availability and security to avoid the targeting of high reputation vehicles in attacks. The encryption scheme used would require that CAVs be authorized to install updates, and all of the updates would be secured via the use of smart contracts. Smart contracts can only operate via the use of blockchain technology and, as such, are known to be incredibly secure and well-used by people in electronically conducted transactions today. Blockchain technology serves as the basis for the scheme as a whole, again heightening its relevance and role in the future development of CAV operation, security, and safety.

Another method of providing secure software updates via the use of blockchain technology continues to use a cloud-based structure [13]. Wireless Remote Software Updates are, instead of tasked to several deliverers to distribute, available via cloud storage of a car manufacturer or software provider. In order to ensure the security of the update, the software provider begins a transaction, using its private key and a signature constructed via the signed hash of the software binary maintained inside of the cloud structure. Using this signature, the transaction is verified by overlay nodes within, and the manufacturer then signs the transaction. Following this, the overlay block managers supervising the public blockchain broadcast the transaction by checking the signature and ensuring that the manufacturer used the set private keys. Finally, the overlay block managers send out the transaction to all members of their clusters, and all of the connected devices can verify and download the update from the cloud storage. Figure 8 illustrates this process, providing security through the use of extensive checks instead of using outside entities to deliver essential updates [13].

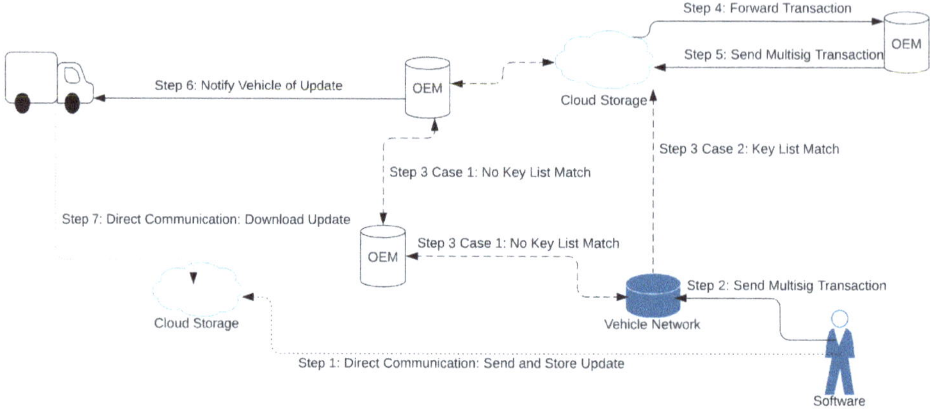

Figure 8. A diagram showing how software updates are created, verified, and delivered to CAVs across the network through a cloud-based system, inspired by [13].

On top of securing user personal data, it is necessary to provide the security for the messages that is sent between vehicles. Vehicle-to-vehicle communications are necessary to keep both of the entities updated in traffic conditions and, in turn, mitigate accidents on the road. However, if these communications are compromised, so too is the network in the event of a malicious user carrying out attacks over these communication channels. Because of the huge risk that is involved in allowing vehicle-to-vehicle communication at this time and the significant benefit that this feature would provide to CAV capabilities, finding a solution that provides security while allowing these necessary interactions is a priority of many developers.

In response to this known issue, Rowan et al. [15] looked into a potential solution that incorporated many different technologies, including blockchain, to secure communication channels in CAV networks. The proposed solution outlined an advanced security system that made use of visible light and acoustic audio-based side-channel encoding to permit secure communications between two distinct vehicles, as well as using a PKI heavily based on blockchain technology to allow communication between unverified, potentially untrustworthy vehicles. In order to defend the use of blockchain, the researcher team expanded on its past capabilities in remaining secure and stable when faced with a heavy attack, as well as its proposed ability to verify information that is sent by untrusted vehicles through the use of available distributed hash tables maintained by each machine involved. Blockchain technology was incorporated for communication security and overall system stability and reliability. The implementation of the side-channels would focus primarily on the physical security of the transactions that were carried out by checking for and maintaining the identity and location information of a vehicle being communicated with. These numerous protections would allow immunity to RF channel jamming as well as physical attacks on the side channels through the double-check system, implementing blockchain on top of side channels to provide a backup if one of the systems is compromised through the use of the other one [15].

Singh and Kim [66] also brought up how the use of blockchain could serve as the solution for communication-based security vulnerabilities: citing its frequent use in similar security-based applications, as well as its high level of user trust that stems from its past reliability and flexibility between different fields. The proposed system is one that makes use of two types of blockchains: the main blockchain and a local dynamic blockchain, which each work together to store communications between vehicles. The local blockchain works in maintaining communications that are sent to it, sending any that it deems to be strange or out of the ordinary to the main blockchain, which will hold onto the strange data for longer than the local one is able to. This way, the local blockchain can continue quickly storing and maintaining understood data, while the rest is held for a longer analysis period by the main blockchain. By this method, the high power use concerns of blockchain technology are mitigated, allowing for proper network function and the security benefits provided by the blockchain. Overall, the proposed communication network would stay secure and reliable without consuming excessive data or compromising network operation.

Michelin et al. [17] suggested the use of blockchain architecture based on smart city operation, whose infrastructure connects through electronic means, allowing the city and entities operating within it to act in a more efficient, beneficial way. Smart cities need to account for all aspects of traditional cities and, thus, have to heavily consider how roadways and traffic will be controlled and maintained. Today, this is generally done through the use of advanced sensors that routinely monitor traffic conditions and let vehicles make fast, consistent, and correct decisions, helping all the parties operating in, around, or with vehicles [17].

However, the amount of data generated and analyzed by vehicles and the systems they operate over may have some unintended drawbacks. In fact, Michelin et al. noted that in the future, it would likely be possible for vehicles to create around 4000 GB of data per day [17]. Clearly, this much data requires an extensive, reliable system that is well-structured and built to expand easily when faced with greater content production, while still providing for expected security and privacy features that users need. While many systems have been considered to provide for this, there still fails to be one that provides for all necessary qualities of such a system. Eventually, blockchain technology was viewed more extensively in regards to this problem, and it was decided that it would be the best technology to use for system implementation.

Currently, SpeedyChain is a system under heavy consideration, providing measures for intelligent vehicles, elements that are present in smart cities in regards to traffic control (such as traffic lights), Service Providers (SPs), and Roadside Infrastructure Units (RSIs). Common blockchain activities, including transaction and block verification, are all controlled by system users that have significantly higher available computing power and resources than other entities, as well as those that work directly with the smart city to ensure proper operation [17]. The process of introducing new entities, like vehicles and similar network-oriented devices, as depicted in Figures 9 and 10, is well-outlined and customized, especially for blockchain-aided implementation to prevent any coordination problems. When vehicles are first added, they need to undergo a validation process conducted by an RSI, with the validation methods being performed by both the new vehicle and the city-based node [17]. In exchange, the vehicle gets its own block, which has its creation dictated by the blockchain itself. Following the block's creation, the vehicle can use it to request and respond to transactions, with all of the transactions being visible to the system.

Under the system defined, vehicles are able to create data for control or service-based operations. Vehicle control data are uploaded for the use of concerned parties, like traffic management or other vehicles, who wish to minimize the level of traffic in a certain area of the city. However, service data are used instead by certain verified parties who want to sell certain vehicle-related services to users, as well as letting users view statistics on their vehicles to ensure their proper operation maintenance [17]. However, as it stands, there are still numerous problems with vehicle-oriented blockchain technology, with a major problem being the concept of dynamism. To explain, when a vehicle is active, it is moving nearly constantly, so they hardly ever stay in the same exact place over a certain time period. Because location-based data are needed for the system to work, the researcher team came up with a possible solution: the system RSIs and SPs would both take on the role of providing for blockchain implementation, a role that suits them well due to their inherent rigidity and known reliability [17]. Whenever a node is first added to the vehicle system, the RSI, as well as all surrounding vehicles, need to guarantee that the node is valid. This process is called location-based trust establishment [17]. When performed, the process serves to validate the node's operation, which prevents the possible infiltration of malicious vehicles.

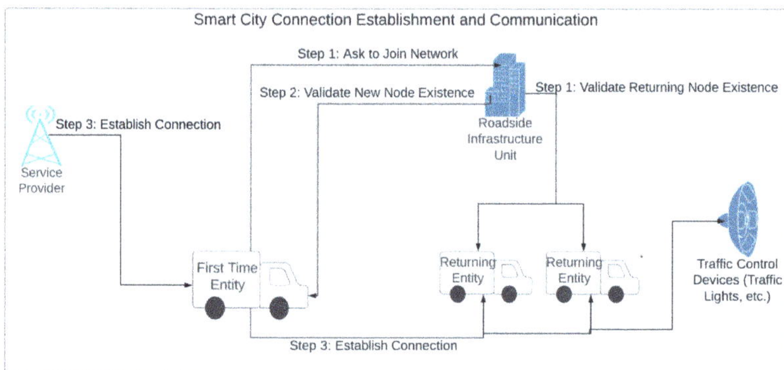

Figure 9. An example of a smart city infrastructure where all nodes, or entities, are capable of processing information and contributing resources to the overall network.

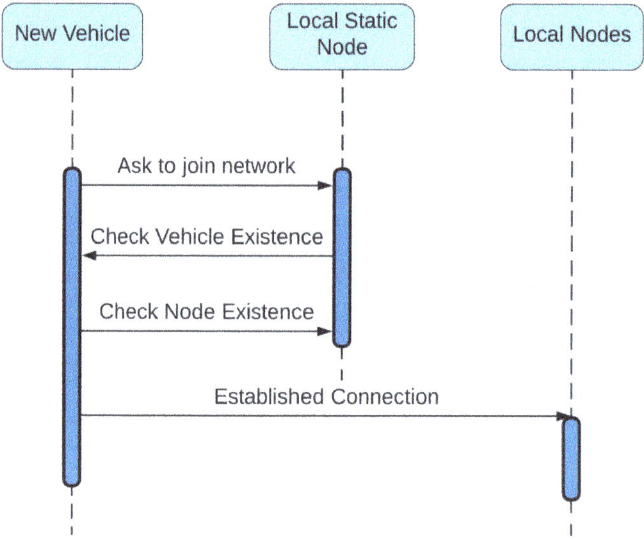

Figure 10. An example UML of a new vehicle entering a Smart City such as the one shown in Figure 9.

Smart cities have been noted to have great potential in improving their security through the use of blockchain technology in their device management system by Gong et al. [67]. Under this method, all devices, including CAVs, within the smart city network are managed under a blockchain-based system that facilitates safe and reliable updates, device control, and access only by approved parties that oversee smart city function and, overall, a secure and trusted network. Additionally, the blockchain implementation ensures the use of a peer-to-peer based system as opposed to one that is based on a single device, both aiding the reliability and protecting against data overflow and the security concerns that it could bring. Different protocols for use allow this system to communicate between any of the device types that are present within the smart city, which aids in communication, as well as device security and maintenance. In securing the network that oversees CAV operation via blockchain, the security and operation of CAVs are maintained by extension, showing that blockchain has a number of potential uses that can directly or indirectly help them function.

Through the previous discussion, it is evident that blockchain has extensive application to CAV security, as noted by several researcher teams and companies. However, all of these implementations differ tremendously, each with their own specific benefits and drawbacks. Table 3 shows a more complete discussion of each method of implementation discussed.

3.2. Financial Transactions and Enhanced Services

With many well-known uses of blockchain centering around securing financial transaction-related data [68–72], it is not surprising that researchers are already looking ahead at how this can benefit CAVs. Beyond proposed use in strictly driving and user safety security measures, blockchain technology has been identified as having immense potential in conducting financial transactions between users driving a vehicle and another entity offering a service to the user [23,73–76]. While it seems like a somewhat outlandish use upon first glance, after considering the number of transportation-related payments that need to be made by people on a day-to-day basis, it is clear that the extension of blockchain technology to this area could provide users with a number of new benefits that could save them time, as well as provide further accessibility for such features to users who may not have access to the type of payment that is required while using more traditional methods on hand [23]. Parking payments, insurance, tolls, and car rentals are just a few examples of common transactions that could be simplified and secured with less time, effort, and user confusion via the implementation of common blockchain technologies [23].

Because CAVs are an emerging technology, they will likely cost too much for average users. Similar to previous emerging technologies, commercial entities will probably be among the first to use CAVs, with prices dropping enough for general consumer use much later. Interested parties that first use this technology will likely offer services, like ride-sharing, to customers that are interested in seeing which industries the technology will head to first; Saranti et al. [18] studied the previous uses of blockchain in CAV systems, taking note of affected areas and those that might be next.

As mentioned, there is a lot of promise for CAVs in ride-sharing, in which companies save time and money by offering unmanned transport to customers. Saranti et al. then moved on to investigate another blockchain-oriented application, in which the technology was used to handle transactions between vehicles, facilitating communication through the system.

The system network is composed of several CAVs, all of which can send messages to a user directly through their phone. A user can access certain information through a mobile application, more specifically, the locations of nearby cars, as well as relevant information on the vehicle currently being driven. This consistent access to accurate information promotes user safety, with the added benefit of building trust between the user and their vehicle's security. In a case where a user wants to use the application to carry out financial transactions, users are also required to enter their own data and, if desired, make a coin ledger.

This coin ledger handles payments between the passenger and CAVs. The vehicle will be able to withdraw the required amount throughout the trip, cost per kilometer, or all at once at the end. When completed, the blockchain-aided transaction is transmitted to the network.

Viewing previous blockchain applications, like its earliest use in Bitcoin, its extension into the financial field surrounding vehicle-based transactions would be relatively easy to apply. As an added benefit, blockchain is most well-known for its widespread influence in securing payments between network-connected entities, and it already has a solid reputation among millions of satisfied users. As such, its implementation could also promote user acceptance of CAVs, allowing for further development in the field and a greater number of studies and system outlines focusing on its future applications.

Table 3. A comparison of selected security methods in terms of advantages, drawbacks, and type of blockchain used.

Model Proposed	Blockchain Type	Advantages	Drawbacks
BARS [10]	Certificate-based blockchain (CerBC) and revoked public key blockchain (RevBC)	Security and trust, authentication, low time/space overhead, anonymity and privacy.	Few results and performance analysis from few experiments and no large implementation.
Self-managed VANET [63]	Ethereum	Decentralized network, fast and reliable operation, no single failure point, secure communications, allowed user applica-tion use.	Customers charged fees for commu-nication and app use in Ethereum-gas, they pay for network.
Intelligent Vehicle Framework Model using Blockchain [64]	-	Security and privacy, IV-TP for speed. Records of communication, tamper-resistant, well-defined layering system for model.	No tests carried out, beyond system analysis, no results or space or time analysis.
Blockchain-based Connected CAV Framework [20]	-	Appropriate data hiding, transparency, data verification. Improvement in attack mitigation from previous methods due to blockchain.	Results from simulations only. No time/space analysis. Approach works well only after certain time interval. Many attackers can still compromise network.
Elliptic Curve Digital Signature Algorithm Based Approach [65]	Exonum	ECDSA securely inputs and validates information, blockchain secures vehicle state data.	No testable system, vague imple-mentation. ECDSA relies on users, but users can be unreliable.
Proof-of-Event VANET [12]	-	Higher success rate in attack detection and decision speed. Secured and pri-vate communications, but verifiable from transparency.	Performance is compromised when few vehicles are in an area. Physical tests not done.
Blockchain-Based Firmware Update Scheme [14]	Ethereum	Update scheme is peer-to-peer and hard to attack, many sources for fast updates, users are rewarded to maintain network. Effective and fast security, validity. Over- load prevented through peer-to-peer ar-chitecture.	No working model, so no tests done to see performance in real world. Users may not have update soon: some far from distributors.
Blockchain-Based Cloud Update Scheme [13]	Lightweight Scalable Blockchain (LSB)	Access to updates through cloud storage, updates verified for security and integrity, providers safely send updates to users, privacy, hash function prevents malicious node access. DDoS attacks are impossible.	Large overhead through various operations to check update security before download, cloud provides all software, which may compromise. No tests or implementation.

Table 3. *Cont.*

Model Proposed	Blockchain Type	Advantages	Drawbacks
Side-Channel Blockchain- Based Security [15]	Bitcoin	Side-channels with blockchain have many protections, if one is compromised, other is used. Physical security allows direct communications with nearby vehicles, can be applied to ensure features between vehicles, like proper spacing. Side-channel protects against wireless transmission interception.	Side-channels lost from outside conditions, causes insecurity. Low performance and speed. No extensive system testing. Vague implementation details.
Blockchain-Based Vehicle to Vehicle and Vehicle to Infrastructure Communications [66]	Branch-Based (Local dynamic and main blockchain combination)	Security in communications, user trust promotion by blockchain use. Branching is more lightweight than most blockchain, vehicles operate in real-time.	Test results are simulated. Vague implementation structure.
SpeedyChain [17]	SpeedyChain	Security with blockchain, integrity through hash functions and condition records. User privacy through timed key changes. System is immune to Sybil attacks and data tampering with SpeedyChain. Performance higher than that in Bitcoin blockchains, and others reliant on PoW.	Limited testing, no space and time analysis. Speed increases a lot with transactions when many vehicles are in a system.
Blockchain-Based Device Management Framework for Smart Cities [67]	Private Blockchain, similar to Ethereum	Data integrity and scalability through blockchain. Management system is applicable to any device type so each is secured. Energy use lowered with proof-of-stake to manage smart contracts. Outline detailed and accounts for many attacks.	No testing or estimated performance in real-world.

As Miller explained in an article that focused on the future adoption of blockchain into several different fields, blockchain technology, with its ability to allow and manage transactions securely and their details as permanent records, directly encourages the possibility of allowing CAVs to carry out transport-related financial transactions [77]. Miller went on to add that different transaction types, such as refueling or vehicle repairs, would be treated differently, with transaction rules differing depending on the exact type of service that is being provided to the user. Figure 11 shows a basic diagram of the model, which indicates the specialization of each individual process to make all transactions as simple as possible [77].

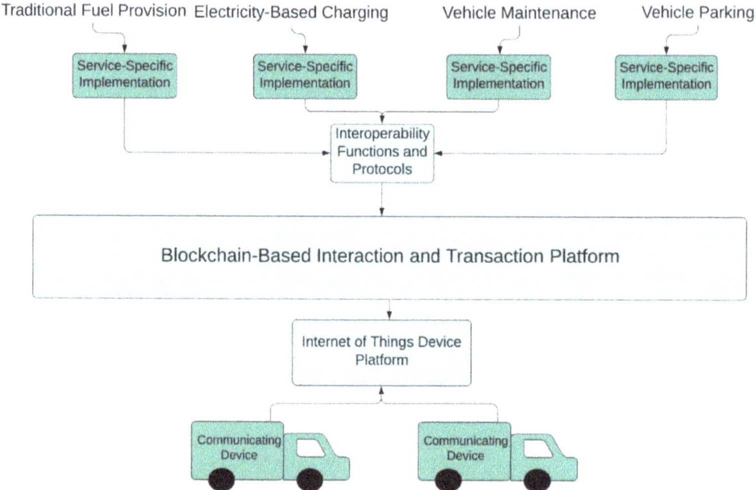

Figure 11. A model showing how transactions can be carried out and recorded through the use of blockchain in CAVs, inspired by [77].

A similar approach is brought up by Yuan and Wang, which also makes use of blockchain in meeting its goal [78]. In their paper, the potential of CAVs acting with the ability to get certain goods or services, like wi-fi or timed parking spots, through the use of stored cryptocurrency powered by blockchain is discussed in detail [78]. Via this method, it may be possible for people to make payments through the use of intelligent agents that act on their behalf, with these agents having all the specific information, like rules and algorithms, to actually carry out requested user transactions. This would greatly ease the process of making transport-related purchases, ensuring that users are never at a loss in how to carry them out, and CAVs are always equipped with the ability to facilitate financial transactions when desired. Because of this, the integration of blockchain-based cryptocurrency is certainly a promising concept and one that could serve to expand on the capabilities of CAVs greatly.

Companies have not ignored the potential for financial applications of CAVs using an underlying blockchain framework. In fact, IBM has begun contributing its own current blockchain-based architecture to promote the development of a system to access such applications [79]. Through this addition, the main goal is to provide security to the system, as well as the ability to create, end, and manage financial transactions with as much user ease of access as possible, as well as allow users to see statistics on their own vehicles in order to guarantee that everything is functioning properly [79]. The Car eWallet system, initially developed by ZF Friedrichshafen AG, is one that cites a huge number of benefits as the reason for the addition of blockchain, including its security, low processing power needs, validation of transactions, the constant and secured maintenance of transaction records, and the consistent availability of such information to only verified parties [80].

Easy user access to beneficial driving-related services, as well as ease of deploying such services to users, are both major reasons for more companies and users to use this network, and, with its broad applications to CAV technology, it could very well be a primary driving force towards user acceptance of CAVs as a whole [80]. Figure 12 presents a visual model of the transaction process carried out by this Car eWallet system.

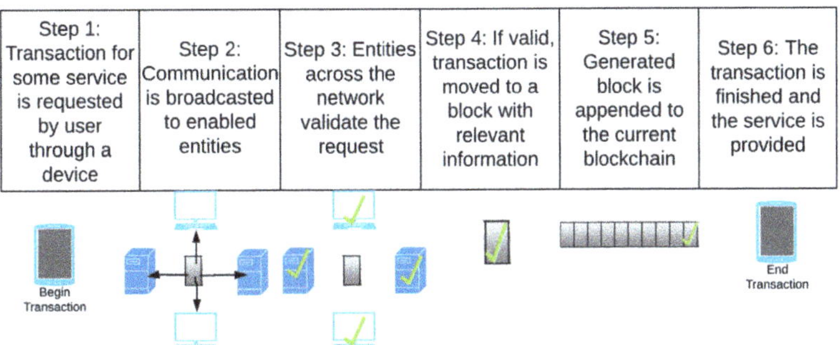

Figure 12. A diagram of the transaction-establishment and completion process, inspired by [81].

Transportation, while used on a day-to-day basis by most people, has a number of implementation-based flaws that limit its range of accessibility. In areas where public transportation is not an option, anyone who is unable to afford or drive their own vehicle is forced to rely on person-based transport, such as Uber or Lyft, which may not be desired due to concerns of safety in trusting complete strangers to drive them to their destination, or impossible in cases where no drivers live nearby. With the high level of activity that is required in the lives of many people today to go to work, run errands, and access certain services, consistently available and accessible transportation is an inviting concept. While working towards this goal, CAVs may face many challenges in moving forward, depending on their exact implementation. As companies jump on new CAV technology to provide their own businesses in on-demand autonomous transport, there is an inherent risk in this promoting a single transport platform, as has been seen in a number of other industries that are controlled by a few well-known giants that limit competition and promote centralization of the network and its data [21]. In such a case, if one of the major few platforms is compromised, an unacceptably large number of users will be affected, so a major goal of CAV system applications must be to promote many different transport options to discourage single points of failure.

Though many may argue that such a centralized setup is unavoidable in business, measures have been discussed further to promote decentralization, including the adaptation of blockchain technology. A concept that was presented by Catapult Transport Systems alongside the University of Sheffield elaborating on the benefits of decentralizing the transportation industry focused on how blockchain lends itself to use as a factor promoting this push on the industry [21]. Unlike many other systems that promote client-server network architecture, blockchain uses an architecture much more similar to peer-to-peer systems in enabling communication between entities. Making use of this system in transport-as-a-service applications would remove controlling entities, allowing for a collection of transport operators to moderate its use instead of putting all public trust in single entities [21]. This necessary application of CAV technology could be made more robust and reliable through these methods, allowing for its enhanced growth and development without compromising the desired decentralization of the CAV system and its use. In particular, blockchain has already shown itself to be well-trusted and used to promote decentralized networks in a number of other industries with great success across fields, so its use in more specific CAV applications would likely be met with approval from the public as well as researchers.

Similarly, Yuan et al. [82] emphasize the importance of having a decentralized Intelligent Transportation System, as well as the role of blockchain technology in ensuring it. The team also presents an outline for how such an architectural design would work, proposing a seven-layer model covering all aspects of the blockchain implementation that would be used. Unlike basic blockchain, this design is especially oriented towards an Intelligent Transportation System approach, which it accomplishes by providing a number of distinct layers. The physical layer is the first of these layers, which includes and secures IoT-enabled entities. Following it is the data layer, providing data blockchains and the ability to operate with or on them. Next is the network layer, which outlines the procedures in forwarding and verifying data and participating on the network. After the network layer is the consensus layer, which keeps track of and provides all necessary consensus algorithms and decides on the most appropriate one for any interaction. Layer five is the incentive layer, which works by motivating the network to keep up data verification efforts through the use of money-based rewarding blockchains that are granted to contributing nodes. The contract layer is used to maintain activating entities for specific blockchains, like algorithms and smart contracts. The final layer, the application layer, keeps track of different scenarios and use cases in the system. When combined, these layers form a comprehensive blockchain structure that allows for the construction and maintenance of an Intelligent Transportation System, aiding in the decentralization of CAV technology and applications [82].

When blockchain technology is considered in terms of CAV services, there are a number of different approaches that are opened up by its addition. For example, Liu et al. [83] has proposed the use of blockchain in enabling EVCE, electric vehicles cloud, and edge networks. This architectural model focuses on allowing the needed, easily opened and ended, communications between network-connected entities, while also allowing for certain other exchanges to occur. When vehicles communicate amongst each other across the network, they cannot only share information, but unused resources from a shared pool that can aid in the speed of operation and provide energy to vehicles that need it. In turn, providing vehicles are rewarded with energy or data coins, depending on what they have provided for other vehicles. These coins allow certain benefits, like greater access to the resource pool between vehicles, or lower prices for energy, which encourages users to contribute their unused resources more often. Roadside units serve as communication providers for the system, working in information exchange, and they serve to validate the information and enable transactions, as shown in Figure 13. Local aggregators have slightly different roles, both facilitating information exchange and serving as an intermediate body between an energy-providing power grid and the requesting vehicles to provide access to energy-exchange features, using available batteries to accomplish the latter [83].

The VANETs previously discussed have been noted to be particularly oriented towards the adoption of a financial transaction method, as they primarily act as huge networks of different user-owned vehicles, all of which may have different needs that can be met by suppliers. In their research paper, Benjamin Leiding and William V. Vorobev build on this, outlining a potential transaction-focused network architecture for use in VANETs, citing the number of areas in such networks that goods and services would be desired by users [84]. While similar networks are already in place after implementation by specific companies, they noted that these networks all have different requirements, standards, and methods of operation, which makes interoperability extremely difficult [84]. In order to remedy this problem, they proposed a single unified platform that allows interactions between vehicles and any other enabled devices, which makes the process of carrying out transactions much easier for users [84]. In addition, the proposed platform outlined the foundations of a system for auctioning such goods and services in cases where an agreement needs to be made on a price between a user and seller, providing greater flexibility in the types of purchases that can be made through CAVs [84]. The proposed method was also built on blockchain technology, culminating in a final outline for a network that is able to support full interaction between user vehicles and service-providing devices, letting them interact to their fullest potential by making purchases whenever necessary, regardless of manufacturer or exact object or service being bought [84]. Figure 14 shows a diagram of this system.

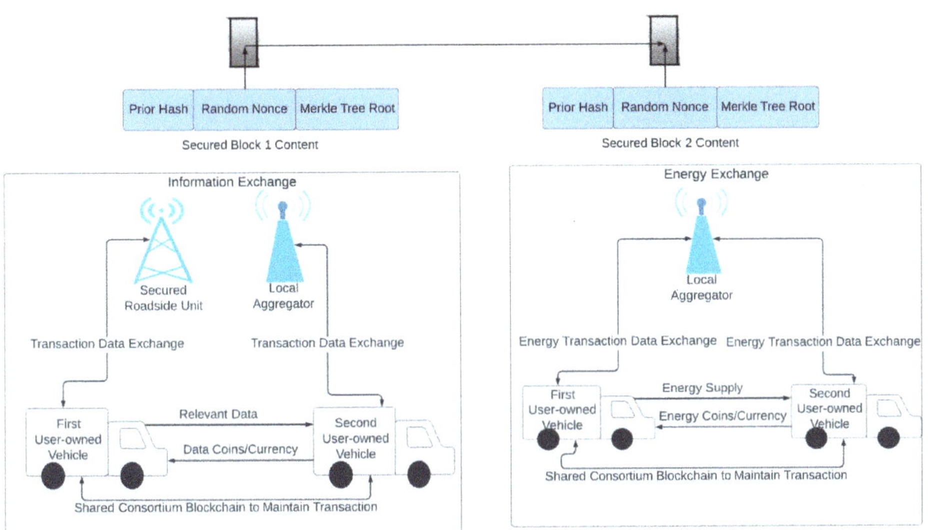

Figure 13. A model showing how energy and information is exchanged over the network using blockchain, inspired by [83].

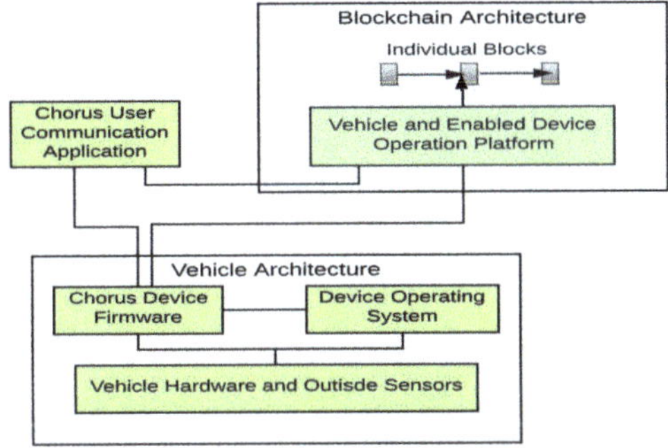

Figure 14. VANETs setup via blockchain, inspired by [84].

Blockchain has been noted as being particularly useful in carrying out financial transactions, which has led to this described influx of different proposed models for how exactly it can be applied to allow for CAV transactions. While they all seek to accomplish similar goals, they all differ in the manner of implementation, and the exact provisions offered, which makes them all unique. Table 4 shows a comprehensive discussion of each method of implementation discussed.

Table 4. Comparison of selected financial transaction methods in terms of advantages, drawbacks, and type of blockchain used.

Model Proposed	Blockchain Type	Advantages	Drawbacks
Blockchain-Aided Transport Transaction System [18]	-	Public, accessible transport via CAV through an app, high availability in any location. Blockchain eases transactions, and can be applied to parking and tolls. Blockchain can also add security to transactions, peer-to-peer based.	Vague description, no real system details. Mentioned privacy, security, and ethical concerns with no exact solution. No time/space analysis.
Autonomous Transaction System [77]	-	A structure outlining a unique approach for each feature integration, ensuring no lax measures in optimizing each. Features contribute to ease of operating vehicles and increased accessibility to users who are not confident driving or performing actions relating to vehicles.	Very vague outline, no implementation detail. No time/space analysis, only base features.
Vehicle Transaction System [78]	-	Defined, comprehensive layered blockchain to cover all functions, blockchain allows new functions, like financial transactions for parking or wi-fi.	Vague outline, no implementation detail. No time/space analysis, just potential.
Car eWallet [79,80]	Hyperledger	Transactions are possible, service and good detection by vehicle, actions can be performed with little user effort, present use, testing, and implementation with great success. Security and records through blockchain.	The technology has not been very widely implemented yet, and many of its promised features have yet to be incorporated.
Decentralized Transport Network [21]	-	Comprehensive discussion of past implementation, different types of blockchain, and use in securing and allowing transport-based transactions. Predictions for future use and how it could be implemented, focus on decentralization.	Vague implementation detail, more a list of desired features, future growth, and potential based on past use.
Blockchain-Based ITS [82]	Ethereum	Intensive model to handle all operation of blockchain, discussion of past blockchain use and application, like decentralization, trust, and network device security. Decentralization greatly benefits potential services offered.	No actual implementation of model, no tests, and no time/space analysis. More focus on potential and under-lying blockchain than how it will be implemented with CAVs.
Blockchain-Aided EVCE [83]	Consortium	Defined resource-sharing models to further peer-to-peer transaction availability between cars. Encouragement of user contribution of data and energy. Decentralized, enhanced services, multiple security outlines.	No tests or implementation, and no time/space analysis.
Chorus V2X Model [84]	Ethereum	Prototype implementation and testing. Blockchain use for security, blockchain enables service and good transactions, like transport or maintenance. Unified platform insures compatibility between networks. Flexibility in transaction details and execution. Requirement analysis.	No full system implementation, no time/space analysis, no testing carried out.

3.3. Driving Record Maintenance

In most areas of the world today, records of actions performed and their results are maintained and analyzed to come to important conclusions, such as how much damage was done in an accident, how much money is owed, and what must be done in reaction to certain events. In driving, this no different: traffic accidents are both extremely common and, in many cases, remarkably lengthy and challenging to document [85–89]. The introduction of CAVs with no event-recording measures will not necessarily help to remedy this flaw and, as such, many researchers have come to blockchain technology and similar technologies as a possible mechanism to overcome it instead. Providing evidence through accurate and secured records maintained in a database could greatly ease the current accident documentation process and issue insurance claims, determine who is responsible, and outline how reparations should be made [90–94].

With growing concern over the accountability assigned in traffic accidents involving one or more CAVs, there must be measures to avoid collisions and measures of tracking and securely maintaining information on crashes in the case that one still occurs. Accidents are not always recorded and documented effectively and, even when they are, it can be difficult to determine which parties were at fault. However, in the case of traffic incidents, this information is needed to determine what actions need to be taken to resolve the situation. From these needs, a system description from the authors of [95] emerged, focusing on the application of an approach that is based heavily on blockchain technology in maintaining records of vehicle statistics and decisions made to get a clearer view of events leading up to and following accidents to determine which party was at fault.

While not involving blockchain directly, it was the inspiration for a proposed method to take records of data, maintain these records securely, and promote the safety of CAVs and the people surrounding them [95].

Aste et al. noticed the same potential, saying that the use of unified databases to hold information are inefficient and ineffective when these databases have different information relating to the same case [96]. With the peer-to-peer based model, in which all vehicles present share gathered information, such concerns are mitigated, and the proof is ensured to be provided in a faster, more accurate, and reliable manner. Blockchain was chosen by them to be an incredibly effective method to underly such a system as well, citing its high security, ability to provide proof of existence, and maintained, up-to-date, and readily available information on transactions to verified parties. Data transparency, a decentralized network layout, and the adaption of blockchain technology to a storage facility for the information provided by a number of different relevant devices were all also reasons for its incorporation. Extensive prior use of blockchain in verifying and validating different entities before allowing them to interact is another benefit of the technology. This approach needs further consideration in the adaption of such record-maintaining structures to CAV systems [96].

Oham et al. outlined a similar approach to analyzing accident data for blame attribution [97]. This system manages to avoid several common errors in ensuring that parties are not wrongfully accused or let off for roles in accidents. The model described is resilient, relying on not one, but several parties, and it makes use of blockchain to validate provided evidence and enable only parties directly involved to present information. When accidents occur on the road, parties that are close to or involved in the accident can present evidence, information that they witnessed regarding the accident, including information that is related to time and location, and other factors gathered through visual and auditory sensors in vehicles. Blockchain, they explained, is a natural choice due to its inherent security and heavy use in data validation and decentralized networks. The peer-to-peer based system, in which all entities are able to present, agree on, and invalidate evidence presented makes it harder to miss specific crash details and, in turn, greatly eases the process of determining exactly what happened, who should be blamed, and who must be compensated. Figure 15 shows a basic model for this approach.

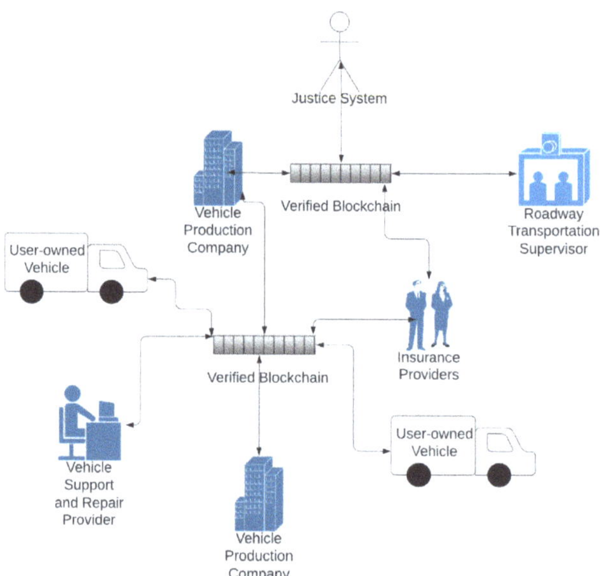

Figure 15. A model outlining the basic proposed operation of a blockchain-based information-gathering system to validate and collect accident details, inspired by [97].

Cebe et al. [98] also outlined a method of using blockchain to record events and information in the event of accidents, with this approach making use of a type of permission-oriented blockchain-based architecture, which guarantees the access to and use of vehicle-collected data only when necessary to analyze accidents, and only by parties with certain permissions. To continue providing the anonymity and security necessary in information sharing, aliases are assigned to any users participating in the blockchain, so their information will not be compromised. Information gathered can then be used to assess the entire event and, eventually, assign blame to guilty parties and offer reparations to those harmed. Because of heightened overhead in a variety of similar applications, the method accounts for preserving speed and processing capabilities by only requiring the hash values of data provided to be stored and shared in favor of the entire ledger. Figure 16 [98] outlines parties assigned permission to add, analyze, and access accident-based information.

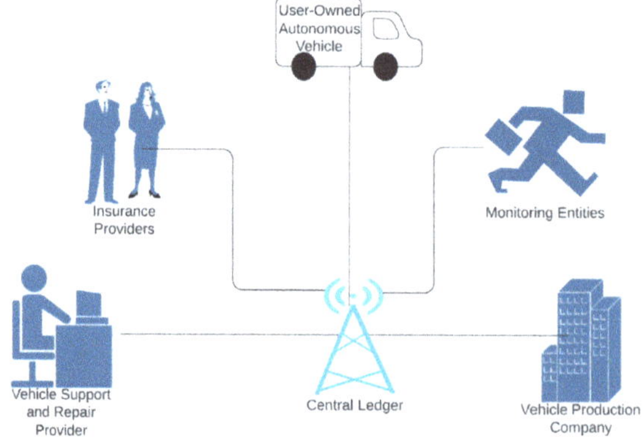

Figure 16. Different types of users who may have permission to operate on or with accident-related data, inspired by [98].

In terms of event-recording in piecing together accident-based information, the method used must be both understandable and efficient in determining case-sensitive details. As mentioned, it can be difficult to prove what has happened in the case of traffic incidents, which is why blockchain has been viewed as a potential solution due to its extensive background in maintaining and validating records. However, when CAVs are involved, considerations need to be made for both the user and their vehicle in accident assessment. In other words, how much of the accident was due to the user, and how much was due to flaws inherent in the CAV itself?

M. Ugwu et al. proposed a new type of permissioned blockchain-based system to address concerns on how accidents can be assessed in such a model, with this system operating in a series of two tiers to promote the separation of different data types to their respective tiers [99]. As in permissioned blockchain networks, each entity involved in the accident assessment and recovery process has its own distinct permissions to add, edit, and view data, being allowed and restricted based on their defined roles. Each tier is made up of three classes of objects: those who send data, those who validate sent data, and those who monitor overall functionality and activity on their tier. Tier one deals primarily in exactly how responsible each party is for the crash, determined through communications occurring at this level. Following the full assessment at this level, each party's distinct roles are known, and the known information is moved to the second tier. Next, at tier two, the presented information is viewed and used in determining how exactly each party should be held responsible for their roles. This two-tiered approach is extremely organized and easy to view due to the clear roles, players involved, and task isolation in each area, improving the overall system's efficiency and activity. Figure 17 shows a model of behaving entities in the system [99].

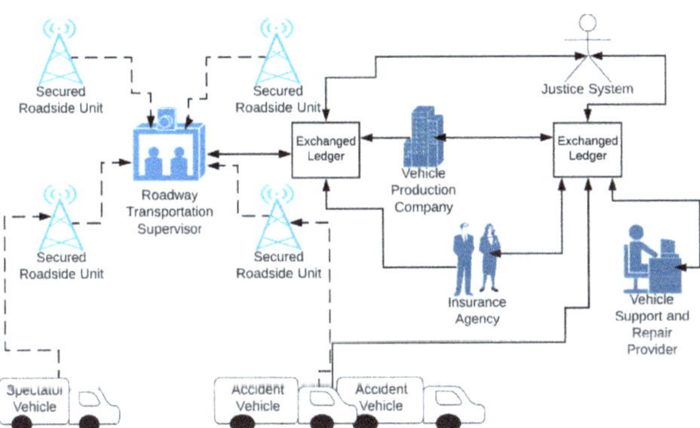

Figure 17. Different types of users who may have permission to operate on or with accident-related data under a tiered model, inspired by [99].

The implementation of a tiered system has shown promising results through its many benefits. In fact, it has shown numerous benefits over similarly proposed systems, like Block4Forensics, in its enhanced abilities in proving the existence of participating entities and their behaviors, entity involvement, and the ongoing activities of the blockchain underlying transaction control and validation. By these new capabilities, it is clear that blockchain is even more applicable to event-recording and validation in CAV systems than previously thought. If more thought is put into how exactly CAVs and the entities they work with specifically can benefit from various applications of blockchain technology, they can reach greater capabilities than ever before. Blockchain is applicable to CAV systems and outside systems that regularly interact with CAVs, so it must be thought of in terms of how it can be applied to both, not simply how it can be directly used in CAVs.

Insurance is often difficult for people to manage, and companies tend to have great difficulty in maintaining accurate information that is related to the driving history of an individual, especially in the case of undocumented traffic incidents and history of reckless driving. Claims that a driver has suffered some form of personal or property damage must be backed by extensive evidence, which is not always readily available due to the absence of constant surveillance among all roadways. Even today, when technology and electronic record keeping exists all around us, driving records do not necessarily have every single instance of traffic-related wrongdoing committed by an individual, but such information can prove itself to be extremely valuable in the case of damaging traffic accidents in assigning blame and determining future insurance rates for individuals responsible, as well as how individuals that are hurt can be properly helped by their insurance plan.

The use of blockchain in CAV insurance cannot be overlooked to remedy these flaws. To understand why, consider the primary function of blockchain: maintaining, securing, and allowing transactions. These transactions are maintained and they can be extended to recording specific conditions, like those that are internal or external to the vehicle in question in the event of a traffic accident. Automobile insurance, including that for future CAVs, is an area that could face great improvement from the addition of blockchain technology, as mentioned by Wang [100]. A consistent history of records with up to date information on vehicle actions, conditions, and location that is secured and accessible only by verified insurance-providing parties means that users can rest assured that they will be able to quickly and effectively claim benefits if they suffer harm or damages in an incident, and those insurance providers will have a simpler time accessing reliable user vehicle data and shifting prices based on user reliability. Kudwa has also assessed the future of blockchain in CAV insurance, who elaborates on just how far blockchain technology can be extended to such an approach [101]. His focus is primarily on the simplification of several auto insurance-related processes, including the ease of users in issuing claims, including those that are based on vehicle and personal injuries or damages. Figure 18 shows a proposed model of how blockchain would be used to assist in general claims [101]. As shown, the prior issue of gathering, verifying, and analyzing extensive proof is made significantly less taxing on all parties that are involved by the addition of blockchain technology due to its known accuracy, tamper-resistant nature, and overall security in the information recorded.

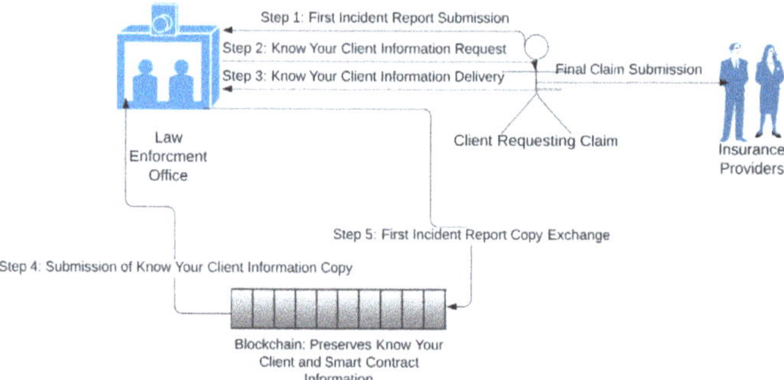

Figure 18. Information on how insurance claim processing and analysis can be benefited through blockchain, inspired by [101].

The potential use of blockchain in CAVs record-maintenance to resolve known problems in providing accurate insurance provisions is also noted in [102]. The constant maintenance of internal and external vehicle conditions and operations that are used by blockchain technology would be able to keep a collection of records, with these records being accessible if needed to present evidence of wrongdoing or traffic violation. In addition, this would greatly shorten the amount of time that is required to file claims and appropriately charge individuals responsible. Blockchain, above other existing means of maintaining such information, has shown great success in maintaining security and resistance to tampering of information, which would allow for access as needed as well as absolutely ensuring that no change has been made to the data recorded.

Consistent driving records that are reliable and accessible to verified parties is also essential in more specific operations. For example, as discussed in an article presented by Deloitte, blockchain-aided recording of delivery vehicles can provide companies or individuals with up to date information on when they can expect shipments to be delivered, as well as the current and prior states of this vehicle to inform them if something is amiss [103]. Additionally, in the case that a product is harmed through transport, internal vehicle conditions can be analyzed to determine so, proving that the customer had no responsibility for the poor state of the product and guaranteeing that they will be compensated instead of blamed for the damages. Similarly, customers will not be able to claim that a product that was damaged wholly by themselves was faulty upon arrival, as the internal conditions of the vehicle will prove that the package was secure for the duration of the trip. This allows for transparency in product condition, delivery status, and accountability in regard to damages, greatly aiding in the return process, as well as maintaining user satisfaction and company reliability.

Continuous vehicle status records also provide for a number of new features regarding vehicle rentals. In terms of leasing company benefits, the company can be sure of the past actions and track records of users prior to setting a rental price and allowing them to request certain vehicles. This setup also directly provides for the protection of safe CAVs operators in the case that a vehicle they were renting was damaged by an outside entity, which ensures that they will not receive full blame for the event in question. In such a way, responsible users and the companies that serve them benefit from this system.

The use of blockchain in transactions has already been noted as a primary reason for the technology by many, and this known benefit can be applied across most industries, including CAVs in the case of enabling and securing the generation and acceptance of contracts. The system that was outlined by [19] demonstrates the potential use of blockchain technology for allowing users to easily and securely make transportation-related agreements, letting them carpool, call autonomous taxis, and charge their vehicles without needing to perform such operations by more traditional, time-consuming means. This system is designed as a full-scale charging system, securing transactions and providing services directly between an CAVs and charging station, gathering essential user input with as little difficulty as possible.

Many people are still skeptical of the use of autonomous technology and so-called smart vehicles in maintaining user security, trust, and immunity to attackers. However, the adoption of such technology could meet these criteria and improve on existing measures in place in terms of availability and ease of access and use. Blockchain technology maintains records of all the transactions that take place, and its adoption in Bitcoin has already proven its safety and security measures. By traditional means, there is often a concern of what kind of currency a given service will require, with carpooling services ranging in whether or not they will accept electronically-made payments and, for some users, entering and exiting a vehicle to refuel or charge it can be a difficult or time-consuming task. Through the implementation of blockchain technology in allowing traditional transportation-based financial transactions to take place between machines with less direct user input, such actions can be simplified, allowing for user ease of use, access, and overall support of CAVs technology.

With the amount of attention blockchain has received in keeping secure and well-documented records of important data for later use, the focus on how this function can be applied to CAVs is not surprising. As a result, many different system outlines are available to study, test, and consider for further use, each differing in their exact specifications. Table 5 shows a table providing an overview of each of the outlines discussed previously in terms of advantages and disadvantages.

are crossed during operation [31]. Through the tests conducted, Westerlund was able to demonstrate the high level of security, reliable and correct operation, and acceptable time and space use provided by the proposed system [31].

Another major source of error for CAVs today is the logic surrounding their lane change operation, as the act of switching between lanes is an inherently complex operation due to its dependence on a variety of internal and external factors [26]. Information transfer is greatly hindered by this, as the sheer amount of information that a system needs to track can be overwhelming, and even the slightest mistake or delay can lead to a collision. The data collected can also be excessive and even dangerous in the eyes of users who do not want information on their location and driving habits made known to hackers or other malicious parties.

While blockchain technology has certainly been very promising, it also relies on the use of extensive record maintenance and ledger updates to operating effectively, which presents problems in the energy use and response delay of CAVs [116]. Energy use is a huge hindrance for many vehicles and other devices today, so it cannot be ignored when present on such a major scale in an entire class of developing technology. It is not enough for CAVs to be secure and safe for users: their energy consumption must also be a factor that is considered before their full implementation. In addition, the significant number of transactions occurring at any given time over the network has dangerous implications for the network as a whole. The potential for network overload contributes to overall instability, as well as providing an opening for attackers to misdirect drivers and compromise user information. In such a case, regardless of the security and safety benefits provided by the implementation of blockchain technology into CAVs systems, it would be far too risky to use. However, at the same time, the number of transactions cannot simply be reduced without further thought. This would risk eliminating any of the beneficial aspects that are provided by blockchain technology, which makes its use pointless. Because it is not possible to prioritize either of these aspects without drastically compromising the network and devices on it, there is no truly secure way to implement blockchain technology into CAVs systems until its excessive energy-consumption and rate of transactions can be resolved.

Responding to energy-use concerns, Sharma came up with a method of drastically reducing the number of transactions that are carried out over blockchain without compromising its many benefits. Figure 19 shows a model of the approach, labeling several points where energy can be further conserved to allow improved efficiency. By his method, the number of transactions and overall energy-use would be reduced via the use of his designed distributed clustering model, with calculated 40.16% energy conservation on average and an 82.06% reduction in the number of transactions. The model does this by utilizing the optimal slots to update blockchain ledgers instead of choosing any slot indiscriminately, which is found via the use of an optimal transaction model selected from Cluster Heads. As noted, this potential reduction in energy use and the number of transactions carried out across the CAVs network would greatly increase the efficiency, speed, and overall operation of these devices, which makes it an extremely promising method to consider for blockchain implementation.

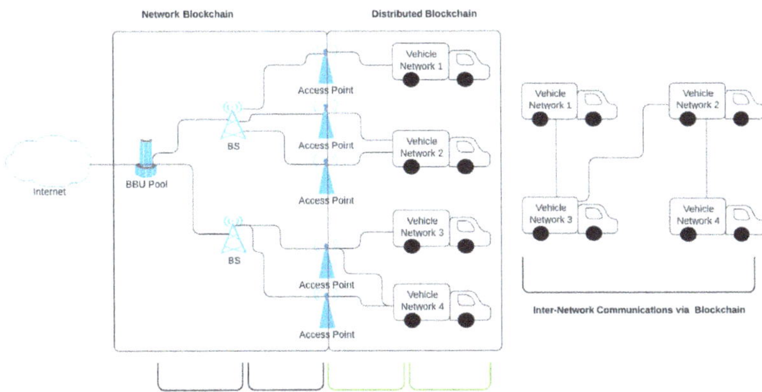

Figure 19. A model showing the structure of the areas of the CAVs network where energy use can be improved, inspired by [116].

Charging systems put in place to provide for consistent and reliable energy provisions in the case of requested transactions must also be assessed before CAVs systems as a whole can function effectively. In order to understand why, keep in mind how many vehicles may be completely autonomous in the future, and how this may open opportunities for the overload of charging stations and their resources. In such a case, the network itself may become compromised, and vehicles in need of energy may be unable to access it, compromising the operation of vehicles across the network. Energy-allocating transactions need to be readily available, presented via a scalable and resilient system, and immune to overload.

Looking to find a way to ensure this functionality, Jin et al. noted that blockchain technology is not only useful in CAVs themselves: it has a variety of traits that would benefit the operation of charging stations. Using it, energy can be provided to vehicles via a decentralized network, with ensured security and operation, very closely mirroring the architectural layout of CAVs systems as a whole. Inherent scalability, flexibility, and security present in blockchain technology have made it very well suited for such use, so it is certainly worth further consideration in terms of this application.

Charging system optimization is, as mentioned, a strong factor in determining the optimization of CAVs systems as a whole, since by improving their efficiency, CAVs will be able to operate more efficiently while still providing for the safety and health of our world in terms of environmental impact. To address concerns about the current capabilities of energy-providing units, Su et al. [117] designed a comprehensive system that works to provide a charge in a specialized, user-specific way, which guarantees that any given car will receive the optimal treatment and that charge-distributors can function with greater effect and efficiency. With the incorporation of blockchain in the system, the security level is guaranteed to carry out needed transactions, mitigating energy use concerns by implementing a permission-based model that avoids the use of an outside entity to overlook transactions. In addition, the authors discussed the possibility of using a consensus algorithm, called a delegated Byzantine fault tolerance algorithm, which heavily cuts the amount of energy that is used in carrying out transactions. Transactions are similarly carried out via blockchain to ensure the reliable, secure, and decentralized approach that is desired, with records of transactions being secured and always maintained. Figure 20 shows the results from analyzing the proposed model and previous methods for utility relative to the cleanness and environmental-consciousness of energy in use [117].

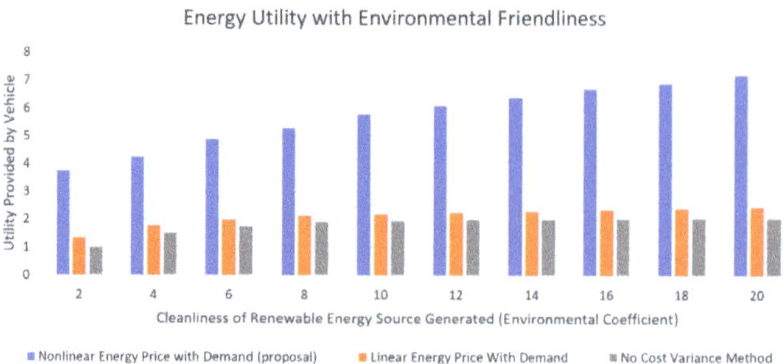

Figure 20. A model showing the environmental impact and utility of charging stations behaving under different methods, adopted from [117].

The concept of energy exchange between distinct CAVs and defined charging units or other charge providers has faced significant improvement from blockchain technology use. In an article that was presented by Javed et al. [118], an entirely new framework to charge vehicles, called a Mobile-Vehicle-to-Vehicle method, was outlined, which facilitates decentralized charging between network entities, and showed marked improvement in data security, transaction speed, energy cost, and ease in locating and accessing charge providing units for users. THe benefits listed are accomplished through, among other factors like improved algorithms, the incorporation of blockchain technology. An example of its benefits in relation to the costs of vehicles traveling overall based on the number of vehicles moving is shown in Figure 21 [118]. Underlying blockchain was used primarily in security, but also ensured user trust and approval through its extensive past use in known technologies and marked reliability, as well as its transparency in letting users view ongoing transactions in terms of who is sending and receiving information. Its implementation is an absolute necessity in this case, as, without the stability and security it provides, this charging system would be dangerous for its users, and, thus, not a feasible option for implementation [118].

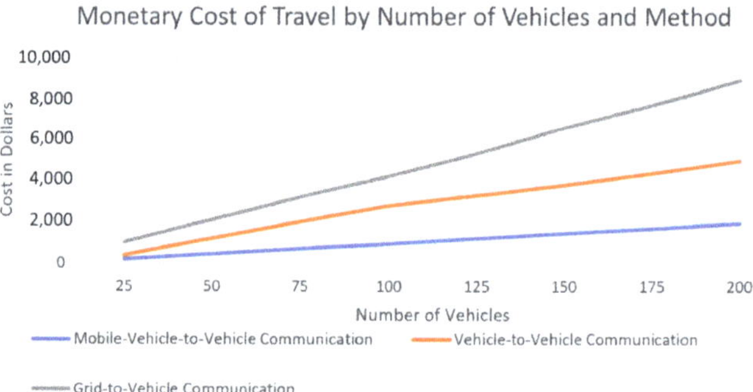

Figure 21. A graph showing the cost to transport CAVs based on method, adopted from [118].

Another example of the use of blockchain-based scheduling optimization in benefiting the energy-exchange process is shown in an article that was presented by Huang et al. [119]. In this case, a consortium blockchain-based architecture is used to define the presented charge schedule, which aims to optimize the efficiency and utility of charge stations and their users. Similar to the method that is discussed above, blockchain is used primarily in its decentralized network promotion, as well as its inherent security and promotion of user trust and transparency. Figure 22 shows a base outline of the discussed approach to charging hybrid EVs, and, as pictured, involves a number of different entities that behave in distinctly specified ways. Under the hybrid-based system, many different charging methods can be used, such as vehicle-to-vehicle and mobile charging vehicle-to-vehicle style transactions. By this method, the charge would be available to vehicles, regardless of type, model, and individual vehicle capabilities through the offer of a variety of charging methods instead of expecting all to follow a singular style best.

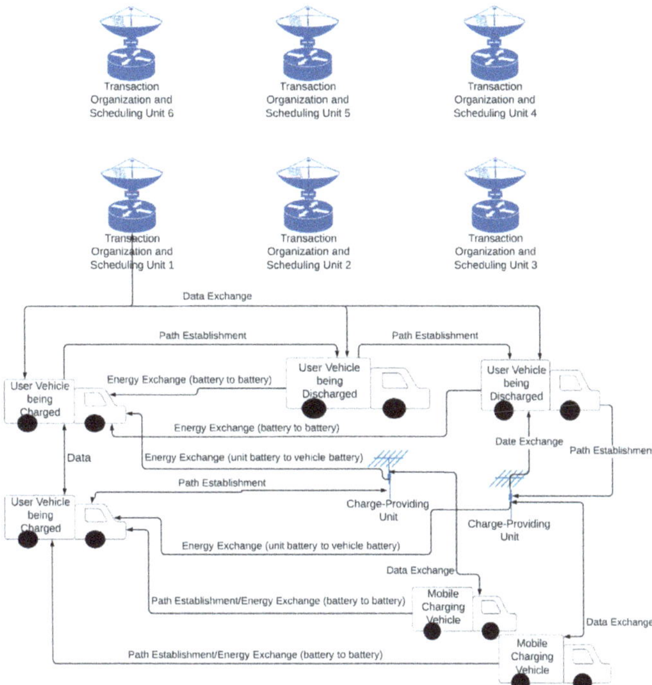

Figure 22. A diagram presenting the overall layout of the hybrid-based vehicle system, inspired by [119].

Open-source, usable networks for charging electrically-powered vehicles have not only been proposed; some have already been put into place. The first example of such a network was provided by the Energy Web Foundation (EWF), which put forward the energy-based blockchain system, called the Energy Web Chain [120]. With its network, the EWF hopes that energy can be provided to vehicles via an efficient, cost-effective, and overall accessible and optimized process that, in turn, contributes to the efficiency of the vehicles in using themselves. German-based company Share&Charge has also implemented its own network, which is called the Open Charging Network (OCN), to provide electric vehicles with fuel via the use of an expansive, mobile network [121]. Among other methods and technologies, blockchain technology was a huge basis for the design of the OCN, its security, mobility, and broad scope of applications causing it to have significant attention in such areas.

With its heavy recent growth and success, it is extremely likely that this technology will soon also be applied to CAVs. In this case, the optimization of the charging system needs to be given just as much thought as that of the CAVs system, since charging will be a central part of the proper operation of CAVs. Open, decentralized networks to provide energy must be available, reliable, and presented in a user-friendly manner, while not compromising the speed expected in carrying out energy-based transactions.

The difficulty faced by CAVs attempting to change lanes safely has already been noted and studied by a number of individuals and organizations, all rushing to find a way to combat this flaw. In response, one study, which was conducted by Fu et al., decided to implement blockchain technology into a more generally used machine-learning approach in an attempt to expedite the process [26]. In this implementation, they made use of vehicular blockchain in tandem with a deep reinforcement learning model, which acted to secure and protect data collected as it related to the user. This information gathering style allowed for users to remain anonymous and completely secure without sacrificing the educational benefits offered by having access to numerous sources of information relating to CAVs progress and behavior. Using this method, it was expected that users would have fewer privacy and security concerns about CAVs, be more willing to participate in data collection on their vehicles for research purposes, and that the information gathered would be shared between CAVs more quickly and effectively, all of which would lead to a shorter learning period for CAVs, and in turn, fewer accidents.

In order to test this idea, the group assessed the security of two separate groups: one using a collective learning approach without blockchain technology, and the other implementing blockchain technology to aid privacy and security. From the security tests conducted, it was shown that, when compared to the collective learning approach, the CAVs following the blockchain-based approach had a notably higher success ratio in changing lanes as malicious nodes were added to produce faulty and harmful data. The results showed that blockchain technology improved the safety and privacy of user information, supporting the use of blockchain in further development in the lane-changing capabilities of CAVs [26].

The use of blockchain has been applied and proposed for use in a variety of systems aiming to benefit its potential applications, as well as how well resources that are involved in such operations are maintained and utilized. As discussed above, these system and method outlines, while often similar, are never the exact same in how they are applied or what problem they aim to solve, and each has its own specific strengths and weaknesses. Table 6 shows a comprehensive discussion of each method mentioned above and how they relate to each other.

Table 6. A comparison of the selected previously discussed blockchain-based improvement and energy use outlines in terms of advantages, drawbacks, and type of blockchain used.

Model Proposed	Blockchain Type	Advantages	Drawbacks
Hyperledger-Based Intersection-Traversal Algorithm [114]	Hyperledger	Working logic to determine priority and order, peer-to-peer structure to avoid overload or single failure point. Secure transactions and communications.	Operation time is too slow for real-time response, and unusable as a result.
Multi-Agent Autonomous Intersection Management System [115]	Private Hyperledger	Secured communications between vehicles and other vehicles or infrastructure-governing devices, in-depth explanation of how intersections are controlled to avoid collisions.	No tests or working implementation yet, no time/space analysis with roadway conditions.
Blockchain-Based Collision Avoidance [31]	Ethereum	Extensive testing and traffic reduction, and in turn, greater safety for drivers. Relatively high security and acceptable time and space use.	No real-life testing, only results are from simulations. Plaintext private keys, and anonymity is not ensured.
Blockchain-Based Collective Learning for Lane Changing [26]	-	Secure communications and group-based information retrieval, high malicious node detection and information reliability, higher lane change success rate, improved execution time and space.	There is greater difficulty in finding malicious nodes when they comprise a larger part of the network, so performance slightly degrades. All tests are simulated, there have been no real-world tests. No full execution time/space analysis.
Energy-Optimized Blockchain System [116]	-	Improved energy use, and reduced transaction number with no performance loss. Less network strain and high security in transactions. Extensive resource-use and performance consideration.	Limited testing, all results thus far are simulated, wide-scale testing is needed.
Blockchain-Based Energy Trading Network	Ethereum	Well-outlined energy management and exchange system outline and simulation testing. Ability for several different transaction types for energy, like bidding and offering in auctions. Decentralized network to avoid overload and single failure point. High security and trust.	All testing done through simulations, no real-world condition test. No considera-tion for how full network will operate, just individual energy-providing units in it.
Permissioned Energy Blockchain [117]	-	Improved energy management through use and demand analysis, decentralized network. High security, consensus, and information reliability through reputation analysis. Optimized charge station utility, and overall optimized use in turn.	Little testing has been carried out on this model, no real-world implementation, all results are simulated so far.
Energy-Exchange System [118]	Consortium	Decentralized system, reduced cost, and improved utility of charging systems through optimized charge schedule. Blockchain ensures security and trust, rewards given for active participation in network, and lowest distance from vehicle to charging system is calculated when needed.	All tests are simulated, may be inaccurate to real-world results. The model is not optimized, and presents concerns of high resource use that may drain the network or compromise security. Security can still be improved, as mentioned.

Table 6. *Cont.*

Model Proposed	Blockchain Type	Advantages	Drawbacks
Scheduled Charge System [119]	Consortium	Improved operation of charge stations through charge schedule algorithm. High security and privacy, user benefits through demand and location considerations in determining a price, as well as hybrid architecture allowing for charge for all vehicle types.	All testing is done through simulation, no real-world analysis. No space analysis for proposed system.
Blockchain-Based Energy Network [120]	Ethereum	Implemented energy system, high security, detection of malicious nodes, energy efficiency and high scalability, accessible and cost-optimized energy for users. Open-source implementation for increased user access.	Limited implementation to a few company systems, not available to the public on a wide scale. Fairly new, so there is limited testing and performance analysis.
Blockchain-Based Open Charge Network [121]	Ethereum	Implemented network to charge user vehicles, open-source and highly available to users. High security, scalability, and lack of unnecessary additional middle parties in carrying out transactions. Fairly large-scale public implementation.	Limited use and test results due to it being a relatively new technology.

4. Application of Blockchain in Collective Decision

Before their full deployment, autonomous vehicles must have both individual and group operations that are assured to be secure. After all, the majority of accidents that occur require more than one party, so there must be a way to allow for autonomous vehicles to decide what maneuvers to make with other parties in proximity to avoid accidents. While many solutions have been posed, all with their advantages and drawbacks, blockchain has been applied rather extensively in this area through a variety of applications, highlighting its ability to allow the safe and consistent operation of a variety of related parties within a group.

There are not many methods of ensuring well-timed, secured communication between distinct devices operating in a group, as mentioned previously. Many researchers have noted that while devices may work adequately when operating alone, incorporating a group is a much more difficult topic, as a great deal of the rules underlying their operation are based upon human cues that are not readily understood by machines. As a potential solution to such group-based operation problems, blockchain has emerged and been employed in common group-based exercises to test its ability to govern such interactions. For example, an experiment that was conducted by Moran Cerf, Sandra Matz, and Aviram Berg, which incorporated its use into a Public Goods game, showed that its use of Smart Contracts allowed operating entities to understand and take opportunities that yielded better results [122]. Under different rule sets, this logical operation could be applied to vehicle systems as well, allowing them to avoid collisions and operate optimally on the road, securing the safety of all group entities and their passengers.

Cooperative decision making is essential to a variety of different systems, including swarm robotics, in which blockchain has already been extremely successfully applied [123]. The secure and consistent control of these autonomous devices was one of the main features keeping it back, but, with the application of blockchain technology in the field, many of the underlying problems were overcome due largely to the security, flexibility, and scalability of blockchain, as well as its low resource use when using its Proof-of-Authority algorithm [123–125]. Through an experiment implementing it into the decision-making process, it was shown to excel in this area, allowing for different parties in the swarm to communicate seamlessly, and thus avoid colliding and allow work in an intelligent manner [123].

Similarly, this use of blockchain to encourage cooperation within a system is maintained by studies conducted by Malavika Nair, and Daniel Sutter [126]. Starting by outlining the history of blockchain, its more common uses today, and its predicted growth, their paper discusses the potential impact it will have on other group-oriented applications [126] Through past uses of blockchain, it is evident that, as it continues to be implemented into such problems, it will greatly contribute to entities' ability to communicate quickly and securely over a network, making it a promising choice in future studies regarding AV group operation. This expectation is due to its strengths in allowing crowd-based applications to thrive, in addition to it consistently providing users with expected security and privacy needs [126].

Blockchain, by its very nature, lends itself to use in ensuring the cooperation of autonomous entities, as emphasized by researchers in [127]. Their study, which begun with the goal of finding an efficient, scalable, and effective way to allow the control of large groups of robotic entities, arrived at the conclusion that blockchain would serve as an effective means [127]. This conclusion was reached due to the known benefits blockchain has above similar methods in privacy, security, and decentralized network incorporation [127]. Through its incorporation, it is reasonable to expect that its incorporation may also gain such benefits into AVs, which would be a significant step forward towards their wide-scale implementation and acceptance.

Even more specifically, blockchain has had its use proposed in ensuring the safety of vehicles operating in a platoon, a group of closely positioned operating vehicles. One project involved its use in maintaining these platoon vehicles' safety needs while simultaneously making them secure from information-stealing and attacks [128]. Both of these traits, while being necessary to AV operation, are often viewed as mutually exclusive, since the increase of one tends to imply the relapse of the other, but, through the use of blockchain, it is expected that both of the requirements may be met [128]. Such advanced operation would allow for the hastened and safer deployment of large-scale AV structures, making blockchain incredibly desirable incorporation.

LIPS (Leadership Incentives for Platoons) is another such method that is built around blockchain to ensure the appropriate operation of connected AVs across a network [129]. Similarly, the method proposed is primarily put forth to aid AV platoons, this time increasing their ability to form dynamically by providing certain benefits, often in the form of payment, to vehicles willing to lead [129]. As proposed, this payment method would be carried out over blockchain architecture, a natural choice due to its origin in Bitcoin, ensuring secure and consistently available payment between entities [129]. In tests carried out to test this proposal, it was shown to be effective in fulfilling its goals, with a number of future areas for improvement in platoon operation and capabilities [129].

Studies of how blockchain can be applied to help AV platoons have extended far beyond simply ensuring their security; however, such technology has been extended to allow for the enhanced operation of such platoons in automated group toll payment for charging. In fact, such an application was tested in an experiment conducted by Zuobin Ying, Longyang Yi, and Maode Ma [130]. In the case of platoon operation, while it allows for the eased navigation of a group to retrieve fuel with as little wasted time as possible, there is the possibility of vehicles trying to sneak through without paying or providing incorrect information to lie to a governing distribution authority [130]. For a time, this was a pressing issue, significantly delaying its full implementation. However, it was discovered, through the test that was carried out, that blockchain could provide the perfect mechanism to allow such operation through its smart contract feature and noted security prowess [130].

In addition to approaches that are entirely reliant on the blockchain, its flexibility has allowed it to be considered an option to complement others based on differing technology. For example, it is highlighted as an excellent choice of architecture to support platoons' intelligence while navigating difficult intersections, as mentioned by a team of researchers proposing potential solutions to such problems [131]. Blockchain is viewed as a natural choice to support platoon navigational and operative techniques has also been mentioned by Emanuel Regnath and Sebastian Steinhorst, who, in their research on how to supervise platoons of AVs, noted the applicability of blockchain to verify parties that are involved in the group [132].

Based on increased interest and study and growing capabilities, blockchain is expected to provide numerous benefits in the area of group supervision, in which case its application to AVs to benefit platoon operation would be a natural next step. Prior results in implementation testing have shown to allow great strides forward already, providing a promising look at AVs' future capabilities if blockchain is to be utilized.

5. Future Research Directions, Challenges and Barriers

Bockchain has a lot of potential to be used in different areas, particularly in the field of Autonomous Vehicles, as discussed through this paper. However, before this can be deployed on a large scale, several problems surrounding the technology must be examined and repaired to ensure that it meets necessary requirements for energy consumption, resource use, and response time [24,25].

Nevertheless, the general form of Blockchain is quite inefficient in terms of consuming computation resource. Lightweight applications of Blockchain have shown a great deal of progress in reducing computational resource strain, and, even now, research is ongoing to improve on this current drawback. Blockchain technology itself also does not have an inherent flaw in computational resource consumption: this is instead linked directly to its PoW algorithm, which has been studied, with some proposals being put forth to reduce resource consumption [24,133]. As alternative algorithms to PoW are studied, Blockchain will likely be able to overcome its current high degree of resource use.

The high cost of implementation for Blockchain is the next challenge, which could be expensive in applications, such as CAVs. There are not many works discussing the expected cost for a large implementation, but examples of costs to develop other common blockchain applications are discussed across several papers [134,135]. However, as with many technologies, this cost can be expected to reduce significantly over time as more people study it, improve its efficiency, and become familiar with its structure and implementation enough to increase the number of workers who are able to work on implementing large-scale networks as needed.

Another challenge with using Blockchain in CAVs application is the fact that it requires high energy consumption. However, that depends on the exact type of blockchain type, there may be a much lower drain on energy use, as in non PoW models in use today [136]. While a good deal of blockchain applications today are fairly energy-intensive, a number of papers have also thought of how to resolve this issue, coming to a variety of possible solutions depending on the type of application desired [25]. In addition, Blockchain has shown that it provides a platform for users to interact more directly with their energy use and obtainment, as in proposed energy exchanging mechanisms between vehicles, as discussed by papers in the past, which could make up for higher rates of consumption in allowing more user control of how much energy they obtain at a time and from where they can access it [137].

Another main drawback of Blockchain is the responding time when more users are connected to the network, which makes them no suitable for safety applications. Delays are a known problem for Blockchain today, and as such, there is currently a lot of discussion on how to proceed when working to resolve the issue, as discussed in several papers [138]. Although there is not currently an agreed-upon solution to be used in all blockchain implementations, many incorporations of blockchain technology today still use their own methods to account for the issue of delay when under use by many people. For example, proposed methods like parallel proof of work [139] have shown to allow significant improvement in this area already. More lightweight blockchain applications have also been implemented, as with Block4Forensic, to improve the speeds of Blockchain when several users are in a network. With these examples, it's clear that these flaws are already noted and are currently being examined in a variety of ways to come to a resolution. In time, this flaw will most likely be mitigated, at which point blockchain technology will be closer to being ready for full use.

6. Conclusions

The blockchain technology contributes many advantages to network architecture such as scalability, security, and user safety which results in heavily being considered for future applications in CAVs systems. From its conception, blockchain technology was designed with several key features in mind: secured and maintained transactions, the use of a decentralized network, and user data protection, all of which could be extremely beneficial to future CAVs growth in a variety of ways. CAVs systems may not adequately protect against malicious user interference, leading to numerous accidents that may result in the loss of life. Thus, it still takes time to completely embrace the CAVs technology by public communities, such as engineers, mankind's scholars, legal intellectuals, social scientists, and moral philosophers. We need to take step forward to incorporate novel technologies, such as blockchain, to alleviate these concerns, as shown through experiments, outlines, and increased researcher interest in the idea. For the reasons that are discussed above, it is clear that the potential relationships between blockchain and CAVs technology need to be studied further in order to promote further development in both fields. By making use of a reliable, trusted technology to serve as a backbone for future CAVs systems, developers could be more certain in the safety and security of their products, as could users. In this way, the full potential of CAVs technology would be realized, which would contribute to a safer, more secure, and more accessible world through the mitigation of traffic accidents.

Author Contributions: Conceptualization, N.K. and G.T.; methodology, N.K. and G.T.; validation, A.S.; writing—review and editing, N.K., G.T., and A.S.; visualization, G.T. and N.K.; supervision, N.K. and A.S.; project administration, N.K. and A.S.; funding acquisition, A.S. All authors have read and agreed to the published version of the manuscript.

Funding: Partial support of this research was provided by the National Science Foundation under Grant No. CNS-1919855. Any opinions, findings, and conclusions, or recommendations expressed in this material are those of the author(s) and do not necessarily reflect the views of the sponsoring agency.

Conflicts of Interest: The authors declare no conflict of interest.

References

1. Madrigal, A.C. 7 Arguments Against the Autonomous-Vehicle Utopia. 2018. Available online: theatlantic.com (accessed on 1 September 2020).
2. Siddiqui, F. Silicon Valley Pioneered Self-Driving Cars. But Some of Its Tech-Savvy Residents Don't Want Them Tested in Their Neighborhoods. 2019. Available online: washingtonpost.com (accessed on 1 September 2020).
3. Top 5 Dangers of Self-Driving Cars. 2019 Available online: technology.org (accessed on 1 September 2020).
4. Top 3 Possible Dangers of Self-Driving Cars. Available online: vesttech.com (accessed on 1 September 2020).
5. Deign, J. Why Self-Driving Cars Might Make Traffic Worse. 2020. Available online: https://www.greentechmedia.com/articles/read/why-a-world-with-self-driving-cars-might-not-be-such-a-great-idea (accessed on 1 September 2020).
6. Ahram, T.; Sargolzaei, A.; Sargolzaei, S.; Daniels, J.; Amaba, B. Blockchain technology innovations. In Proceedings of the 2017 IEEE Technology & Engineering Management Conference (TEMSCON), Santa Clara, CA, USA, 8–10 June 2017; pp. 137–141.
7. Daniel, J.; Sargolzaei, A.; Abdelghani, M.; Sargolzaei, S.; Amaba, B. Blockchain Technology, Cognitive Computing, and Healthcare Innovations. *J. Adv. Inf. Technol.* **2017**, *8*, 194–198. [CrossRef]
8. Zheng, Z.; Xie, S.; Dai, H.N.; Wang, H. Blockchain challenges and opportunities: A survey. *Int. J. Web Grid Serv.* **2018**, *14*, 352–375. [CrossRef]
9. Kyriakidis, M.; Happee, R.; de Winter, J.C. Public opinion on automated driving: Results of an international questionnaire among 5000 respondents. *Transp. Res. Part F Traffic Psychol. Behav.* **2015**, *32*, 127–140. [CrossRef]
10. Lu, Z.; Wang, Q.; Qu, G.; Liu, Z. Bars: A blockchain-based anonymous reputation system for trust management in vanets. In Proceedings of the 2018 17th IEEE International Conference On Trust, Security And Privacy in Computing And Communications/12th IEEE International Conference On Big Data Science And Engineering (TrustCom/BigDataSE), New York, NY, USA, 1–3 August 2018; pp.98–103.
11. Cao, Y.; Morley Mao, Z. Autonomous Vehicles can be Fooled to 'See' Nonexistent Obstacles. Available online: https://theconversation.com/autonomous-vehicles-can-be-fooled-to-see-nonexistent-obstacles-129427#:~:text=BystrategicallyspoofingtheLiDAR,blockingtrafficorbrakingabruptly (accessed on 1 September 2020).
12. Al-Ali, M.S.; Al-Mohammed, H.A.; Alkaeed, M. Reputation Based Traffic Event Validation and Vehicle Authentication using Blockchain Technology. In Proceedings of the 2020 IEEE International Conference on Informatics, IoT, and Enabling Technologies (ICIoT), Doha, Qatar, 2–5 February 2020; pp. 451–456.
13. Dorri, A.; Steger, M.; Kanhere, S.S.; Jurdak, R. Blockchain: A distributed solution to automotive security and privacy. *IEEE Commun. Mag.* **2017**, *55*, 119–125. [CrossRef]
14. Baza, M.; Nabil, M.; Lasla, N.; Fidan, K.; Mahmoud, M.; Abdallah, M. Blockchain-based Firmware Update Scheme Tailored for Autonomous Vehicles. In Proceedings of the 2019 IEEE Wireless Communications and Networking Conference (WCNC), Marrakesh, Morocco, 15–18 April 2019; pp. 1–7.
15. Rowan, S.; Clear, M.; Gerla, M.; Huggard, M.; Goldrick, C.M. Securing Vehicle to Vehicle Communications using Blockchain through Visible Light and Acoustic Side-Channels. *arXiv* **2017**, arXiv:1704.02553.
16. Nakamoto, S. Bitcoin: A peer-to-peer electronic cash system. Available online: https://bitcoin.org/en/bitcoin-paper (accessed on 1 September 2020).
17. Michelin, R.A.; Dorri, A.; Lunardi, R.C.; Steger, M.; Kanhere, S.S.; Jurdak, R.; Zorzo, A.F. SpeedyChain: A framework for decoupling data from blockchain for smart cities. *arXiv* **2018**, arXiv:1807.01980.
18. Saranti, P.G.; Chondrogianni, D.; Karatzas, S. Autonomous Vehicles and Blockchain Technology Are Shaping the Future of Transportation. In *The 4th Conference on Sustainable Urban Mobility*; Springer: Berlin/Heidelberg, Germany, 2018; pp. 797–803.
19. Ranchal-Pedrosa, A.; Pau, G. ChargeltUp: On Blockchain-based technologies for Autonomous Vehicles. In Proceedings of the 1st Workshop on Cryptocurrencies and Blockchains for Distributed Systems, Munich, Germany, 15 June 2018; pp. 87–92. [CrossRef]
20. Rathee, G.; Sharma, A.; Iqbal, R.; Alogaily, M.; Jablan, N.; Kumar, R. A Blockchain Framework for Securing Connected and Autonomous Vehicles. *Sensors* **2019**, *19*, 3165. [CrossRef] [PubMed]

21. Carter, C.; Koh, L.D. Blockchain Disruption in Transport: Are You Decentralised Yet? Available online: https://trid.trb.org/view/1527923 (accessed on 1 September 2020).
22. Fadhil, M.; Owenson, G.; Adda, M. A Bitcoin Model for Evaluation of Clustering to Improve Propagation Delay in Bitcoin Network. In Proceedings of the 2016 IEEE Intl Conference on Computational Science and Engineering (CSE) and IEEE Intl Conference on Embedded and Ubiquitous Computing (EUC) and 15th Intl Symposium on Distributed Computing and Applications for Business Engineering (DCABES), Paris, France, 24–26 August 2016; pp. 468–475. [CrossRef]
23. Horwitz, L. Data Center-Impact of Driverless Cars Could Broaden with Blockchain. Available online: https://www.cisco.com/c/en/us/solutions/data-center/blockchain-driverless-cars.html (accessed on 1 September 2020).
24. Juričić, V.; Radošević, M.; Fuzul, E. Optimizing the Resource Consumption of Blockchain Technology in Business Systems. *Bus. Syst. Res. J.* **2020**, *11*, 78–92. [CrossRef]
25. Ghosh, E.; Das, B. A Study on the Issue of Blockchain's Energy Consumption, In *Proceedings of International Ethical Hacking Conference 2019, eHaCON 2019, Kolkata, India, 2020*; Springer: Singapore, Singapore; pp. 63–75. [CrossRef]
26. Fu, Y.; Li, C.; Yu, F.R.; Luan, T.H.; Zhang, Y. An Autonomous Lane Changing System with Knowledge Accumulation and Transfer Assisted by Vehicular Blockchain. *IEEE Internet Things J.* **2020**, *7*, 1–14. [CrossRef]
27. Choncholas, J.; Bhardwaj, K.; Gavrilovska, A. GeoENS: Blockchain-based Infrastructure for Service Discovery at the Edge. Available online: https://www.usenix.org/conference/hotedge20/presentation/choncholas (accessed on 1 September 2020).
28. Anatomy of Autonomous Vehicles: Is GIS Really Under the Hood of Self-Driving Cars? Available online: https://gisgeography.com/autonomous-vehicles-gis-self-driving-cars/ (accessed on 1 September 2020).
29. Petrovskaya, A.; Thrun, S. *Model Based Vehicle Tracking for Autonomous Driving in Urban Environments*; MIT Press: Cambridge, MA, USA, 2008. [CrossRef]
30. Sharp, C.; Schaffert, S.; Woo, A.; Sastry, N.; Karlof, C.; Sastry, S.; Culler, D. Design and Implementation of a Sensor Network System for Vehicle Tracking and Autonomous Interception. In Proceeedings of the Second European Workshop on Wireless Sensor Networks, Istanbul, Turkey, 31 January–2 February 2005; pp. 93–107.
31. Westerlund, R. Decentralized Reservation of Spatial Volumes by Autonomous Vehicles: Investigating the Applicability of Blockchain and Smart Contracts. Available online: https://www.diva-portal.org/smash/get/diva2:1437488/FULLTEXT01.pdf (accessed on 1 September 2020).
32. Zadobrischi, E.; Cosovanu, L.M.; Dimian, M. Traffic Flow Density Model and Dynamic Traffic Congestion Model Simulation Based on Practice Case with Vehicle Network and System Traffic Intelligent Communication. Available online: https://www.mdpi.com/2073-8994/12/7/1172 (accessed on 1 September 2020).
33. Van Arem, B.; Van Driel, C.J.; Visser, R. The Impact of Cooperative Adaptive Cruise Control on Traffic-Flow Characteristics. *IEEE Trans. Intell. Transp. Syst.* **2006**, *7*, 429–436. [CrossRef]
34. Yakub Abualhoul, M. Visible Light and Radio Communication for Cooperative Autonomous Driving: Applied to Vehicle Convoy. Ph.D. Thesis, Mines ParisTech, Paris, France, 2016. [CrossRef]
35. Kent, T.; Pipe, A.; Richards, A.; Hutchinson, J.; Schuster, W. A Connected Autonomous Vehicle Testbed: Capabilities, Experimental Processes and Lessons Learned. *Automation* **2020**, *1*, 17–32. [CrossRef]
36. Sichitiu, M.; Kihl, M. Inter-vehicle communication systems: A survey. *IEEE Commun. Surv. Tutorials* **2008**, *10*, 88–105. [CrossRef]
37. Luo, J.; Hubaux, J.P. A Survey of Inter-Vehicle Communication. 2004. Available online: core.ac.uk (accessed on 1 September 2020).
38. Jameel, F.; Awais Javed, M.; Zeadally, S.; Jantti, R. Efficient Mining Cluster Selection for Blockchain-based Cellular V2X Communications. *IEEE Trans. Intell. Transp. Syst.* **2020**, 1–9. . [CrossRef]
39. Jawhar, I.; Mohamed, N.; Zhang, L. Inter-vehicular Communication Systems, Protocols and Middleware. In Proceedings of the 2010 IEEE Fifth International Conference on Networking, Architecture, and Storage, Macau, China, 15–17 July 2010; pp. 282–287.
40. Takatori, Y.; Hasegawa, T. Quantitative Performance Evaluation of Predictive Collision Warning System based on Inter-Vehicle Communication. *Int. J. ITS Res.* **2007**, *4*, 13–20.
41. Lin, F.; Wang, K.; Zhao, Y.; Wang, S. Integrated Avoid Collision Control of Autonomous Vehicle Based on Trajectory Re-Planning and V2V Information Interaction. *Sensors* **2020**, *20*, 1079. [CrossRef]
42. Reichardt, D.; Shick, J. Collision Avoidance in Dynamic Environments Applied to Autonomous Vehicle Guidance on the Motorway. In Proceedings of the Intelligent Vehicles '94 Symposium, Paris, France, 24–26 October 1994; pp. 74–78. [CrossRef]
43. Funke, J.; Brown, M.; Erlien, S.M.; Gerdes, J.C. Collision Avoidance and Stabilization for Autonomous Vehicles in Emergency Scenarios. *IEEE Trans. Control. Syst. Technol.* **2017**, *25*, 1204–1216. [CrossRef]
44. Wang, P.; Gao, S.; Li, L.; Sun, B.; Cheng, S. Obstacle Avoidance Path Planning Design for Autonomous Driving Vehicles Based on an Improved Artificial Potential Field Algorithm. *Energies* **2019**, *12*, 2342. [CrossRef]
45. Gupta, R.; Tanwar, S.; Kumar, N.; Tyagi, S. Blockchain-based security attack resilience schemes for autonomous vehicles in industry 4.0: A systematic review. *Comput. Electr. Eng.* **2020**, *86*, 106717. [CrossRef]
46. Chattopadhyay, A.; Lam, K.Y.; Tavva, Y. Autonomous Vehicle: Security by Design. *IEEE Trans. Intell. Transp. Syst.* **2020**, 1–15. [CrossRef]

47. Zelle, D.; Rieke, R.; Plappert, C.; Kraus, C.; Levshun, D.; Chechulin, A. SEPAD-Security Evaluation Platform for Autonomous Driving. In Proceedings of the 2020 28th Euromicro International Conference on Parallel, Distributed and Network-Based Processing (PDP), Västerås, Sweden, 11–13 March 2020; IEEE Computer Society: Los Alamitos, CA, USA, 2020; pp. 413–420. [CrossRef]
48. Ferdowsi, A.; Challita, U.; Saad, W.; Mandayam, N.B. Robust Deep Reinforcement Learning for Security and Safety in Autonomous Vehicle Systems. In Proceedings of the 2018 21st International Conference on Intelligent Transportation Systems (ITSC), Maui, HI, USA, 4–7 November 2018; pp. 307–312.
49. Amoozadeh, M.; Raghuramu, A.; Chuah, C.N.; Ghosal, D.; Michael Zhang, H.; Rowe, J.; Levitt, K. Security Vulnerabilities of Connected Vehicle Streams and Their Impact on Cooperative Driving. *IEEE Commun. Mag.* **2015**, *53*, 126–132. [CrossRef]
50. Pokhrel, S.; Choi, J. A Decentralized Federated Learning Approach For Connected Autonomous Vehicles. In Proceedings of the IEEE Wireless Communications and Networking Conference Workshops (WCNCW), Seoul, Korea, 6–9 April 2020. [CrossRef]
51. Lam, A.Y.; Leung, Y.W.; Chu, X. Autonomous-Vehicle Public Transportation System: Scheduling and Admission Control. *IEEE Trans. Intell. Transp. Syst.* **2016**, *17*, 1210–1226. [CrossRef]
52. Llewellyn, P. Cybersecurity and Autonomous Vehicle Technology. Available online: aertech.com (accessed on1 September 2020).
53. Autonomous Vehicles: What Are the Security Risks? Available online: opentext.com (accessed on 1 September 2020).
54. Butcher, L. Increased Security for Autonomous and Connected Vehicles. Available online: autonomousvehicleinternational.com (accessed on 1 September 2020).
55. Raiyn, J. Data and Cyber Security in Autonomous Vehicle Networks. *Transp. Telecommun. J.* **2018**, *19*, 325–334. [CrossRef]
56. Blanco, S. Data Security For Autonomous Vehicles Can And Should Be Treated With Respect. Available online: forbes.com (accessed on 1 September 2020).
57. Jones, M. Tesla Personal Data Oversight Highlights Autonomous Vehicle Data Privacy Issue. Available online: techhq.com (accessed on 1 September 2020).
58. Henrique Ruffo, G. Tesla Data Leak: Components With Personal Info Find Their Way On eBay. Available online: insideevs.com (accessed on 1 September 2020).
59. Greenberg, A. Hackers Remotely Kill a Jeep on the Highway-With Me in It. Available online: wired.com (accessed on 1 September 2020).
60. Evans, S. Unsafe at Any Connection: Autonomous Vehicles Lacking in Privacy, Security Protections. Available online: sharaevans.com (accessed on 1 September 2020).
61. Bowles, J. Autonomous Vehicles and the Threat of Hacking. Available online: cpomagazine.com (accessed on 1 September 2020).
62. Hartenstein, H.; Laberteaux, K. *VANET: Vehicular Applications and Inter-Networking Technologies*; John Wiley & Sons: Hoboken, NJ, USA, 2009; Volume 1.
63. Leiding, B.; Memarmoshrefi, P.; Hogrefe, D. Self-managed and blockchain-based vehicular ad-hoc networks. In Proceedings of the 2016 ACM International Joint Conference on Pervasive and Ubiquitous Computing, Heidelberg, Germany, 12–16 September 2016; pp. 137–140. [CrossRef]
64. Singh, M.; Kim, S. Blockchain Based Intelligent Vehicle Data sharing Framework. *arXiv* **2017**, arXiv:1708.09721.
65. Narbayeva, S.; Bakibayev, T.; Abeshev, K.; Makarova, I.; Shubenkova, K.; Pashkevich, A. Blockchain Technology on the Way of Autonomous Vehicles Development. *Transp. Res. Procedia* **2020**, *44*, 168–175. [CrossRef]
66. Singh, M.; Kim, S. Branch Based Blockchain Technology in Intelligent Vehicle. *Comput. Netw.* **2018**, *145*, 219–231. [CrossRef]
67. Gong, S.; Tcydenova, E.; Jo, J.; Lee, Y.; Park, J. Blockchain-Based Secure Device Management Framework for an Internet of Things Network in a Smart City. *Sustainability* **2019**, *11*, 3889. [CrossRef]
68. The Impact of Blockchain on Banks & Financial Institution. Available online: asiablockchainreview.com (accessed on 1 September 2020).
69. Spilka, D. Blockchain and the Unbanked: Changes Coming to Global Finance. Available online: ibm.com (accessed on 1 September 2020).
70. Schlapkohl, K. Central Bank Digital Currency Explained. Available online: ibm.com (accessed on 1 September 2020).
71. Blockchain in Financial Services. Available online: ey.com (accessed on 1 September 2020).
72. Sullivan, M. The Future of Blockchain in Financial Services. Available online: blocktelegraph.io (accessed on 1 September 2020).
73. Joshi, N. Autonomous Vehicles and Blockchain. Available online: bbntimes.com (accessed on 1 September 2020).
74. Zambon, A. Autonomous Vehicles and Blockchain. Available online: octotelematics.com(accessed on 1 September 2020).
75. Fenech, G. The Link Between Autonomous Vehicles and Blockchain. Available online: forbes.com (accessed on 1 September 2020).
76. Blockchain @ Auto Finance: How Blockchain Can Enable the Future of Mobility. Available online: deloitte.com (accessed on 1 September 2020).
77. Miller, D. Blockchain and the Internet of Things in the Industrial Sector. *IT Prof.* **2018**, *20*, 15–18. [CrossRef]
78. Yuan, Y.; Wang, F.Y. Blockchain and Cryptocurrencies: Model, Techniques, and Applications. *IEEE Trans. Syst. Man Cybern. Syst.* **2018**, *48*, 1421–1428. [CrossRef]
79. Car Ewallet: Make Your Car a Wallet. Available online: chainstep.com (accessed on 1 September 2020).
80. We built Car eWallet on blockchain to securely facilitate machine-to-machine transactions. Available online: chainstep.com (accessed on 1 September 2020).
81. Mine, A. Blockchain based car wallet Car eWallet. Available online: gaiax-blockchain.com (accessed on 1 September 2020).

82. Yuan, Y.; Wang, F.Y. Towards blockchain-based intelligent transportation systems. In Proceedings of the 2016 IEEE 19th International Conference on Intelligent Transportation Systems (ITSC), Rio de Janeiro, Brazil, 1–4 November 2016; pp. 2663–2668.
83. Liu, H.; Zhang, Y.; Yang, T. Blockchain-Enabled Security in Electric Vehicles Cloud and Edge Computing. *IEEE Netw.* **2018**, *32*, 78–83. [CrossRef]
84. Leiding, B.; Vorobev, W.V. Enabling the Vehicle Economy Using a Blockchain-Based Value Transaction Layer Protocol for Vehicular Ad-Hoc Networksguo. 2018. Available online: uploads-ssl.webflow.com (accessed on 1 September 2020).
85. Steps You Can Take to Correct a Mistake in the Police Report After Your Car Accident. Available online: braunslaw.com (accessed on 1 September 2020).
86. Goguen, D. Checklist of Records to Gather After a Car Accident. Available online: nolo.com (accessed on 1 September 2020).
87. Landers, D. Tips for Settling a Car Accident Claim. Available online: nolo.com (accessed on 1 September 2020).
88. Ways to Investigate the Cause of a Car Accident. Available online: dolmanlaw.com (accessed on 1 September 2020).
89. Who is Responsible for Your Car Accident. Available online: dolmanlaw.com (accessed on 1 September 2020).
90. Huckstep, R. Four Ways Autonomous Vehicles Will Change Auto Insurance. Available online: the-digital-insurer.com (accessed on 1 September 2020).
91. Carlson, D. The Autonomous Vehicle Revolution: How Insurance Must Adapt. Available online: marsh.com (accessed on 1 September 2020).
92. Lyne, A. Driverless Cars and the Insurance Industry. Available online: capsicumre.com (accessed on 1 September 2020).
93. Barnett, D. Autonomous Cars and Auto Insurance: What's Going to Happen. Available online: atlantainsurance.com (accessed on 1 September 2020).
94. Notte, J. How Do Self-Driving Safety Features Affect Your Car Insurance? Available online: thesimpledollar.com (accessed on 1 September 2020).
95. Guo, H.; Meamari, E.; Shen, C.C. Blockchain-inspired Event Recording System for Autonomous Vehicles. In Proceedings of the 2018 1st IEEE International Conference on Hot Information-Centric Networking (HotICN), Shenzhen, China , 15–17 August 2018; pp. 218–222.
96. Aste, T.; Tasca, P.; Di Matteo, T. Blockchain Technologies: The Foreseeable Impact on Society and Industry. *Computer* **2017**, *50*, 18–28. [CrossRef]
97. Oham, C.; Kanhere, S.S.; Jurdak, R.; Jha, S. A Blockchain Based Liability Attribution Framework for Autonomous Vehicles. *arXiv* **2018**, arXiv:1802.05050.
98. Cebe, M.; Erdin, E.; Akkaya, K.; Aksu, H.; Uluagac, S. Block4Forensic: An Integrated Lightweight Blockchain Framework for Forensics Applications of Connected Vehicles. *arXiv* **2018**, arXiv:1802.00561.
99. Ugwu, M.C.; Okpala, I.C.; Oham, C.I.; Nwakanma, C.I. A Tiered Blockchain Framework for Vehicular Forensics. *Int. J. Netw. Secur. Its Appl. IJNSA* **2018**. [CrossRef]
100. Wang, A. The Future of Blockchain in Insurance. Available online: genre.com (accessed on 1 September 2020).
101. Kudwa, A.S. Blockchain: Life and Vehicle Insurance. Available online: mindtree.com (accessed on 1 September 2020).
102. Oham, C.; Jurdak, R.; Kanhere, S.S.; Dorri, A.; Jha, S.K. B-FICA: BlockChain based Framework for Auto- Claim and Adjudication. In Proceedings of the IEEE International Conference on Internet of Things (iThings) and IEEE Green Computing and Communications (GreenCom) and IEEE Cyber, Physical and Social Computing (CPSCom) and IEEE Smart Data (SmartData), Halifax, NS, Canada, 30 July–3 August 2018; pp. 1171–1180.
103. Accelerating Techonlogy Disruption in the Automotive Market: Blockchain in the Automotive Industry. Available online: deloitte.com (accessed on 1 September 2020)
104. Autonomous Car Crashes: Who or What-Is to Blame. Available online: knowledge.wharton.upenn.edu (accessed on 1 September 2020).
105. Goh, B.; Shirouzu, N. Chinese Man Blames Tesla Autopilot Function for Son's Crash. Available online: reuters.com (accessed on 1 September 2020).
106. Shepardson, D. Google Says it Bears 'Some Responsibility' After Self-Driving Car Hit Bus. Available online: reuters.com (accessed on 1 September 2020).
107. Yadron, D.; Tynan, D. Tesla Driver Dies in First Fatal Crash While Using Autopilot Mode. Available online: theguardian.com (accessed on 1 September 2020).
108. Baldwin, R. Driver in Fatal Tesla Model X Crash Had Complained About Autopilot. Available online: caranddriver.com (accessed on 1 September 2020).
109. Yamauchi, M. How Will Autonomous Vehicles Charge Themselves? Available online: pluglesspower.com (accessed on 1 September 2020).
110. Frangoul, A. VW Subsidiary to Help with Pilot of Robotic Charging Stations for Self-Driving Vehicles. Available online: cnbc.com (accessed on 1 September 2020).
111. EZ-Linck Beats Tesla with Its Completely Cable-Less Charging. Available online: autonomousevcharging.com (accessed on 1 September 2020).
112. Redefining Charging: Automatic Conductive Connection Device. Available online: volterio.com (accessed on 1 September 2020).
113. Morris, C. What's the Best Way to Grow Electric Vehicle Charging Infrastructure? Available online: evannex.com (accessed on 1 September 2020).

114. Buzachis, A.; Filocamo, B.; Fazio, M.; Ruiz, J.A.; Sotelo, M.; Villari, M. Distributed Priority Based Management of Road Intersections Using Blockchain. In Proceedings of the 2019 IEEE Symposium on Computers and Communications (ISCC), Barcelona, Spain, 29 June–3 July 2019; pp. 1159–1164.
115. Buzachis, A.; Celesti, A.; Galletta, A.; Fazio, M.; Villari, M. A Secure and Dependable Multi-Agent Autonomous Intersection Management (MA-AIM) System Leveraging Blockchain Facilities. In Proceedings of the IEEE/ACM International Conference on Utility and Cloud Computing Companion (UCC Companion), Zurich, Switzerland, 17–20 December 2018; pp. 226–231. [CrossRef]
116. Sharma, V. An energy-efficient transaction model for the blockchain-enabled internet of vehicles (IoV). *IEEE Commun. Lett.* **2018**, *23*, 246–249. [CrossRef]
117. Su, Z.; Wang, Y.; Xu, Q.; Fei, M.; Tian, Y.C.; Zhang, N. A Secure Charging Scheme for Electric Vehicles With Smart Communities in Energy Blockchain. *IEEE Internet Things J.* **2018**. [CrossRef]
118. Javed, M.U.; Javaid, N.; Aldegheishem, A.; Alrajeh, N.; Tahir, M.; Ramzan, M. Scheduling Charging of Electric Vehicles in a Secured Manner by Emphasizing Cost Minimization Using Blockchain Technology and IPFS. *Sustainability* **2020**, *12*, 5151. [CrossRef]
119. Huang, X.; Zhang, Y.; Li, D.; Han, L. An optimal scheduling algorithm for hybrid EV charging scenario using consortium blockchains. *Future Gener. Comput. Syst.* **2018**, *91*. [CrossRef]
120. Energy Web Foundation Launches Worlds First Public, Open-Source, Enterprise-Grade Blockchain Tailored to the Energy Sector. Available online: energyweb.org (accessed on 1 September 2020).
121. Burgahn, C. Launch of the Open Charging Network. Available online: shareandcharge.com (accessed on 1 September 2020).
122. Cerf, M.; Matz, S.; Berg, A. Using Blockchain to Improve Decision Making That Benefits the Public Good. *Front. Blockchain* **2020**, *3*, 13. [CrossRef]
123. Singh, P.; Singh, R.; Nandi, S.; Ghafoor, K.; Rawat, D.B.; Nandi, S. An Efficient Blockchain-Based Approach for Cooperative Decision Making in Swarm Robotics. *Internet Technol. Lett.* **2019**, *3*, e140. [CrossRef]
124. Abbaspour, A.; Mokhtari, S.; Sargolzaei, A.; Yen, K.K. A Survey on Active Fault-Tolerant Control Systems. *Electronics* **2020**, *9*, 1513. [CrossRef]
125. Mokhtari, S.; Abbaspour, A.; Yen, K.K.; Sargolzaei, A. A Machine Learning Approach for Anomaly Detection in Industrial Control Systems Based on Measurement Data. *Electronics* **2021**, *10*, 407. [CrossRef]
126. Nair, M.; Sutter, D. The Blockchain and Increasing Cooperative Efficacy. *Indep. Rev.* **2018**, *22*, 529–550.
127. Khan, A.T.; Cao, X.; Li, S.; Milosevic, Z. Blockchain Technology with Applications to Distributed Control and Cooperative Robotics: A Survey. *Int. J. Robot. Control* **2019**, *2*, 36. [CrossRef]
128. Hexmoor, H.; Alsamaraee, S.; Almaghshi, M. BlockChain for Improved Platoon Security. *Int. J. Inf.* **2018**, *7*, 1–6.
129. Ledbetter, B.; Wehunt, S.; Rahman, M.A.; Manshaei, M.H. LIPs: A Protocol for Leadership Incentives for Heterogeneous and Dynamic Platoons. In Proceedings of the 2019 IEEE 43rd Annual Computer Software and Applications Conference (COMPSAC), Milwaukee, WI, USA, 15–19 July 2019; Volume 1, pp. 535–544. [CrossRef]
130. Ying, Z.; Yi, L.; Ma, M. BEHT: Blockchain-Based Efficient Highway Toll Paradigm for Opportunistic Autonomous Vehicle Platoon. *Wirel. Commun. Mob. Comput.* **2020**, *2020*, 1–13. [CrossRef]
131. Saiáns-Vázquez, J.V.; Ordóñez-Morales, E.F.; López-Nores, M.; Blanco-Fernández, Y.; Bravo-Torres, J.F.; Pazos-Arias, J.J.; Gil-Solla, A.; Ramos-Cabrer, M. Intersection Intelligence: Supporting Urban Platooning with Virtual Traffic Lights over Virtualized Intersection-Based Routing. *Sensors* **2018**, *18*. [CrossRef] [PubMed]
132. Regnath, E.; Steinhorst, S. CUBA: Chained Unanimous Byzantine Agreement for Decentralized Platoon Management. In Proceedings of the 2019 Design, Automation Test in Europe Conference Exhibition (DATE), Florence, Italy, 25–29 March 2019; pp. 426–431. [CrossRef]
133. Pashayev, I. Reducing Computational Waste : Space and Usefulness. Available online: https://crypto.stanford.edu/cs359c/17sp/projects/IskandarPashayev.pdf (accessed on 1 September 2020).
134. How to Determine the Cost of Blockchain Implementation? Available online: https://www.leewayhertz.com/cost-of-blockchain-implementation/ (accessed on 1 September 2020).
135. Blockchain Solution Implementation. Available online: https://www.tpptechnology.com/blog/blockchain-solutions-implementation-how-much-does-it-cost-in-2020/ (accessed on 1 September 2020).
136. Sedlmeir, J.; Buhl, H.; Fridgen, G.; Keller, R. The Energy Consumption of Blockchain Technology: Beyond Myth. *Bus. Inf. Syst. Eng.* **2020**, *62*. [CrossRef]
137. K, A.; Verma, P.; Southernwood, J.; Massey, B.; Corcoran, P. Blockchain in Energy Efficiency: Potential Applications and Benefits. *Energies* **2019**. [CrossRef]
138. Zhou, Q.; Huang, H.; Zheng, Z. Solutions to Scalability of Blockchain: A Survey. *IEEE Access* **2020**. [CrossRef]
139. Hazari, S.; Mahmoud, Q. Improving Transaction Speed and Scalability of Blockchain Systems via Parallel Proof of Work. *Future Internet* **2020**, *12*, 125. [CrossRef]

MDPI

Article

A Secure Control Design for Networked Control Systems with Linear Dynamics under a Time-Delay Switch Attack

Mauro Victorio [1,†], Arman Sargolzaei [2,*,†] and Mohammad Reza Khalghani [1,†]

1 Department of Electrical and Computer Engineering, Florida Polytechnic University, Lakeland, FL 33805, USA; mvictorio@floridapoly.edu (M.V.); Khalghani@ieee.org (M.R.K.)
2 Mechanical Engineering Department, Tennessee Technological University, Cookeville, TN 38505, USA
* Correspondence: a.sargolzaei@gmail.com
† These authors contributed equally to this work.

Abstract: Networked control systems (NCSs) are designed to control and monitor large-scale and complex systems remotely. The communication connectivity in an NCS allows agents to quickly communicate with each other to respond to abrupt changes in the system quickly, thus reducing complexity and increasing efficiency. Despite all these advantages, NCSs are vulnerable to cyberattacks. Injecting cyberattacks, such as a time-delay switch (TDS) attack, into communication channels has the potential to make NCSs inefficient or even unstable. This paper presents a Lyapunov-based approach to detecting and estimating TDS attacks in real time. A secure control strategy is designed to mitigate the effects of TDS attacks in real time. The stability of the secure control system is investigated using the Lyapunov theory. The proposed TDS attack estimator's performance and secure control strategy are evaluated in simulations and a hardware-in-the-loop environment.

Keywords: time-delay switch attack; networked control systems; secure control design; Lyapunov theory; attack estimation; hardware-in-the-loop testing

Citation: Victorio, M.; Sargolzaei, A.; Khalghani, M.R. A Secure Control Design for Networked Control Systems with Linear Dynamics under a Time-Delay Switch Attack. *Electronics* **2021**, *10*, 322. https://doi.org/10.3390/electronics10030322

Academic Editor: Qusay H. Mahmoud
Received: 28 December 2020
Accepted: 25 January 2021
Published: 30 January 2021

1. Introduction

A networked control system (NCS) is a type of control system in which the control and feedback data packets are exchanged through communication channels between agents and the controller. NCS systems are used to enhance the efficiency and reliability of the control systems [1–4]. The simplicity and efficiency of NCSs for constructing networks among multi-agent systems have received significant attention over the past years. Although leveraging communication channels in an NCS can control and supervise the system more efficiently and reliably, NCSs are prone to cyber disruptions—either inherent or intentional ones, such as cyberattacks. The most known cyberattacks are denial of service (DoS), which disables access to the system information or a service [5], false data injection (FDI), which intentionally manipulates the exchange of data [6], replay attack, which maliciously repeats valid data transmissions [7], and a newly found attack, the time-delay switch (TDS) [3]. The number and intensity of these cyber manipulations have grown in recent years. One of these cyber attacks was the 2015 Ukraine Blackout event, in which about 225,000 customers lost their electricity for several hours. This cyber incident was a successful FDI on an actual power grid [8]. The 2019 Venezuela blackout resulted from a cyberattack on energy supplies in eighteen states that affected two-thirds of the country [9,10]. These cyber disruptions determine the criticality of studying cyber attacks on NCSs, like power grids.

A TDS attack is made by inserting time delays into communication channels of NCSs [11]. Since NCSs are time-sensitive and require updated measurement signals, a TDS can be highly destructive [11,12]. Time delays can occur purposefully or inherently in a wide range of engineering systems [12–14]. In general, time delays are common in control systems and can influence the stability of control systems. Even worse delays can occur when an adversary injects random TDS attacks into NCSs, making the systems inefficient or even unstable. This circumstance stems from the fact that the controller needs to receive the measurement values in real time to be able to generate the control signals. NCSs transmit the sensor measurements from agents to a centralized control unit through communication channels, and injecting TDS attacks will result in instability in NCSs. Therefore, it is crucial to design a secure NCS that is robust to both natural delays and TDS attacks [3,15].

Even though it has been shown in the literature that TDS attacks can cause instability in NCSs [11], only a few studies have focused on detecting TDS attacks in real time, and none have investigated the compensation of TDS attacks by designing a secure controller. A neural network (NN) approach was developed in [16] as a tool for estimating a time delay in industrial communication systems with nonlinear dynamics, but the stability of this controller has not been investigated. Another NN-based approach was introduced in [17] to estimate the state of the system in real time. The aforementioned proposed algorithms require offline training and cannot detect TDS attacks in real time. Although machine learning techniques have been utilized for cyber attack detection in NCSs [18], various susceptible and erroneous detection results have been reported in the literature [19,20]. These machine learning methods are prone to maliciously altering the training or test data and cause disastrous operation issues and system instability [20]. A robust controller was introduced in [21] for systems with nonlinear dynamics. The proposed approach can compensate for the effects of TDS attacks in real time without detecting them. However, the proposed controller can only mitigate small amounts of TDS attacks. The approach proposed in [22] uses a neural-network-based detection algorithm to detect and estimate the TDS attack in real time. However, the proposed approach can estimate the TDS attacks accurately, but it cannot mitigate the effects of TDS attacks in real time. To mitigate the effect of a TDS attack, Ref. [12] proposed an adaptive control algorithm that estimates TDS attacks introduced into measurement signals. This work was able to detect and mitigate TDS attacks in real time. However, the stability of the proposed method was not investigated due to the nature of the controller design. Furthermore, the convergence of attack detection and estimation requires further investigation. Table 1 summarizes the advantages and disadvantages of approaches in the literature.

Table 1. The advantages and disadvantages of the approaches in the literature.

Approach	Advantage	Disadvantage
Machine-learning-based approaches [16–20]	These approaches do not require the dynamic model of the system	Stability analysis of neural network (NN)-based approaches is complex; offline learning time is required
Robust controller-based approach [21]	There is no need to detect attacks in real time	The system is not efficient due to its robustness to potential faults, failures, and attacks
NN-based detection approach [22]	Ability of detection and estimation of time-delay switch (TDS) attacks	Stability analysis requires further investigation; it cannot mitigate the effects of TDS attacks
Least-mean-square-based approach [12]	Accurate estimation of TDS attacks with a linear model of the system	Stability analysis is complex

The contributions of this papers are as follows:

- It develops a novel secure control strategy to estimate and compensate TDS attacks in real time for NCSs with linear dynamics to address the current barriers in the detection and compensation of TDS attacks in the literature.
- It designs a secure controller that is able to mitigate the effects of TDS attacks using the proposed model-based algorithm.
- The controller and estimator in this paper are designed based on the Lyapunov theory to guarantee the stability of an NCS under a TDS attack. The proposed method is compared with an NN-based method [22] to show the efficacy of the proposed technique in the presence of the TDS attack.

To summarize, the contribution of the paper is its proposal of a novel TDS attack detection technique along with its design of a secure Lyapunov-based controller for NCSs with linear dynamics under TDS attacks. The proposed algorithm is a model-based method that is able to detect and compensate for TDS attacks in real time with low computational complexity compared with learning-based methods.

The rest of the paper is organized as follows. Section 2 presents a general model for an NCS under a TDS attack. The proposed controller with a Luenberger observer is described in Section 3. Section 4 verifies the controller's stability. The case study is described in Section 5. Simulations and results are demonstrated in Section 6, and the conclusion is presented in Section 7.

2. Networked Control Systems

Despite all of the benefits of NCSs, including that they are fast, reliable, and have remote capability, NCSs are vulnerable to cyber disruptions and threats, since NCSs are highly dependent on information, communication, and cyber interfaces [1–3]. NCSs are used in critical infrastructures, and their security is categorized as one of the important concerns in the nation. Therefore, the vulnerability to external interruptions and attacks must be prevented. Since these systems heavily rely on networked communications, this dependency introduces delays, data packet drops, bandwidth allocations, and other cybersecurity issues in the data transfer process [11,23,24].

Specifically, delays can cause catastrophic failures in the systems that are under control. Measured data and control commands must be transferred within a limited period; otherwise, introducing delays to these signals makes the system unable to trace its operating conditions and causes an undesired response [3,25].

2.1. Dynamic Model of an NCS

A general diagram of an NCS system is presented in Figure 1. The communication between the plant and the controller takes place through a networked communication channel. Commands from the controller to the actuators and sensing signals from measurement sensors, which are vulnerable points of NCSs, must flow through the network. The NCS is mathematically represented by the following state-space model, assuming that the NCS has J agents:

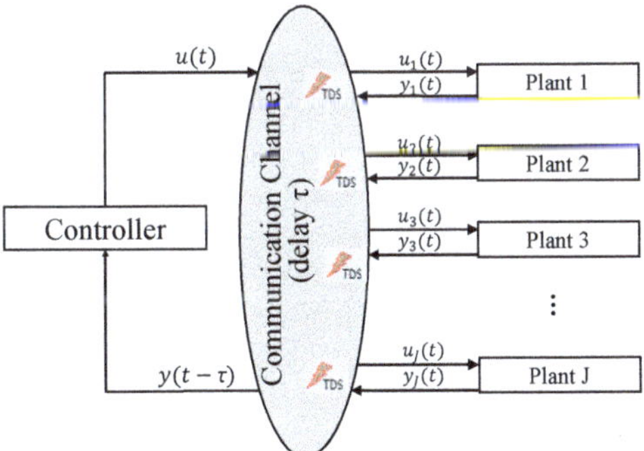

Figure 1. General diagram of a networked control system (NCS) with multiple plants under time-delay switch (TDS) attacks.

$$\begin{cases} \dot{x}(t) = Ax(t) + Bu(t) \\ y(t) = Cx(t), \end{cases} \tag{1}$$

where $x(t)$ indicates the states of NCS and can be described as

$$x(t) = \begin{bmatrix} x_1(t)^T & x_2(t)^T & \cdots & x_J(t)^T \end{bmatrix}^T \tag{2}$$

and x_i is the state-space vector of the i_{th} agent of the NCS.

In a similar manner, $u(t)$ is the vector of inputs and $y(t)$ is the output, the aggregated state measurements of the system, which are described as

$$u(t) = \begin{bmatrix} u_1(t)^T & u_2(t)^T & \cdots & u_J(t)^T \end{bmatrix}^T \tag{3}$$

$$y(t) = \begin{bmatrix} y_1(t)^T & y_2(t)^T & \cdots & y_J(t)^T \end{bmatrix}^T. \tag{4}$$

Each agent in the NCS can have a different number of inputs $u_i(t)$, outputs $y_i(t)$, and states $x_i(t)$. This means that each vector in $x(t), u(t)$, and $y(t)$ has its own dimension.

$$x_i(t) = \begin{bmatrix} x_{i,1}(t) & x_{i,2}(t) & \cdots & x_{i,n_{xi}}(t) \end{bmatrix}^T \tag{5}$$

$$u_i(t) = \begin{bmatrix} u_{i,1}(t) & u_{i,2}(t) & \cdots & u_{i,n_{ui}}(t) \end{bmatrix}^T \tag{6}$$

$$y_i(t) = \begin{bmatrix} y_{i,1}(t) & y_{i,2}(t) & \cdots & y_{i,n_{yi}}(t) \end{bmatrix}^T, \tag{7}$$

where n_{xi}, n_{ui}, and n_{yi} are the dimensions of each vector of the state-space, inputs, and outputs, respectively, for the i_{th} agent.

The matrix A is described as:

$$A = \begin{bmatrix} A_{11} & A_{12} & \cdots & A_{1J} \\ A_{21} & A_{22} & \cdots & A_{2J} \\ \vdots & \vdots & \ddots & \vdots \\ A_{J1} & A_{J2} & \cdots & A_{JJ} \end{bmatrix}. \tag{8}$$

Here, each sub-conjunct A_{ii} has the dimension $n_{xi} \times n_{xi}$ of the i_{th} agent in the system. A_{ab} represents the mutual dependency between agents.

If all the agents are independent of each other, the matrix changes to the following:

$$A = \begin{bmatrix} A_{11} & 0 & \cdots & 0 \\ 0 & A_{22} & \cdots & 0 \\ \vdots & \vdots & \ddots & \vdots \\ 0 & 0 & \cdots & A_{JJ} \end{bmatrix}. \tag{9}$$

In the same way, the matrix B is defined as

$$B = \begin{bmatrix} B_1 & 0 & \cdots & 0 \\ 0 & B_2 & \cdots & 0 \\ \vdots & \vdots & \ddots & \vdots \\ 0 & 0 & \cdots & B_J \end{bmatrix}. \tag{10}$$

Therefore, any single B_i has the dimension $n_{xi} \times n_{ui}$ based on the number of terms in the state vector (n_{xi}) and the number of terms in the input vector (n_{ui}) of the i_{th} agent in the NCS.

Matrix C is composed of C_i for each agent in the system. The dimension of each C_i is $n_{yi} \times n_{xi}$. Matrix C is defined as:

$$C = \begin{bmatrix} C_1 & 0 & \cdots & 0 \\ 0 & C_2 & \cdots & 0 \\ \vdots & \vdots & \ddots & \vdots \\ 0 & 0 & \cdots & C_J \end{bmatrix}. \tag{11}$$

2.2. NCS under a TDS Attack

The controller of an NCS is considered to be an optimal controller, which is described as

$$u(t) = -Kz(t), \tag{12}$$

where z is the signal measured by the centralized controller.

Since delays are introduced to the communication channel, the signals received by the controller from each agent in the NCS are not the $y_i(t)$ originally observed from the system output. This paper assumes that an adversary cannot access control signals and only injects the TDS attacks into the measurement signals. The measured signal under a TDS attack can be described as

$$z_i(t) = y_i(t - \tau_i), \quad i \in \{1, 2, ..., J\}, \tag{13}$$

where τ_i is the delay of the i_{th} element of the NCS.

3. Controller Design: Lyapunov Compensation with a Luenberger Observer

3.1. Observer Design

A Luenberger observer is designed such that the error between the system measurement and estimated output converges to zero:

$$\begin{cases} \dot{\hat{x}} = Ax + Bu + L(y - \hat{y}) + \psi_2 \\ \hat{y} = C\hat{x} \end{cases}, \tag{14}$$

where the vectors $\hat{x}$, $\hat{u}$, and $\hat{y}$ are the state-space, the inputs, and the outputs of the observer, respectively. The compensator signal ψ_2 will be designed based on the subsequent stability analysis. The Luenberger gain L is a constant scalar number that multiplies the error between the output of the system $y = Cx$ and the estimated value from the observer $\hat{y} = C\hat{x}$.

Substituting the y and $\hat{y}$ elements in (14) yields

$$\begin{cases} \dot{\hat{x}} = A\hat{x} + B\hat{u} + LC(x - \hat{x}) + \psi_2 \\ \hat{y} = C\hat{x} \end{cases}. \tag{15}$$

Grouping the elements with $\hat{x}$, the system representation can be written as

$$\begin{cases} \dot{\hat{x}}(t) = (A - LC)\hat{x} + B\hat{u} + LCx + \psi_2 \\ \hat{y} = C\hat{x} \end{cases}. \tag{16}$$

The last term in Equation (16) is the response received by the observer from the plant through the delayed channel, which is actually $x(t - \tau)$. Given this variable, the state-space representation is changed to the following:

$$\begin{cases} \dot{\hat{x}}(t) = (A - LC)\hat{x} + B\hat{u} + LCx(t - \tau) + \psi_2 \\ \hat{y} = C\hat{x} \end{cases}. \tag{17}$$

3.2. Controller Design

The proposed control diagram is presented in Figure 2.

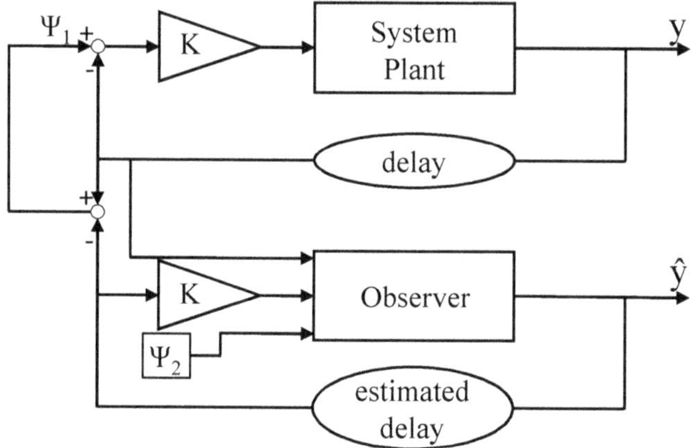

Figure 2. The proposed robust observer-based controller.

The delay presented in the superior feedback loop is the attack delay inserted into the system. The feedback controller is the same for both the plant and observer environment (K matrix). The terms of K must be defined to implement an optimal response to the system. It was used as a linear quadratic regulator (LQR) method to create the proper K matrix with the individual gains for each state in the system [26].

The delay in the system data is estimated at the feedback loop of the observer. This estimated delay signal is applied in the controller to adjust the control commands based on the inserted delay. The system model is set with an initial condition vector and the reference signal, which is a zero vector. It is replicated on the observer, and because of that, the inputs are $u = -Kx$ and $\hat{u} = -K\hat{x}$.

As seen in (18) and (19), the input signal of the system is affected by the delay in the communication channel. In the same way, the estimate delay is injected into the observer feedback response to also emulate the delay at the observer.

$$u = -Kx(t - \tau) + \psi_1 \tag{18}$$

$$\hat{u} = -K\hat{x}(t - \hat{\tau}) \tag{19}$$

As Figure 2 shows, ψ_1 compensates for the delay attack in the control input, and ψ_2 compensates for the delay attack in the observer process. These two terms will be further elaborated upon in the next section.

4. Proposed Delay Detection Method

In this section, ψ_1 and ψ_2 are designed using the Lyapunov approach. Furthermore, this section illustrates how TDS attacks are estimated in real time.

Consider the Lyapunov function described as

$$V_c = \frac{1}{2}\tilde{x}\tilde{x}^T + \frac{\alpha}{2}\tilde{\tau}\tilde{\tau}^T, \tag{20}$$

where α is a positive gain, $\tilde{x} \triangleq x - \hat{x}$ is the state estimation error, and $\tilde{\tau} \triangleq \tau - \hat{\tau}$ is the delay estimation error.

In (21), the derivative of the Lyapunov function is taken to obtain the required parameters and design a controller that is resilient against TDS attacks. It should be noted that time-delay attacks here target the input signals u_{real} and u_{obs}. The output data y from the plant also have the delay attack shift in this model.

$$\dot{V}_c = \tilde{x}\dot{\tilde{x}}^T + \alpha\tilde{\tau}\dot{\tilde{\tau}}^T \tag{21}$$

Substituting $\dot{\tilde{x}} = (\dot{x} - \dot{\hat{x}})$ and $\dot{\tilde{\tau}} = (\dot{\tau} - \dot{\hat{\tau}})$ into (21), we obtain:

$$\dot{V}_c = \tilde{x}\{\dot{x} - \dot{\hat{x}}\}^T + \alpha\tilde{\tau}\{\dot{\tau} - \dot{\hat{\tau}}\}^T. \tag{22}$$

Then, $\dot{x}$ and $\dot{\hat{x}}$ are substituted according to (1) and (14), respectively. The equation changes into:

$$\begin{aligned} \dot{V}_c =& \tilde{x}\{Ax + Bu - ((A - LC)\hat{x} + B\hat{u} \\ &+ LCx(t - \tau) + \psi_2)\}^T + \alpha\tilde{\tau}(\dot{\tau} - \dot{\hat{\tau}})^T. \end{aligned} \tag{23}$$

Substituting the signals u and $\hat{u}$ from (18) and (19), the equation becomes the following:

$$\begin{aligned} \dot{V}_c = \tilde{x}\{&Ax + B(-Kx(t - \tau) + \psi_1) \\ &- (A\hat{x} + B(-K\hat{x}(t - \hat{\tau}) \\ &+ LCx(t - \tau) + \psi_2)\}^T + \alpha\tilde{\tau}(\dot{\tau} - \dot{\hat{\tau}})^T. \end{aligned} \tag{24}$$

Considering a constant delay in the channel, even if it happens over short periods of time, the derivative of the delay is going to be null ($\dot{\tau} = 0$). Using a Taylor series, the delayed signal is modeled approximately up to the first derivative term as $x(t - \tau) = x - \dot{x}\tau$.

$$\begin{aligned} \dot{V}_c = \tilde{x}\{&(A - LC)\tilde{x} \\ &- BK(x(t - \tau) - \hat{x}(t - \hat{\tau})) + B\psi_1 \\ &+ LC\dot{x}\tau - \psi_2\}^T - \alpha\tilde{\tau}(\dot{\hat{\tau}})^T \end{aligned} \tag{25}$$

Assuming that the second and third terms in (25) can cancel out each other, ψ_1 can be obtained as follows:

$$\psi_1 = K(x(t - \tau) - \hat{x}(t - \hat{\tau})). \tag{26}$$

After this, the derivative of the Lyapunov function is simplified as in (27).

$$\begin{aligned} \dot{V}_c = \tilde{x}\{&(A - LC)\tilde{x} \\ &+ LC(\dot{\hat{x}} + \dot{\tilde{x}})(\hat{\tau} + \tilde{\tau}) - \psi_2\}^T - \alpha\tilde{\tau}(\dot{\hat{\tau}})^T \end{aligned} \tag{27}$$

Applying the distributive property, Equation (27) is extended to (E:dem4):

$$\begin{aligned} \dot{V}_c = \tilde{x}\{&(A - LC)\tilde{x} \\ &+ LC(\dot{\hat{x}}\hat{\tau} + \dot{\hat{x}}\tilde{\tau} + \dot{\tilde{x}}\hat{\tau} + \dot{\tilde{x}}\tilde{\tau}) - \psi_2\}^T - \alpha\tilde{\tau}(\dot{\hat{\tau}})^T. \end{aligned} \tag{28}$$

Simplifying (27), we obtain the following equations, which will be used to find the estimated delay $\hat{\tau}$:

$$\tilde{x}LC(\dot{\hat{x}} + \dot{\tilde{x}})\tilde{\tau}^T = \alpha\tilde{\tau}(\dot{\hat{\tau}})^T \tag{29}$$

$$\dot{\hat{\tau}} = \frac{LC}{\alpha}(\dot{\hat{x}} + \dot{\tilde{x}})\tilde{x}. \tag{30}$$

The element α is a parameter for characterizing the system to make it possible to run the computation of the estimated delay value. Unfortunately, there is not a previous relation between the physical parameters of the system and the value of α. The moment when the attack is deployed and the amount of delay injected have different effects on the system. The method used to get α is based on the peak value observed when tracking the error signal—the difference between the plant response and the observer response—and defining the α value. Then, some series of tests must be performed under known TDS attack conditions to validate the definition of α. An offset may be applied according to the essay results to guarantee more precise estimates.

ψ_2 is obtained below:

$$\psi_2 = LC(\dot{\hat{x}} + \dot{\tilde{x}})\hat{\tau}. \quad (31)$$

The Lyapunov theory is applied to guarantee that the system remains stable at different operating points. The Luenberger gain L must satisfy the following equation:

$$\tilde{x}\{(A - LC)\tilde{x}\}^T < 0. \quad (32)$$

The derivative $\dot{\hat{x}}$ can be taken from the observer. However, the error signal and its derivative $\dot{\tilde{x}}$ must be deduced—because it is not possible to measure data at the plant output before the delayed channel—by checking the error $x(t - \tau) - \hat{x}(t - \hat{\tau})$.

The Taylor series approximation must be considered; neglecting the higher-order terms, the delayed signal is represented by the following:

$$x(t - \tau) = x - \dot{x}\tau. \quad (33)$$

Similarly, the estimated states are obtained below:

$$\hat{x}(t - \hat{\tau}) = \hat{x} - \dot{\hat{x}}\hat{\tau}. \quad (34)$$

Subtracting Equations (33) from (34) will result in (36):

$$x(t - \tau) - \hat{x}(t - \hat{\tau}) = \tilde{x} - \dot{x}\tau + \dot{\hat{x}}\hat{\tau} \quad (35)$$

$$x(t - \tau) - \hat{x}(t - \hat{\tau}) = \tilde{x} - \dot{x}(\hat{\tau} + \tilde{\tau}) + \dot{\hat{x}}\hat{\tau}. \quad (36)$$

The term $\tilde{\tau}$ can be neglected because it is expected to go to zero due to the proper function of the proposed estimator. In this case, the Equation (36) becomes:

$$x(t - \tau) - \hat{x}(t - \hat{\tau}) = \tilde{x} - \dot{\tilde{x}}\hat{\tau}. \quad (37)$$

Considering that the system must be stable because the chosen value of L must guarantee the Lyapunov stability criteria, the Luenberger observer must bring estimation error signals in the form of $\tilde{x} = a.e^{-kt}$. Then, $\dot{\tilde{x}} = -k\tilde{x}$. Taking $k = 1$ (the effects of this k can be absorbed by α) to simplify the calculations, Equation (37) becomes

$$x(t - \tau) - \hat{x}(t - \hat{\tau}) = \tilde{x}(t)(1 + \hat{\tau}) \quad (38)$$

$$\tilde{x}(t) = \frac{x(t - \tau) - \hat{x}(t - \hat{\tau}))}{1 + \hat{\tau}}. \quad (39)$$

This way, it is possible to deduce $\tilde{x}(t)$ and its derivative $\dot{\tilde{x}}(t)$ used in Equations (30) and (31). They are obtained with the difference between consecutive samples in the digital readings.

5. NCS Case Study: Load Frequency Control Design for Power Grids

Load variation may create frequency oscillations and ultimately cause power grid instability. When the the grid load increases, the frequency decreases, and vice versa. Once a rapid load change is sensed, the electrical torque output is changed, which leads to a mismatch between the electrical and mechanical torques. This torque mismatch results in turbine speed changes that, in turn, cause frequency oscillations in the grid [27].

The load frequency control (LFC) of power systems is crucial in sustaining the grid frequency within a predefined range. LFC guarantees that the grid generators can properly maintain a balance between load and energy supply using the regulation of generators' set-points [27].

The mathematical model for an NCS applied to the power grid with several agents was developed in [12]. The mathematical model presented in Equation (1) with the x_i vector is defined as the following:

$$x_i(t) = [\Delta f_i(t)\ \Delta P_{g_i}(t)\ \Delta P_{tu_i}(t)\ \Delta P_{pf_i}(t)\ e_i(t)]^T, \tag{40}$$

where $\Delta f_i(t)$ is the frequency deviation, $\Delta P_{g_i}(t)$ is the generated power deviation, $\Delta P_{tu_i}(t)$ is the turbine position, $\Delta P_{pf_i}(t)$ is the tie-line power flow, and $e_i(t)]^T$ is the control error given by $e_i(t) = \int_0^t (\beta_i \Delta f_i + \Delta P_{pf_i})dt$.

Therefore, the equations are going to have $x_i(t)$, $u_i(t)$, and $y_i(t) \in \mathbb{R}^5$, where $u_i(t)$ is the control input vector, $x_i(t)$ is the state vector, and $y_i(t)$ is the output of the i_{th} agent in the system.

In this study, multiple grids are connected to the same tie-line within the same NCS. This means that agents can communicate with each other and exchange power among themselves through the grid. The matrix A will be similar to that presented in (8) with the following elements: Each A_{ii} will be

$$A_{ii} = \begin{bmatrix} \frac{-\mu_i}{J_i} & \frac{1}{J_i} & 0 & \frac{-1}{J_i} & 0 \\ 0 & \frac{-1}{T_{tu,i}} & \frac{1}{T_{tu,i}} & 0 & 0 \\ \frac{-1}{\omega_i T_{g,i}} & 0 & \frac{-1}{T_{g,i}} & 0 & 0 \\ \sum_{i=j,j=1}^{2} 2\pi T_{i,j} & 0 & 0 & 0 & 0 \\ \beta_i & 0 & 0 & 1 & 0 \end{bmatrix} \tag{41}$$

and A_{ij} will be

$$A_{i,j} = \begin{bmatrix} 0 & 0 & 0 & 0 & 0 \\ 0 & 0 & 0 & 0 & 0 \\ 0 & 0 & 0 & 0 & 0 \\ -2\pi T_{i,j} & 0 & 0 & 0 & 0 \\ 0 & 0 & 0 & 0 & 0 \end{bmatrix}. \tag{42}$$

The parameters presented in (41) and (42) are described as follows:

J_i – generator moment of inertia
β_i – frequency bias factor
ω_i – speed-drop coefficient
μ_i – damping coefficient
$T_{g,i}$ – governor time constant
$T_{tu,i}$ – turbine time constant
$T_{i,j}$ – stiffness constant between i_{th} and j_{th} agents

Similarly to (10), the matrix B is defined below:

$$B_i = \begin{bmatrix} 0 & 0 & \frac{1}{T_{g,i}} & 0 & 0 \end{bmatrix}^T. \tag{43}$$

Since vectors $y(t)$ and $x(t)$ are the same as those shown in Equation (1), the matrix C presented in Equation (1) is an identity matrix, and finally, the matrix D is considered null in this study.

6. Simulation Results

This section illustrates the performance of the proposed secure control design along with the TDS attack detection technique through the case study that was introduced in Section 5. On top of the simulation, the method was implemented in a hardware-in-the-loop (HIL) environment to show that the proposed method is practical. For the HIL testing, we used a DS1104 Controller Board in connection with the MATLAB Simulink software. The LFC model was implemented via MATLAB Simulink 2019a, and it was converted into a C program to be loaded into the Dspace board. The results were observed using the Dspace console in real time. To show that the proposed method can perform well under measurement noise, we added zero-mean Gaussian noise during the HIL performance testing. As shown in Figure 3, the NCS, including two agents, was subjected to TDS attacks. Table 2 illustrates the parameter values that were used in the simulation.

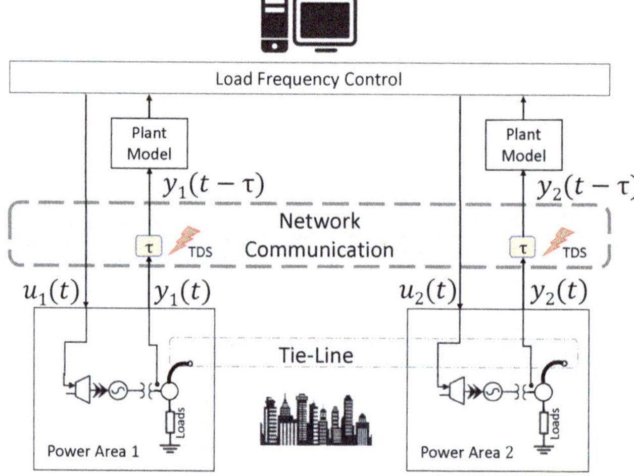

Figure 3. Two-agent load frequency control (LFC) system.

Table 2. Simulated Power Areas' Parameter Values

Description	Symbol	Power Area 1	Power Area 2
Generator Moment of Inertia	J_1	10	12
Frequency Bias Factor	β_1	21.5	21
Speed-Drop Coefficient	ω_1	0.05	0.05
Damping Coefficient	μ_1	1.5	1
Governor Time Constant	$T_{g,1}$	0.12 s	0.18 s
Turbine Time Constant	$T_{tu,1}$	0.2 s	0.45 s
Stiffness Constant *i-j*	$T_{1,2}$	0.198 pu/rad	0.198 pu/rad

The delay attacks were simulated based on the following assumptions:

Assumption 1.1. *The TDS attack takes place at a certain moment and remains constant.*

Assumption 1.2. *The TDS attack affects all the states. The starting moment and the delay period are the same for all the states.*

Assumption 1.3. *From the moment the attack is launched, it persists until the end of the simulation.*

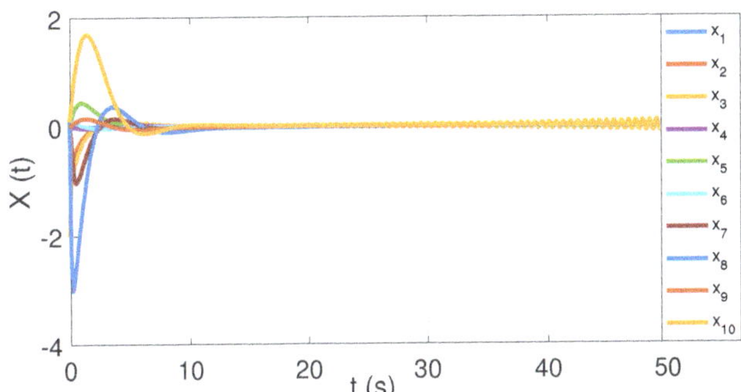

Figure 4. The NCS system under a TDS attack of $\tau = 0.19$ s at the instant $t_0 = 1$ s.

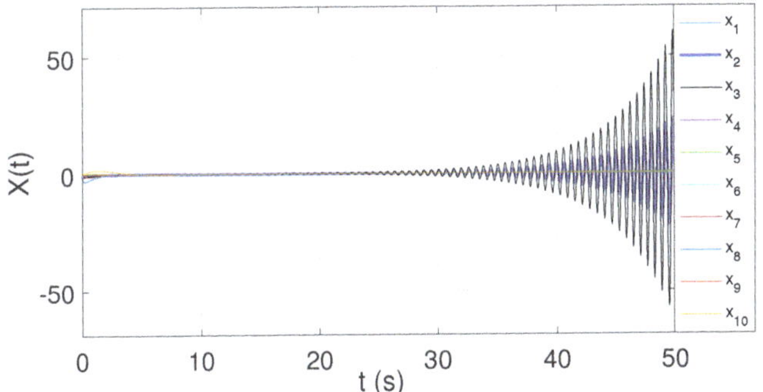

Figure 5. The NCS system under evaluation submitted to a TDS attack of $\tau = 0.20$ s at the instant $t_0 = 1$ s.

6.1. Vulnerability Analysis

When running the NCS in an optimal control situation, with no compensation algorithm to correct the delays, it can be observed that the system becomes unstable for attacks of $\tau = 190$ ms or higher. Figures 4 and 5 show the response of the system under a TDS attack. Both attacks were started at $t_0 = 1$ s, but the intensity of the attack is different. In the first case, $\tau = 0.19$ s, and in the other one, $\tau = 0.2$ s. It can be observed that the delay of 0.2 s had an aggressive effect on the system, leading the response to a divergent behavior with a highly increasing rate. Figure 4 represents a threshold condition, where the system started facing unstable responses due to the delay. It is possible to observe that the instability started around the instant at 30 s and the oscillations grew at a low rate. However, it was an unstable condition, and the system could not work under this attack. To detect and mitigate the effects of an attack, a resilient controller must be applied.

6.2. TDS Attack Detection and State Estimation

The observer compensation technique discussed in Section 4 made the system operate correctly, even with TDS attacks of around $\tau = 260$ ms. The same TDS attack as that shown in Figure 5 was simulated, but the system was equipped with the proposed secure controller. As Figure 6 shows, the adverse effect of the TDS attack was improved by the proposed state estimation mechanism.

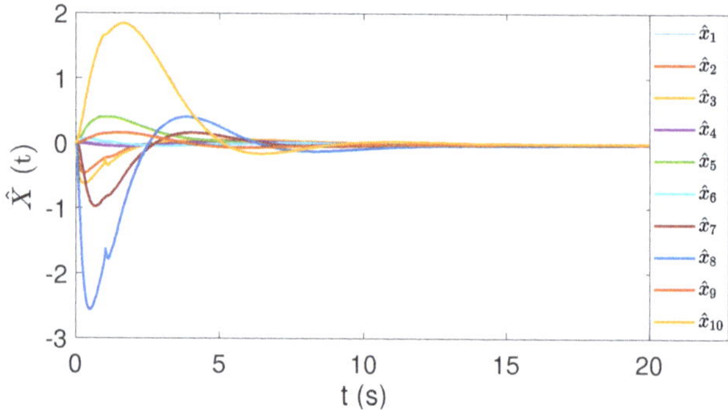

Figure 6. NCS system under a TDS attack of $\tau = 0.2$ s at the instant $t_0 = 1$ s with the Luenberger observer operating to compensate the errors.

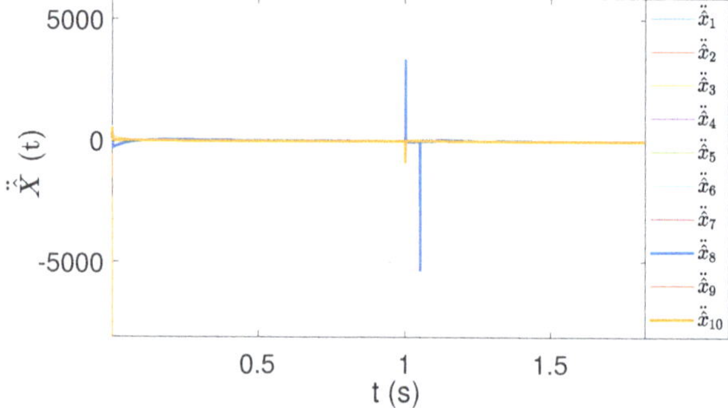

Figure 7. Second derivative of $\hat{x}$ used to detected the instant in time t_0 when the TDS is deployed.

The simulation was repeated with different time delays inserted by the TDS attack into the NCS at different starting instants ($t_0 = 1, 2, ..., 5$ seconds). Table 3 shows the results for three different time delays τ. The results show an Mean Square Error (MSE) around 5%.

Table 3. Delay estimation results—MSE.

τ (s)	$\hat{\tau}$ (s)	MSE (s)
0.15	0.1538	5.26%
0.22	0.2202	4.61%
0.26	0.2540	5.70%

The estimation of the delay depends on the parameter α mentioned in (30). The definition of α is a result of a complex set of essays running under known conditions to evaluate how the system responds to an injected delay depending on the amount of delay and the time instant at which was deployed. When an attack takes place, the error between the actual and the estimated states from the observer increases, and this is also reflected in the variations in $\hat{x}$. Using the second derivative of $\hat{x}$, the time instant of the TDS attack is detected (as shown in Figure 7). Thus, it is deployed, and the error is used to define the value to be chosen for α. The solution for computing α was created by using a constant value for alpha and then comparing the results with the offset that should be needed to get the correct values of $\hat{\tau}$. A polynomial approximation was created by taking the instant values of $\dot{\hat{x}}$; then, it was incorporated in the α computation.

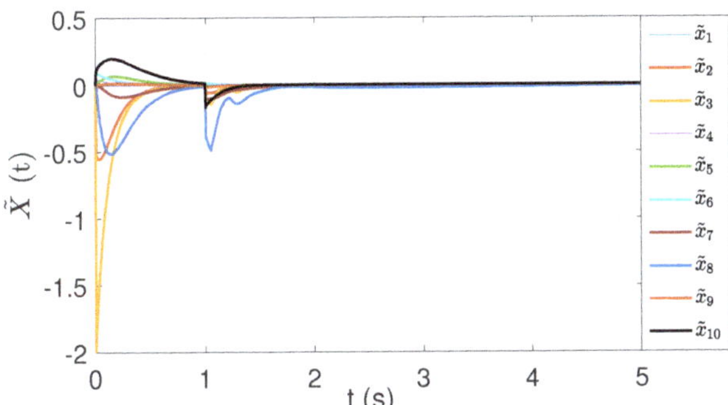

Figure 8. The state estimation error signal $\tilde{x}$ has a peak at the instant at 1 s, indicating that the attack was deployed. This peak value is used to estimate the proper α for $\hat{\tau}$ estimation.

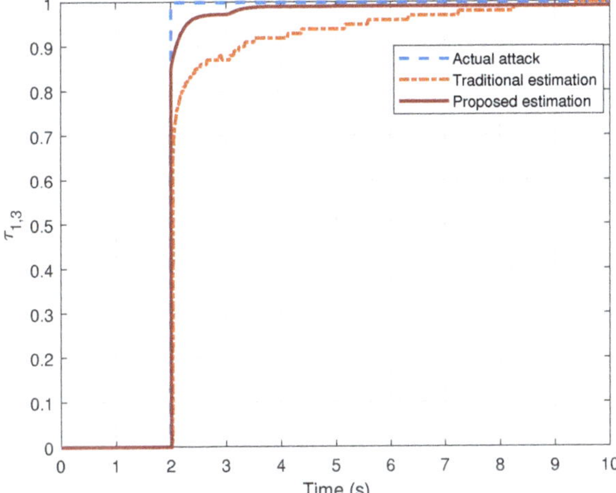

Figure 9. Delay estimation using the proposed method compared with the traditional method.

6.3. Performance Evaluation

The performance of the method proposed in this paper is compared with the performance of the control method presented in [22], which works based on an adaptive NN technique to estimate the time delay to which the NCS is exposed. The simulations were performed based on the following assumptions:

Assumption 2.1. *The delay attack takes place in a certain moment and remains constant.*

Assumption 2.2. *The delay attack targets only the state of* $x_3(t)$.

Assumption 2.3. *The attack persists from the moment it is launched until the end of the simulation.*

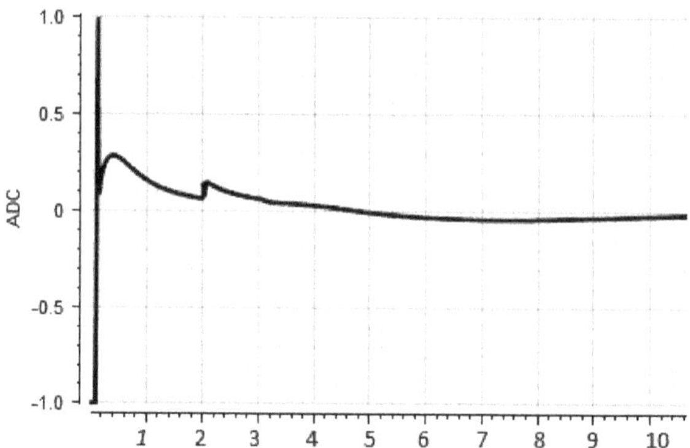

Figure 10. Response of x_3 with controller correction for a $\tau = 1.0$ s delay injected at $t_0 = 2.0$ s. The signal was observed in a Dspace input to simulate the input of the plant received through a feedback loop with a delay.

A delay $\tau = 1$ s was inserted from $t_0 = 2$ s. The state estimation errors are shown in Figure 8. The results show that the observer was able to accurately estimate measurement signals, even when the NCS was under TDS attacks. The results associated with the proposed method are presented in Figure 9. As shown in the figure, it is clear that the proposed TDS attack detection technique is more accurate than the NN-based (traditional) detection technique.

The NN approach allows consecutive computations in order to correct the estimation error and compensate frequency discrepancies in the LFC. The proposed method based on the Lyapunov theory presented in (20) depends on the parameter α, which is a characteristic quantity of the system, but it also depends on the instant the attack takes place. A study using the values of $\dot{\hat{x}}$, $\hat{x}$, and $\tilde{x}$ was used to calculate this parameter. So, the final value of $\hat{\tau}$ was reached more quickly, but a steady-state error can happen. The better the adjustment of the α parameter, the lower the steady-state error.

Another experiment to validate the Lyapunov method for avoiding TDS attack effects was made in a hardware-in-the-loop environment using external hardware equipment, the dSPACE CP1104 platform. A feedback loop was made for the x_3 state through the connection on the board. A time delay of $\tau = 1.0$ s affected the LFC, and the resilient performance of the proposed controller was seen in the results. The signal response on x_3 is shown in Figure 10. This test is important because it shows the effectiveness of the solution in a real-life system, demonstrating how the correction takes place to keep the system working properly, even when it is subjected to a TDS attack.

7. Conclusions

This paper presented an approach to providing adjustments to the input signal in an NCS structure in order to keep a plant working properly, even under a TDS attack. The compensation signal is defined based on the Lyapunov theory and a Luenberger observer. The proposed approach was evaluated through a case study in which a two-agent LFC system was monitored and controlled. The performance of the controller was validated in a hardware-in-the-loop simulation using the dSPACE platform to emulate the feedback loop of the NCS under TDS attacks. It should be noted that the proposed method can be applied to any NCSs, such as cooperative driving systems.

During the transient response of the system, the proposed delay estimation is more precise compared with other techniques in the literature. Furthermore, the controller and estimator were designed based on Lyapunov stability analysis, unlike other traditional techniques. When the most significant variations take place, it is possible to detect the peaks caused by the attack. The final portion of the transient response already has lower variation, and it brings uncertainties to the estimation of $\hat{\tau}$. If the attack is deployed at an instant close to the starting point, the estimation is also affected because the observer will get the actual values from the plant. The variation in a delay over time is also a strong challenge for estimations, but this paper considered only constant delay attacks. Future work will enhance the proposed method to investigate time-variable TDS attacks. Although this paper focused on an NCS with dependent agents, the proposed method is general and can be used for a multi-agent system with independent dynamics.

Author Contributions: Methodology, M.V. and A.S.; Writing – original draft, M.V., A.S. and M.R.K.; Writing – review and editing, M.V., A.S. and M.R.K. All authors have read and agreed to the published version of the manuscript.

Funding: Partial support of this research was provided by the Woodrow W. Everett, Jr. SCEEE Development Fund in cooperation with the Southeastern Association of Electrical Engineering Department Heads and the National Science Foundation under Grant No. CNS-1919855. Any opinions, findings, and conclusions, or recommendations expressed in this material are those of the author(s) and do not necessarily reflect the views of the sponsoring agency.

Conflicts of Interest: The authors declare no conflict of interest.

Abbreviations

NCS	Networked Control System
TDS	Time-Delay Switch
DoS	Denial-of-Service
FDI	False Data Injection
NN	Neural Network
LQR	Linear Quadratic Regulator
LFC	Load Frequency Control
HIL	Hardware-in-the-Loop

References

1. Hespanha, J.P.; Naghshtabrizi, P.; Xu, Y. A Survey of Recent Results in Networked Control Systems. *Proc. IEEE* **2007**, *95*, 138–162. [CrossRef]
2. Yarali, A.; Rahman, S. Smart Grid Networks: Promises and Challenges. *J. Commun.* **2012**, *7*. [CrossRef]
3. Sargolzaei A.; Abbaspour A.; Al Faruque, M.A.; Eddin, A.S.; Yen, K. Security Challenges of Networked Control Systems. *Sustain. Interdepend. Netw. Stud. Syst. Decis. Control* **2018**, *145*, 77–95.
4. Abbaspour, A.; Mokhtari, S.; Sargolzaei, A.; Yen, K.K. A Survey on Active Fault-Tolerant Control Systems. *Electronics* **2020**, *9*, 1513. [CrossRef]
5. Lu, A.; Yang, G. Observer-Based Control for Cyber-Physical Systems Under Denial-of-Service With a Decentralized Event-Triggered Scheme. *IEEE Trans. Cybern.* **2019**, *50*, 4886–4895. [CrossRef] [PubMed]
6. Khalghani, M.R.; Solanki, J.; Solanki, S.K.; Khooban, M.H.; Sargolzaei, A. Resilient Frequency Control Design for Microgrids Under False Data Injection. *IEEE Trans. Ind. Electron.* **2021**, *68*, 2151–2162. [CrossRef]
7. Hosseinzadeh, M.; Sinopoli, B.; Garone, E. Feasibility and Detection of Replay Attack in Networked Constrained Cyber-Physical Systems. In Proceedings of the 2019 57th Annual Allerton Conference on Communication, Control, and Computing (Allerton), Monticello, IL, USA, 22–27 September 2019; pp. 712–717. [CrossRef]
8. Liang, G.; Weller, S.R.; Zhao, J.; Luo, F.; Dong, Z.Y. The 2015 Ukraine Blackout: Implications for False Data Injection Attacks. *IEEE Trans. Power Syst.* **2017**, *32*, 3317–3318. [CrossRef]
9. Long, H.; Wu, Z.; Fang, C.; Gu, W.; Wei, X.; Zhan, H. Cyber-attack Detection Strategy Based on Distribution System State Estimation. *J. Mod. Power Syst. Clean Energy* **2020**, *8*, 669–678. [CrossRef]
10. Li, F.; Yan, X.; Xie, Y.; Sang, Z.; Yuan, X. A Review of Cyber-Attack Methods in Cyber-Physical Power System. In Proceedings of the 2019 IEEE 8th International Conference on Advanced Power System Automation and Protection (APAP), Xi'an, China, 21–24 October 2019; pp. 1335–1339. [CrossRef]
11. Sargolzaei, A.; Yen, K.; Abdelghani, M. Time-Delay Switch Attack on Load Frequency Control in Smart Grid. *Adv. Commun. Technol.* **2013**, *5*, 55–64.
12. Sargolzaei, A.; Yen, K.K.; Abdelghani, M.N. Preventing Time-Delay Switch Attack on Load Frequency Control in Distributed Power Systems. *IEEE Trans. Smart Grid* **2016**, *7*, 1176–1185. [CrossRef]
13. Chaudhuri, B.; Majumder, R.; Pal, B. Wide-Area Measurement-Based Stabilizing Control of Power System Considering Signal Transmission Delay. *Power Syst. IEEE Trans.* **2004**, *19*, 1971–1979. [CrossRef]

14. Wu, H.; Tsakalis, K.; Thomas Heydt, G. Evaluation of Time Delay Effects to Wide-Area Power System Stabilizer Design. *Power Syst. IEEE Trans.* **2004**, *19*, 1935–1941. [CrossRef]
15. Ali, H.; Dasgupta, D. Effects of Time Delays in the Electric Power Grid. In *Critical Infrastructure Protection VI*; Butts, J.; Shenoi, S., Eds.; Springer: Berlin/Heidelberg, Germany, 2012; pp. 139–154.
16. Tan, Y. Time-varying time-delay estimation for nonlinear systems using neural networks. *Int. J. Appl. Math. Comput. Sci.* **2004**, *14*, 63–68.
17. Sadeghzadeh, N.; Afshar, A.; Menhaj, M.B. An MLP neural network for time delay prediction in networked control systems. In Proceedings of the Chinese Control and Decision Conference, Yantai, China, 2–4 July 2008; pp. 5314–5318.
18. Karimipour, H.; Dehghantanha, A.; Parizi, R.M.; Choo, K.R.; Leung, H. A Deep and Scalable Unsupervised Machine Learning System for Cyber-Attack Detection in Large-Scale Smart Grids. *IEEE Access* **2019**, *7*, 80778–80788. [CrossRef]
19. Huang, L.; Joseph, A.D.; Nelson, B.; Rubinstein, B.I.; Tygar, J.D. Adversarial Machine Learning. In *Proceedings of the 4th ACM Workshop on Security and Artificial Intelligence*; Association for Computing Machinery: New York, NY, USA, 2011; pp. 43–58. [CrossRef]
20. Pitropakis, N.; Panaousis, E.; Giannetsos, T.; Anastasiadis, E.; Loukas, G. A taxonomy and survey of attacks against machine learning. *Comput. Sci. Rev.* **2019**, *34*, 100199. [CrossRef]
21. Sargolzaei, A.; Yen, K.K.; Abdelghani, M. Control of Nonlinear Heartbeat Models under Time- Delay-Switched Feedback Using Emotional Learning Control. *Int. J. Recent Trends Eng. Technol.* **2014**, *10*, 85.
22. Abbasspour, A.; Sargolzaei, A.; Victorio, M.; Khoshavi, N. A Neural Network-based Approach for Detection of Time Delay Switch Attack on Networked Control Systems. *Procedia Comput. Sci.* **2020**, *168*, 279–288. [CrossRef]
23. Gupta, R.A.; Chow, M. Networked Control System: Overview and Research Trends. *IEEE Trans. Ind. Electron.* **2010**, *57*, 2527–2535. [CrossRef]
24. Boroojeni, K.G.; Amini, M.H.; Iyengar, S. Overview of the Security and Privacy Issues in Smart Grids. In *Smart Grids: Security and Privacy Issues*; Springer: Berlin/Heidelberg, Germany, 2017; pp. 1–16.
25. Galli, S.; Scaglione, A.; Wang, Z. Power Line Communications and the Smart Grid. In Proceedings of the 2010 First IEEE International Conference on Smart Grid Communications, Gaithersburg, MD, USA, 4–6 October 2010; pp. 303–308.
26. Patarroyo-Montenegro, J.F.; Salazar-Duque, J.E.; Andrade, F. LQR Controller with Optimal Reference Tracking for Inverter-Based Generators on Islanded-Mode Microgrids. In Proceedings of the 2018 IEEE ANDESCON, Santiago de Cali, Colombia, 22–24 August 2018; pp. 1–5. [CrossRef]
27. Elgerd, O. Control of electric power systems. *IEEE Control Syst. Mag.* **1981**, *1*, 4–16. [CrossRef]

 MDPI

Article

Longitudinal Control for Connected and Automated Vehicles in Contested Environments

Shirin Noei [1,*,†], Mohammadreza Parvizimosaed [2,†] and Mohammadreza Noei [3,†]

1 Center for Energy Systems Research, Tennessee Technological University, Cookeville, TN 38505, USA
2 Department of Computer Engineering, K. N. Toosi University of Technology, Tehran 16317-14191, Iran; rezaparvizi@email.kntu.ac.ir
3 Department of Electrical and Computer Engineering, Tarbiat Modares University, Tehran 14117-13116, Iran; mohammadrezanoei@modares.ac.ir
* Correspondence: snoei@tntech.edu; Tel.: +1-931-372-6546
† These authors contributed equally to this work.

Abstract: The Society of Automotive Engineers (SAE) defines six levels of driving automation, ranging from Level 0 to Level 5. Automated driving systems perform entire dynamic driving tasks for Levels 3–5 automated vehicles. Delegating dynamic driving tasks from driver to automated driving systems can eliminate crashes attributed to driver errors. Sharing status, sharing intent, seeking agreement, or sharing prescriptive information between road users and vehicles dedicated to automated driving systems can further enhance dynamic driving task performance, safety, and traffic operations. Extensive simulation is required to reduce operating costs and achieve an acceptable risk level before testing cooperative automated driving systems in laboratory environments, test tracks, or public roads. Cooperative automated driving systems can be simulated using a vehicle dynamics simulation tool (e.g., CarMaker and CarSim) or a traffic microsimulation tool (e.g., Vissim and Aimsun). Vehicle dynamics simulation tools are mainly used for verification and validation purposes on a small scale, while traffic microsimulation tools are mainly used for verification purposes on a large scale. Vehicle dynamics simulation tools can simulate longitudinal, lateral, and vertical dynamics for only a few vehicles in each scenario (e.g., up to ten vehicles in CarMaker and up to twenty vehicles in CarSim). Conventional traffic microsimulation tools can simulate vehicle-following, lane-changing, and gap-acceptance behaviors for many vehicles in each scenario without simulating vehicle powertrain. Vehicle dynamics simulation tools are more compute-intensive but more accurate than traffic microsimulation tools. Due to software architecture or computing power limitations, simplifying assumptions underlying convectional traffic microsimulation tools may have been a necessary compromise long ago. There is, therefore, a need for a simulation tool to optimize computational complexity and accuracy to simulate many vehicles in each scenario with reasonable accuracy. This research proposes a traffic microsimulation tool that employs a simplified vehicle powertrain model and a model-based fault detection method to simulate many vehicles with reasonable accuracy at each simulation time step under noise and unknown inputs. Our traffic microsimulation tool considers driver characteristics, vehicle model, grade, pavement conditions, operating mode, vehicle-to-vehicle communication vulnerabilities, and traffic conditions to estimate longitudinal control variables with reasonable accuracy at each simulation time step for many conventional vehicles, vehicles dedicated to automated driving systems, and vehicles equipped with cooperative automated driving systems. Proposed vehicle-following model and longitudinal control functions are verified for fourteen vehicle models, operating in manual, automated, and cooperative automated modes over two driving schedules under three malicious fault magnitudes on transmitted accelerations.

Keywords: traffic microsimulation tool; cooperative automated driving systems; vehicle powertrain; safety; road capacity; contested environments

Citation: Noei, S.; Parvizimosaed, M.; Noei, M. Longitudinal Control for Connected and Automated Vehicles in Contested Environments. *Electronics* **2021**, *10*, 1994. https://doi.org/10.3390/electronics10161994

Academic Editor: Alexey Vinel

Received: 15 June 2021
Accepted: 9 August 2021
Published: 18 August 2021

1. Introduction

SAE defines six levels of driving automation

- Level 0: drivers perform entire dynamic driving tasks;
- Level 1: driver assistance systems execute either longitudinal or lateral vehicle motion control subtask, and drivers perform all remaining dynamic driving tasks;
- Level 2: driver assistance systems execute both longitudinal and lateral vehicle motion control subtasks, and drivers perform all remaining dynamic driving tasks;
- Levels 3–5: automated driving systems perform entire dynamic driving tasks [1].

Dynamic driving tasks are real-time operational (e.g., longitudinal and lateral vehicle motion control) and tactical (e.g., object and event detection, recognition, classification, and response preparation) functions required to operate a vehicle. Delegating dynamic driving tasks to automated driving systems can eliminate 94% of crashes attributed to driver errors [2].

Cooperative driving automation enables cooperation among road users, intending to enhance dynamic driving task performance, safety, and traffic operations. Cooperative driving automation can prevent 439,000 to 615,000 crashes, save 987 to 1366 lives, reduce 305,000 to 418,000 maximum abbreviated injury scale 1–5 injuries, and eliminate 537,000 to 746,000 property damage only vehicles annually [3]. Vehicles equipped with cooperative automated driving systems can also follow their leaders at shorter gaps and with less variation in acceleration than vehicles dedicated to automated driving systems. SAE defines four classes of cooperative driving automation cooperation: Class A (status-sharing), Class B (intent-sharing), Class C (seeking-agreement), and Class D (prescriptive) [4]. Classes C–D cooperative driving automation cooperation can be achieved at Levels 3–5 driving automation.

Cooperative automated driving systems can be simulated using a vehicle dynamics simulation tool (e.g., CarMaker and CarSim) or a traffic microsimulation tool (e.g., Vissim and Aimsun). Vehicle dynamics simulation tools are mainly used to simulate longitudinal, lateral, and vertical dynamics on a small scale, while traffic microsimulation tools are mainly used to simulate vehicle-following, lane-changing, and gap-acceptance behaviors on a large scale.

- **Verification scale:** Vehicle dynamics simulation tools cannot simulate many vehicles in each scenario.
- **Verification resolution:** Conventional traffic microsimulation tools cannot estimate microscopic (e.g., reduction in distance gaps and time gaps) or macroscopic (e.g., increase in road capacity) benefits associated with driving automation or cooperative driving automation with reasonable accuracy;
- **Vehicle powertrain (i.e., engine, transmission, and driveline):** Conventional traffic microsimulation tools do not simulate vehicle powertrain;
- **Maximum acceleration and maximum deceleration:** Conventional traffic microsimulation tools estimate or use constant maximum accelerations and maximum decelerations. Aimsun considers maximum acceleration of 8.2 ft/s^2 and maximum deceleration of 6.6 ft/s^2 as default [5]. Vissim estimates maximum acceleration as $a_{max}(t) \approx 3.5(1 - v(t)/40)$ and maximum deceleration as $d_{max}(t) \approx 20(1 - v(t)/800)$, where a_{max} is maximum acceleration (m/s^2), v is speed (m/s), and d_{max} is maximum deceleration (m/s^2) [5]—since all units in Vissim User Manual are metric, metric units are preferred to report these regression models with full precision. However, maximum acceleration and maximum deceleration are sensitive to vehicle model, grade, pavement conditions, and traffic conditions;

- **Longitudinal control variables:** Conventional longitudinal control functions (e.g., Adaptive Cruise Control (ACC), Cooperative Adaptive Cruise Control (CACC)) rely on constant distance gaps, time gaps, and controller coefficients, potentially sacrificing safety (i.e., when short gaps are set) or reducing road capacity (i.e., when long gaps are set). Conventional traffic microsimulation tools rely on user inputs for distance gap, time gap, and longitudinal controller coefficients (e.g., proportional, integral, and derivative) to simulate vehicles in a platoon or string. However, distance gap, time gap, and longitudinal controller coefficients are sensitive to driver characteristics, vehicle model, grade, pavement conditions, operating mode, malicious fault magnitude, and traffic conditions.
- **Contested environments:** Onboard sensor measurements and transmitted messages are inherently prone to noise, natural fault, and malicious fault. Minor faults may lead to malfunction or even failure if not responded promptly. A single cyberattack can cost an average original equipment manufacturer $1.6 billion a year, assuming one individual recall costs $800 [6]. From 2010 to 2021, 367 cyberattacks on connected vehicles have been reported [6].
  A cyberattack can exploit one user application's vulnerabilities (e.g., spoofing, data falsification, and replay attacks) or multiple user application vulnerabilities (e.g., denial-of-service attack), leading to severe consequences for vehicle and potentially its operating environment [7]. Spoofing, data falsification, replay, and denial-of-service attacks are common cyberattacks on connected vehicles [8]. Spoofing attack is when hackers steal authentication credentials or use a legitimate vehicle's identity to send unchanged or manipulated messages to other vehicles; data falsification attack is when hackers read, insert, or modify transmitted messages; replay attack is when hackers copy a message stream between two vehicles and repeat that stream to other vehicles; denial-of-service attack is when hackers prevent or interfere with target vehicles from receiving specific messages.
  Conventional fault detection methods are broadly classified into model-driven and data-driven methods [9]. Model-driven methods (e.g., unknown input observer and Kalman filter) require partial plant model; data-driven methods (e.g., neural network) require measured inputs and outputs under normal and faulty conditions to derive plant model. Model-driven methods are more computationally intensive but more accurate than data-driven methods [10].
  Conventional traffic microsimulation tools do not simulate contested environments. A simple strategy is to rely on onboard sensor measurements when there is a significant discrepancy between onboard sensor measurements and transmitted messages [11].

Our traffic microsimulation tool is superior to vehicle dynamics simulation tools and conventional traffic microsimulation tools because it can achieve these objectives

- **Verification scale:** simulate many vehicles in each scenario;
- **Verification resolution:** estimate microscopic and macroscopic benefits associated with driving automation and cooperative driving automation with reasonable accuracy [12,13];
- **Vehicle powertrain:** simulate vehicle powertrain;
- **Maximum acceleration and maximum deceleration:** estimate maximum acceleration and maximum deceleration with reasonable accuracy at each simulation time step, considering vehicle model, grade, pavement conditions, and traffic conditions;
- **Distance gap and time gap:** estimate minimum safe distance gap and minimum safe time gap with reasonable accuracy at each simulation time step for vehicles dedicated to automated driving systems or equipped with cooperative automated driving systems, considering vehicle model, grade, pavement conditions, operating mode, vehicle-to-vehicle communication vulnerabilities, and traffic conditions;

- **Longitudinal controller coefficients:** estimate longitudinal controller coefficients (i.e., proportional, integral, and derivative gains) with reasonable accuracy at each simulation time step for vehicles dedicated to automated driving systems, considering vehicle model, grade, pavement conditions, and traffic conditions;
- **Contested environments:** employ a reduced-order Kalman filter unknown input observer to estimate distance gap, speed, and acceleration with reasonable accuracy at each simulation time step for vehicles dedicated to automated driving systems or equipped with cooperative automated driving systems under noise (e.g., measurement noise and process noise) and unknown inputs (e.g., noise with unknown statistics, natural fault, and malicious fault).

2. Literature Review

Longitudinal control variables are mostly treated as constant parameters (see Tables 1 and 2) or variables estimated using empirical or simplified mechanistic models. However, maximum acceleration, maximum deceleration, minimum safe distance gap, and minimum safe time gap are sensitive to driver characteristics, vehicle model, grade, pavement conditions, operating mode, and traffic conditions.

Akçelik and Besley (2001) empirically estimated maximum acceleration and maximum deceleration based on initial speed and final speed for passenger cars, and based on initial speed, final speed, power-to-weight ratio, and grade for trucks [14]. Ahn et al. (2002) generated a lookup table to identify maximum acceleration over 17 driving schedules (see Table 3) [15]. Fang and Elefteriadou (2005) recommended a maximum acceleration and a maximum deceleration for each vehicle classification (i.e., passenger car and truck), interchange configuration (i.e., Single-Point Urban Interchange (SPUI) and diamond), and traffic microsimulation tool (i.e., Vissim, Aimsun, and CORSIM) (see Table 4) [16]. Kuriyama et al. (2010) considered aerodynamic resistance, rolling resistance, and grade resistance in calculating acceleration and deceleration for electric vehicles [17]. Maurya and Bokare (2012) generated a lookup table to identify maximum deceleration for each vehicle classification at each speed range (see Table 5) [18]. Lee et al. (2013) considered a higher maximum acceleration (13.1 ft/s^2 vs. 10.0 ft/s^2) and a lower maximum deceleration (9.8 ft/s^2 vs. 15.0 ft/s^2) for connected vehicles than the Federal Highway Administration's recommended maximum acceleration and maximum deceleration [19]. Anya et al. (2014) believed vehicle-following, lane-changing, travel time, and queue discharge had an impact on maximum acceleration and maximum deceleration [20]. Song et al. (2015) empirically estimated maximum acceleration based on speed [21]. Bokare and Maurya (2017) generated two lookup tables to identify maximum acceleration and maximum deceleration for each vehicle classification (i.e., diesel car, petrol car, and truck) at each speed range (see Table 6) [22]. Ramezani et al. (2018) generated a lookup table to identify maximum acceleration for trucks in CACC mode at each speed range (see Table 7) [23].

Shladover et al. (2010) identified that drivers maintain 2.2 s, 1.6 s, and 1.1 s time gaps for 31.1%, 18.5%, and 50.4% of their vehicle-following time in ACC mode, respectively, and drivers maintain 0.6 s, 0.7 s, 0.9 s, and 1.1 s time gaps for 57%, 24%, 7%, and 12% of their vehicle-following time in CACC mode, respectively [24]. Willigen et al. (2011) recommended a distance headway and a time headway for each platoon size (i.e., 20 and 30) and operating mode (i.e., ACC, CACC with transmitted accelerations, and CACC with estimated accelerations) (see Table 8) [25]. Horiguchi and Oguchi (2014) calculated distance gap for vehicles in CACC mode based on minimum safe distance gap, follower's speed, leader's speed, maximum acceleration, and maximum deceleration [26]. Flores et al. (2017) calculated time gap based on minimum safe time gap, desired time gap, and speed, and calculated distance gap based on actuator delay, speed, maximum deceleration, and maximum jerk [27]. Askari et al. (2017) calculated distance gap based on minimum safe distance gap, follower's speed, reaction time, leader's speed, maximum acceleration, and maximum deceleration [28]. Flores and Milanés (2018) recommended a time gap for each controller type (i.e., fractional-order proportional derivative and integer-order proportional derivative), desired performance (i.e., ensuring loop bandwidth, phase margin, and string stability), and operating mode (i.e., ACC and CACC) (see Table 9) [29]. Chen et al. (2019) calculated time gap for vehicles in ACC and CACC modes based on jam density, free-flow speed, follower's speed, follower's acceleration, and leader's acceleration [30]. Bian et al. (2019) recommended a time headway for each platoon size (i.e., 1, 3, 10, 20, and 30) and controller type (i.e., linear, nonlinear, and nonlinear subject to communication delay) (see Table 10) [31].

Conventional traffic microsimulation tools (1) should be integrated with a vehicle dynamics simulation tool to simulate vehicle powertrain [32], (2) employ kinematics to estimate quantities associated with motion [33], (3) automatically confine accelerations and decelerations to constant (e.g., Aimsun and MITSIM) or estimated (e.g., Vissim and INTEGRATION) maximum accelerations and maximum decelerations, and (4) rely on constant distance gaps and time gaps to simulate longitudinal control for automated vehicles in a platoon or string. This research proposes a traffic microsimulation tool that can estimate maximum acceleration, maximum deceleration, minimum safe distance gap, and minimum safe time gap with reasonable accuracy at each simulation time step for convectional vehicles, vehicles dedicated to automated driving systems, and vehicle equipped with cooperative automated driving systems, considering driver characteristics (see Section 3.1), vehicle model (see Section 3.2), pavement conditions (see Section 3.2.3), grade (see Section 3.2.3), operating mode (see Section 3.5), traffic conditions (see Section 3.2.3), and vehicle-to-vehicle communication vulnerabilities (see Section 4).

Table 1. Constant distance gaps and time gaps used in literature.

Author	Distance Gap (ft)	Time Gap (s) ACC	CACC
Bu et al. (2010) [34]	-	[1.1–2.2]	[0.6–1.1]
Naus et al. (2010) [35]	-	2.6 *	0.8 *
Shladover et al. (2010); Liu (2018) [24,36]	-	1.1, 1.6, 2.2	0.7, 0.9, 1.1
Ploeg et al. (2011) [37]	-	-	0.7 *
Willigen et al. (2011) [25]	15.8 *, 25.2 *, 26.0 *, 34.5 *, 39.9 *, 57.2 *	0.5 *, 0.6 *	0.2 *, 0.3 *, 0.4 *
Shladover et al. (2012) [38]	-	-	0.5
Zhao and Sun (2013) [39]	-	1.4	0.5
Horiguchi and Oguchi (2014) [26]	-	2.0, 2.0 *	-
Segata et al. (2014) [40]	16.4	0.3 *, 1.2 *	-
Milanés and Shladover (2014) [41]	-	1.1	0.6
Nikolos et al. (2015); Delis et al. (2016) [42,43]	-	1.2	1.0
Wang et al. (2017) [44]	42.7, 46.9, 68.2	-	0.4, 0.5, 0.6 *, 0.7, 0.9 *
Terruzzi et al. (2017) [45]	16.4 *	1.4 *	1.0 *
Zhou et al. (2017) [46]	16.4	0.5, 1.0, 1.5, 2.0	0.5, 1.0, 1.5, 2.0
Askari et al. (2017) [28]	9.8, 13.1	1.1	0.8
Flores and Milanés (2018) [29]	-	0.5, 0.6	0.3
Chen et al. (2019) [30]	-	[0.2–2.2]	-
Bian et al. (2019) [31]	-	-	0.0 *, 0.1 *, 0.2 *, 0.3 *, 0.4 *, 0.6 *, 0.9 *, 1.0 *, 1.1 *

* headway.

Table 2. Constant maximum accelerations and maximum decelerations used in literature.

Author	Max. Acceleration (ft/s^2)	Max. Deceleration (ft/s^2)
Akçelik and Besley (2001) [14]	8.8	10.1
Lemessi (2001) [47]	8.2	8.2
Ahn et al. (2002) [15]	3.4, 3.9, 4.8, 5.0, 5.4, 5.5, 7.3, 7.4, 7.7, 8.3, 8.5, 8.7, 9.3, 10.1	-
Rakha and Ding (2003) [48]	4.9	8.2
Wang and Liu (2005) [49]	8.2	11.5
Fang and Elefteriadou (2005) [16]	4.9, 6.9, 8.2, 9.2, 11.5, 15.1	9.8, 12.1, 15.1
Ossen et al. (2006) [50]	26.2	26.2
Kesting et al. (2007) [51]	4.9	13.1
Kesting and Treiber (2008) [52]	19.7	19.7
Kuriyama et al. (2010) [17]	8.8	9.8
Talebpour et al. (2011) [53]	13.1	26.2
Song et al. (2012) [54]	11.5	13.1
Shladover et al. (2012) [38]	6.6	6.6
Lee and Park (2012); Lee et al. (2013) [19,55]	13.1	9.8
Maurya and Bokare (2012) [18]	-	2.4, 2.5, 2.9, 5.0, 5.1, 5.3
Treiber and Kesting (2013) [56]	1.7, 4.6, 4.8	2.1, 4.8
Anya et al. (2014) [20]	0.7, 1.5, 8.5, 9.8, 11.2, 19.1, 22.0, 25.0	1.6, 3.7, 16.4, 19.7, 23.0, 36.7, 44.0, 51.5
Li et al. (2014) [57]	4.5	11.0
Tang et al. (2014) [58]	-	19.7
Desiraju et al. (2014); Liu et al. (2018) [59,60]	6.6	-
Horiguchi and Oguchi (2014) [26]	5.2	-
Song et al. (2015) [21]	8.8	-
Amoozadeh et al. (2015); Zhou et al. (2017) [46,61]	9.8	16.4
Bokare and Maurya (2017) [22]	2.5, 2.9, 3.1, 3.3, 6.2, 6.5, 7.3, 8.1, 9.4	2.4, 2.5, 2.9, 11.0, 13.0, 14.1, 14.2, 14.8, 16.4
Askari et al. (2017) [28]	2.6, 4.9, 8.2	6.6
Li et al. (2017) [62]	3.3	-
Ramezani et al. (2018) [23]	0.4, 0.5, 0.8, 1.3, 1.6, 1.8, 8.2	9.8

Table 3. Maximum acceleration vs. driving schedule.

Driving Schedule	Max. Acceleration (ft/s^2)
Freeway, High Speed	3.9
Freeway, LOS * A–C	5.0
Freeway, LOS * D	3.4
Freeway, LOS * E	7.7
Freeway, LOS * F & LA92	10.1
Freeway, LOS * G	5.5
Freeway Ramps & Arterial/Collectors LOS * C–D	8.3
Arterial/Collectors LOS * A–B	7.3
Arterial/Collectors LOS * E–F	8.5
Local Roadways	5.4
Non-Freeway Area-Wide Urban Travel	9.3
LA4 & Running 505	4.8
ST01	7.4
New York City Cycle	8.7

* Level of Service.

Table 4. Maximum acceleration and maximum deceleration vs. traffic microsimulation tool, vehicle classification, and interchange configuration.

Specification	Max. Acceleration (ft/s^2)	Max. Deceleration (ft/s^2)
Passenger Car, SPUI, Vissim	11.5	-
Passenger Car, Diamond, Vissim & Aimsun	6.9	-
Truck, SPUI, Vissim	8.2	-
Truck, Diamond, Vissim	4.9	-
Passenger Car, SPUI, Aimsun	9.2	-
SPUI, CORSIM	15.1	9.8
Diamond, CORSIM	6.9	15.1

Table 5. Maximum deceleration vs. vehicle classification.

Vehicle Classification	Speed Range (ft/s)	Max. Deceleration (ft/s^2)
Passenger Car	[83.8–85.7)	5.0
Passenger Car	[85.7–87.5)	5.1
Passenger Car	[87.5–91.1]	5.3
Truck	[18.2–27.3)	2.4
Truck	[27.3–36.5)	2.5
Truck	[36.5–54.7]	2.9

Table 6. Maximum acceleration and maximum deceleration vs. vehicle classification.

Vehicle Classification	Speed Range (ft/s)	Max. Acceleration (ft/s^2)	Max. Deceleration (ft/s^2)
Diesel Car	[62.0–69.3)	6.2	-
Diesel Car	[69.3–76.6)	7.3	-
Diesel Car	[76.6–83.8)	6.5	-
Diesel Car	[83.8–85.7)	-	14.1
Diesel Car	[85.7–87.5)	-	14.2
Diesel Car	[87.5–89.3)	-	16.4
Diesel Car	[89.3–91.1]	-	14.8
Petrol Car	[55.6–65.6)	-	11.0
Petrol Car	[65.6–75.6)	-	13.0
Petrol Car	[72.9–76.6)	7.3	-
Petrol Car	[75.6–82.9]	-	14.2
Petrol Car	[76.6–80.2)	8.1	-
Petrol Car	[80.2–83.8]	9.4	-
Truck	[18.2–27.3)	2.5	2.4
Truck	[27.3–36.5)	3.3	2.5
Truck	[36.5–45.6)	3.1	2.9
Truck	[45.6–54.7]	2.9	2.9

Table 7. Maximum acceleration.

Speed Range (ft/s)	Max. Acceleration (ft/s^2)
[0–14.7)	1.8
[29.3–44.0)	1.3
[44.0–58.7)	0.8
[58.7–73.3)	0.5
Above 73.3	0.4

Table 8. Distance headway and time headway vs. platoon size and operating mode.

Specification	Distance Headway (ft)	Time Headway (s)
20, ACC	34.5	0.5
20, CACC with Transmitted Accelerations	26.0	0.4
20, CACC with Estimated Accelerations	15.8	0.2
30, ACC	57.2	0.6
30, CACC with Transmitted Accelerations	39.9	0.4
30, CACC with Estimated Accelerations	25.2	0.3

Table 9. Time gap vs. controller type, desired performance, and operating mode.

Specification	Time Gap (s)
Fractional-Order Proportional Derivative, ACC	0.5
Integer-Order Proportional Derivative, Loop Bandwidth and Phase Margin, ACC	0.6
Integer-Order Proportional Derivative, Loop Bandwidth and String Stability, ACC	0.5
CACC	0.3

Table 10. Time headway vs. platoon size and controller type.

Specification	Time Headway (s)
1, Linear	0.3, 0.4, 0.6
3, Linear	0.1, 0.2
1, Nonlinear	0.6
3, Nonlinear	0.2
10, Nonlinear	0.1
20 & 30, Nonlinear	0.0
1, Nonlinear Subject to Communication Delay	0.9, 1.0, 1.1
3, Nonlinear Subject to Communication Delay	0.6, 0.7, 0.9, 1.1

3. Proposed Traffic Microsimulation Tool

Our traffic microsimulation tool enables users to customize driver (see Section 3.1), vehicle (see Section 3.2), road (see Section 3.3), cyberattack (see Section 3.4), and operating mode (see Section 3.5) modules separately. Our traffic microsimulation tool contains ten driver types (conservative to aggressive), fourteen vehicle models (i.e., ten passenger car models and four truck configurations), two driving schedules (i.e., US06 and Cycle D), three malicious fault magnitudes (i.e., malicious increases of 1, 3, and 5 ft/s^2 in transmitted accelerations), and three operating modes (i.e., cooperative automated, automated, and manual) as default to simulate many vehicles with reasonable accuracy at each simulation time step under noise and unknown inputs. Vehicles in manual mode require driver, vehicle, and road modules; vehicles in automated mode require vehicle and road modules; vehicles in cooperative automated mode require vehicle, road, and cyberattack modules to be implemented (see Figure 1).

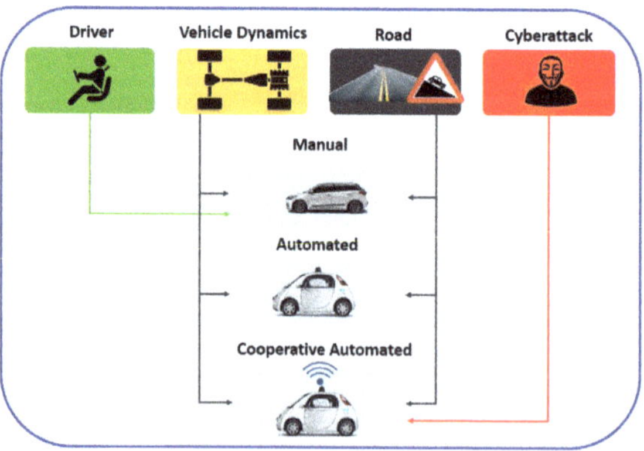

Figure 1. Proposed traffic microsimulation tool.

3.1. Driver Module

Ten driver types are considered as default (based on an assumed value in CORSIM—a traffic microsimulation tool): type 1 is a conservative driver; type 10 is an aggressive driver. Each driver type is associated with a speed multiplier, an acceleration multiplier, a deceleration multiplier, and a percentage included in traffic which follows a normal distribution as default [63].

3.2. Vehicle Module

Fourteen vehicle models are included as default (assumed vehicles in SwashSim—a traffic microsimulation tool): 2006 Honda Civic Si, 2008 Chevy Impala, 1998 Buick Century, 2004 Chevy Tahoe, 2002 Chevy Silverado, 1998 Chevy S10 Blazer, 2011 Ford F150, 2009 Honda Civic, 2005 Mazda 6, and 2004 Pontiac Grand Am, single-unit truck with PACCAR PX-7 engine, intermediate semi-trailer with PACCAR MX-13 engine, interstate semi-trailer with PACCAR MX-13 engine, and double semi-trailer with PACCAR MX-13 engine. Each vehicle is associated with a torque map, a drag coefficient, a width, a height, a weight, a wheelbase length, a wheel radius, a differential gear ratio, a drive axle slippage, a drivetrain efficiency, a transmission gear ratio, shift up speeds, shift down speeds, and a percentage included in traffic which follows a normal distribution as default [63]. Vehicle module contains vehicle generation, reference speed profiles, and vehicle dynamics submodules: Vehicle dispatching model in Section 3.2.1 is intended to generate many vehicles at an assumed entrance under steady-state conditions; platoon leaders are assumed to follow US06 and Cycle D driving schedules (see Section 3.2.2); maximum acceleration, maximum deceleration, minimum safe distance gap, and minimum safe time gap are estimated based on vehicle dynamics (see Section 3.2.3).

3.2.1. Vehicle Generation

Entry headways follow shifted negative-exponential distribution

$$f(h) = \begin{cases} \lambda e^{-\lambda(h-h_{min})}, & h \geq h_{min} \\ 0, & h < h_{min} \end{cases} \quad (1)$$

where f is probability density function, h is entry headway (s/veh), h_{min} is minimum entry headway (s/veh), λ is distribution parameter (veh/s) calculated as $1/(\bar{h} - h_{min})$, $\bar{h}$ is average entry headway (s/veh) calculated as $3600/q$, and q is flow rate (veh/h).

3.2.2. Reference Speed Profiles

US06 and Cycle D driving schedules are used as reference speed profiles. US06 driving schedule is designed to test passenger cars, representing an 8-mile route with average speed of 70.4 ft/s, maximum speed of 117.8 ft/s, maximum acceleration of 12.3 ft/s^2, maximum deceleration of 10.1 ft/s^2, and 600 s duration. Cycle D driving schedule is designed to test trucks, representing a 5.6-mile route with average speed of 27.6 ft/s, maximum speed of 85.1 ft/s, maximum acceleration of 6.4 ft/s^2, maximum deceleration of 6.8 ft/s^2, and 1060 s duration.

3.2.3. Vehicle Dynamics

Conventional longitudinal control functions control accelerations and decelerations using throttle and brake inputs to maintain a constant distance gap in a platoon (e.g., truck platooning) or a constant time gap in a string (e.g., ACC and CACC). Commanded accelerations and decelerations are automatically confined to maximum accelerations and maximum decelerations specific to vehicle model, grade, pavement conditions, and traffic conditions. Longitudinal controller coefficients can be tuned to achieve desired performance. Conventional traffic microsimulation tools require user inputs for maximum acceleration, maximum deceleration, distance gap, time gap, and longitudinal controller coefficients to simulate vehicles in a platoon or string.

Our traffic microsimulation tool follows these steps at each simulation time step to simulate vehicles in a platoon or string: (1) estimating maximum acceleration and maximum deceleration for each vehicle, considering vehicle model, grade, pavement conditions, and traffic conditions, (2) estimating minimum safe distance gap and minimum safe time gap for each vehicle dedicated to automated driving systems or equipped with cooperative automated driving systems, considering vehicle model, grade, pavement conditions, operating mode, vehicle-to-vehicle communication vulnerabilities, and traffic conditions, (3) checking preset distance gaps and preset time gaps with minimum safe distance gaps and minimum safe time gaps, (4) estimating accelerations and decelerations, considering operating mode, and (5) confining accelerations and decelerations to maximum accelerations and maximum decelerations.

Three significant forces against vehicle motion are aerodynamic resistance, rolling resistance, and grade resistance. Aerodynamic resistance can be calculated as

$$R_a[k] \triangleq \frac{\rho}{2} C_D A_f v^2[k], \tag{2}$$

where R_a is aerodynamic resistance (lb), ρ is air density (slugs/ft^3), C_D is drag coefficient (unitless), A_f is vehicle frontal area (ft^2) calculated as vehicle width (ft) $\times$ vehicle height (ft), v is speed (ft/s), and $[k]$ denotes simulation time step. Rolling resistance can be estimated as

$$R_{rl}[k] \approx f_{rl}[k] W, \tag{3}$$

where R_{rl} is rolling resistance (lb), f_{rl} is rolling resistance coefficient (unitless) estimated as $0.01(1 + v[k]/147)$ for vehicles operating on paved surfaces [64], and W is vehicle weight (lb). Grade resistance can be calculated as

$$R_g \triangleq W \sin\theta, \tag{4}$$

where R_g is grade resistance (lb), and θ is grade (unitless). Tractive effort available to overcome resistance and to provide acceleration can be calculated as $F[k] = min(F_{max}[k], F_e[k])$, where F is available tractive effort (lb), F_{max} is maximum tractive effort (lb), and F_e is engine-generated tractive effort (lb). Maximum tractive effort can be calculated as

$$F_{max}[k] \triangleq \begin{cases} \mu W \dfrac{l_r \cos\theta + h f_{rl}[k]}{L + \mu h}, & \text{front-wheel-drive} \\ \mu W \dfrac{l_f \cos\theta - h f_{rl}[k]}{L - \mu h}, & \text{rear-wheel-drive} \\ \mu W \cos\theta, & \text{all-wheel-drive} \end{cases} \tag{5}$$

where μ is road adhesion coefficient (unitless), l_r is distance from rear axle to gravity center (ft), h is vehicle height (ft), L is wheelbase length (ft), and l_f is distance from front axle to gravity center (ft). Engine speed can be calculated as

$$n_e[k] \triangleq \frac{v[k]\epsilon_0[k]}{2\pi r(1-i)}, \tag{6}$$

where n_e is engine speed (revs/s), ϵ_0 is overall gear reduction ratio (unitless), calculated as transmission gear ratio (unitless), selected based on vehicle speed) $\times$ differential gear ratio (unitless), r is wheel radius (ft), and i is drive axle slippage (unitless). Engine power can be calculated as

$$hp_e[k] \triangleq \frac{2\pi M_e[k] n_e[k]}{550}, \tag{7}$$

where hp_e is engine power (hp), and M_e is torque (ft-lb). Engine-generated tractive effort can be calculated as

$$F_e[k] \triangleq \frac{M_e[k]\epsilon_0[k]\eta_d}{r}, \tag{8}$$

where η_d is drivetrain efficiency (unitless). Maximum braking force can be calculated as

$$B_{max}[k] \triangleq \begin{cases} \eta_b \mu W \dfrac{l_r \cos\theta + h f_{rl}[k]}{L - \eta_b \mu h}, & \text{front-wheel-drive} \\ \eta_b \mu W \dfrac{l_f \cos\theta - h f_{rl}[k]}{L + \eta_b \mu h}, & \text{rear-wheel-drive} \\ \eta_b \mu W \cos\theta, & \text{all-wheel-drive} \end{cases} \tag{9}$$

where B_{max} is maximum braking force (lb), and η_b is braking efficiency (unitless). Maximum acceleration can be estimated as

$$a_{max}[k] \approx \frac{F[k] - R_a[k] - R_{rl}[k] - R_g}{m\gamma_m[k]}, \tag{10}$$

where a_{max} is maximum acceleration (ft/s^2), and γ_m is mass factor (untiless) estimated as $1.04 + 0.0025\epsilon_0^2[k]$ [64], accounting for rotational inertia during acceleration. Maximum deceleration can be estimated as [63]

$$d_{max}[k] \approx \frac{B_{max}[k] + R_a[k] + R_{rl}[k] + R_g}{m\gamma_b}, \tag{11}$$

where d_{max} is maximum deceleration (ft/s^2), and γ_b is mass factor (unitless), accounting for rotational inertia during deceleration. Minimum safe distance gap can be estimated as [63]

$$S_{min}[k] \approx \left(\tau_s^{i+1} + \tau_c^{i+1}\right) v_{i+1}[k] + S_{stop}^{i+1}[k] - S_{stop}^{i}[k], \tag{12}$$

where S_{min} is minimum safe distance gap (ft), τ_s is sensing delay (s), τ_c is communication delay (s), subscript/superscript $i+1$ denotes follower, subscript/superscript i denotes leader, and S_{stop} is minimum stopping distance (ft) estimated as

$$S_{stop}[k] \approx \frac{m\gamma_b}{\rho C_D A_f} \ln\left(1 - \frac{R_a[k]}{B_{max}[k] + R_a[k] + R_{rl}[k] + R_g}\right). \tag{13}$$

Minimum safe time gap can be estimated as [63]

$$T_{min}[k] \approx \tau_s^{i+1} + \tau_c^{i+1} + \tau_{lag}^{i+1}[k] - \tau_{lag}^{i}[k], \tag{14}$$

where T_{min} is minimum safe time gap (s), and τ_{lag} is lag in tracking desired deceleration (s) estimated as $v[k]/d_{max}[k]$.

Assumption 1. *Vehicles have constant speeds during sensing delay and communication delay.*

Remark 1. *Proposed longitudinal dynamics has been previously validated for 53,000 lb and 80,000 lb interstate semi-trailers against an industry-standard simulation tool (i.e., TruckSim) [65–67].*

3.3. Road Module

Any desired freeway segment with a single lane can be simulated. Each freeway segment is associated with a grade, a road adhesion coefficient, and a free-flow speed.

3.4. Cyberattack Module

Three malicious fault magnitudes are assumed as default: 1, 3, and 5 ft/s^2 malicious increase in transmitted accelerations. Each malicious fault magnitude is associated with a percentage injected on traffic which follows a normal distribution as default.

3.5. Operating Mode Module

Three operating modes are considered as default: manual, automated, and cooperative automated. Each operating mode is associated with a percentage included in traffic which follows a normal distribution as default. This section proposes a vehicle-following model for vehicles in manual mode and longitudinal control functions for vehicles in automated and cooperative automated modes.

3.5.1. Manual Mode

Levels 1 and 2 automated vehicles are assumed to have a vehicle-following model similar to the Improved Intelligent Driver Model (IIDM)

$$a_{i+1}[k] \triangleq \begin{cases} n \times a_{max}^{i+1}[k] C_s[k] C_v[k], & S[k] \geq S_{min}[k] \\ -q \times d_{max}^{i+1}[k], & \text{else} \end{cases} \tag{15}$$

where n is acceleration multiplier (unitless), C_s is distance gap coefficient (unitless) calculated as $1-(S_{min}[k]/S[k])^{\alpha}$, S is distance gap (ft) calculated as $x_i[k]-x_{i+1}[k]-L_i$, x is front bumper position (ft), C_v is speed coefficient (unitless) calculated as $1-(m \times v_i[k]/FFS)^{\beta}$, m is speed multiplier (unitless), FFS is free-flow speed (ft/s), q is deceleration multiplier (unitless), and α and β are calibration parameters (unitless). IIDM has fewer calibration parameters and demonstrates a more stable performance than Wiedemann model (i.e., vehicle-following model used in Vissim) [68].

Assumption 2. *There are three significant components underpinning a traffic microsimulation tool (i.e., vehicle-following, lane-changing, and gap-acceptance models). This research mainly focuses on vehicle-following models, assuming vehicles drive in a single lane, and there is no lane-change maneuver (i.e., lane-changing and gap-acceptance models are not required). However, a lane-change maneuver can temporarily affect vehicle-following behaviors (e.g., drivers speed up or slow down to align with acceptable gaps in target lanes; drivers temporarily adopt shorter gaps after a lane-change maneuver; drivers temporarily adopt shorter gaps after a vehicle merges in front).*

3.5.2. Automated Mode

When (1) a vehicle dedicated to automated driving systems approaches a vehicle, or (2) a vehicle equipped with cooperative automated driving systems approaches a vehicle not equipped with cooperative automated driving systems, a longitudinal control function similar to ACC is activated [69]

$$a_{i+1}[k] \triangleq max\Big(min\Big(K_{p,a}[k]e_x[k] + K_{d,a}[k]e_v[k], a_{max}^{i+1}[k]\Big), -d_{max}^{i+1}[k]\Big), \tag{16}$$

where $K_{p,a}$ is proportional gain in automated mode (s^{-2}), e_x is distance gap error (ft) calculated as $S_{des}[k]-S[k]$, S_{des} is desired distance gap (ft) calculated as $max(T_{set}, T_{min}[k-1])v_{i+1}[k-1]$, T_{set} is preset time gap (s), $K_{d,a}$ is derivative gain in automated mode (s^{-1}), and e_v is speed error (ft/s) calculated as $v_i[k]-v_{i+1}[k]$. When no leader is detected, a longitudinal control function similar to cruise control is activated

$$a_{i+1}[k] \triangleq max\Big(min\Big(K_{p,cr}[k](FFS - v_{i+1}[k]), a_{max}^{i+1}[k]\Big), -d_{max}^{i+1}[k]\Big), \tag{17}$$

where $K_{p,cr}$ is proportional gain in cruise mode (s^{-1}). $K_{p,a}$ and $K_{d,a}$ should satisfy (18) to maximize road capacity without compromising safety [63]

$$\begin{cases} \left[\dfrac{2\pi\tau_{lag}^{i+1}}{\sqrt{4K_{p,a}(t)\tau_{lag}^{i+1} - (K_{d,a}(t)+1)^2}} - \dfrac{\pi}{2} \cdot \sqrt{\dfrac{\tau_{lag}^{i+1}}{K_{p,a}(t)}} \right] \times \\ \left[e^{-\dfrac{\pi(K_{d,a}(t)+1)}{\sqrt{4K_{p,a}(t)\tau_{lag}^{i+1} - (K_{d,a}(t)+1)^2}}} - v_i(t) \right] \leq S_{min}(t), & \text{during acceleration} \\ K_{d,a}(t) \leq \dfrac{T_{min}(t)}{8\tau_{lag}^{i+1}} - 1. & \text{during deceleration} \end{cases} \quad (18)$$

3.5.3. Cooperative Automated Mode

When a vehicle equipped with cooperative automated driving systems approaches another vehicle equipped with cooperative automated driving systems, a longitudinal control function similar to CACC is activated [69]

$$a_{i+1}[k] \triangleq max\Big(min\Big(K_{p,c}[k]e_v[k] + K_{i,c}[k]e_x[k] + K_{d,c}[k]a_i[k], a_{max}^{i+1}[k]\Big), -d_{max}^{i+1}[k]\Big), \quad (19)$$

where $K_{p,c}$ is proportional gain in cooperative automated mode (s^{-1}), $K_{i,c}$ is integral gain in cooperative automated mode (s^{-2}), and $K_{d,c}$ is derivative gain in cooperative automated mode (unitless).

Remark 2. *All driver characteristics and vehicle powertrain information used in this research are derived from https://www.automobile-catalog.com/ [63,70,71].*

Assumption 3. *Class B cooperative driving automation cooperation is utilized.*

Assumption 4. *x_i and v_i are prone to measurement noise and process noise, and a_i is prone to measurement noise, process noise, natural fault, and malicious fault (i.e., x_i and v_i are known state subvectors, and a_i is an unknown state subvector).*

4. State and Unknown Input Estimation

Consider a state-space model in which unknown inputs can be modeled as an additive term

$$x[k+1] = Ax[k] + Bu[k] + Dd[k] + \xi[k], \quad (20)$$

$$z[k] = Cx[k] + \theta[k], \quad (21)$$

where $x \in \mathbb{R}^n$ is state vector, $A \in \mathbb{R}^{n\times n}$ is state matrix, $B \in \mathbb{R}^{n\times p}$ is input matrix, $u \in \mathbb{R}^p$ is input vector, $D \in \mathbb{R}^{n\times q}$ is unknown input matrix, $d \in \mathbb{R}^q$ is unknown input vector, $\xi \in \mathbb{R}^n$ is process noise, $z \in \mathbb{R}^m$ is measurement vector, $C \in \mathbb{R}^{m\times n}$ is measurement matrix, and $\theta \in \mathbb{R}^m$ is measurement noise.

Assumption 5. *D is full column rank.*

Assumption 6. *rank CD = rank D.*

Assumption 7. *ξ is white noise: $\mathbb{E}(\xi[k]) = 0_n$, $\mathbb{E}(\xi[k]\xi^T[k]) \triangleq \Xi[k]$, $\Xi \in \mathbb{R}^{n\times n}$ is process noise covariance matrix, $\mathbb{E}(\xi[k]\xi^T[j]) = 0_{n\times n}$ $\forall k, j \geq 0, k \neq j$.*

Assumption 8. *θ is white noise: $\mathbb{E}(\theta[k]) = 0_m$, $\mathbb{E}(\theta[k]\theta^T[k]) \triangleq \Theta[k]$, $\Theta \in \mathbb{R}^{m\times m}$ is measurement noise covariance matrix, and $\mathbb{E}(\theta[k]\theta^T[j]) = 0_{m\times m}$ $\forall k, j \geq 0, k \neq j$.*

Assumption 9. $\mathbb{E}(\xi[k]x^T[0]) = 0_{n\times n}$*, and* $\mathbb{E}(\theta[k]x^T[0]) = 0_{m\times n}$ $\forall k \geq 0$.

Assumption 10. *$\hat{x}[0] = \mathbb{E}(x[0])$ is known, where $\hat{x} \in \mathbb{R}^n$ is state estimation.*

Remark 3. *D is full column rank $\rightarrow [N\ D]^{-1}$ exists, where $N \in \mathbb{R}^{n\times(n-q)}$ [72].*

Let us define $\overline{x}[k] \triangleq [N\ D]^{-1}x[k], \overline{A} \triangleq [N\ D]^{-1}A[N\ D], \overline{B} \triangleq [N\ D]^{-1}B, \overline{D} \triangleq [N\ D]^{-1}D$, and $\overline{C} \triangleq C[N\ D]$, where $\overline{x} \in \mathbb{R}^n$, $\overline{A} \in \mathbb{R}^{n\times n}$, $\overline{B} \in \mathbb{R}^{n\times p}$, $\overline{D} \in \mathbb{R}^{n\times q}$, and $\overline{C} \in \mathbb{R}^{m\times n}$.

Remark 4. *$rank\ CD = rank\ D \rightarrow U \triangleq [CD\ Q]^{-1}$ exists, where $Q \in \mathbb{R}^{m\times(m-q)}$ [72].*

Let us define $\overline{x}[k] \triangleq \left[\overline{x}_1^T\ \overline{x}_2^T\right]^T$, $\overline{A} \triangleq \begin{bmatrix}\overline{A}_{11} & \overline{A}_{12}\\ \overline{A}_{21} & \overline{A}_{22}\end{bmatrix}$, $\overline{B} \triangleq \left[\overline{B}_1^T\ \overline{B}_2^T\right]^T$, $U^{-1} \triangleq \left[U_1^T\ U_2^T\right]^T$, where $\overline{x}_1 \in \mathbb{R}^{n-q}$ is known state subvector, $\overline{x}_2 \in \mathbb{R}^q$ unknown state subvector, $\overline{A}_{11} \in \mathbb{R}^{(n-q)\times(n-q)}$, $\overline{A}_{12} \in \mathbb{R}^{(n-q)\times q}$, $\overline{A}_{21} \in \mathbb{R}^{q\times(n-q)}$, $\overline{A}_{22} \in \mathbb{R}^{q\times q}$, $\overline{B}_1 \in \mathbb{R}^{(n-q)\times p}$, $\overline{B}_2 \in \mathbb{R}^{q\times p}$, $U_1 \in \mathbb{R}^{q\times m}$, and $U_2 \in \mathbb{R}^{(m-q)\times m}$.

Assumption 11. *$rank \begin{bmatrix} sI_{n-q} - \overline{A}_{11} & -\overline{A}_{12} \\ CN & CD \end{bmatrix} = n\ \forall s \in \mathbb{C}, Re(s) \geq 0$.*

Let us define $\widetilde{A}_1 \triangleq \overline{A}_{11} - \overline{A}_{12}U_1CN$, $E_1 \triangleq \overline{A}_{12}U_1$, $\widetilde{C}_1 \triangleq U_2CN$, and $\overline{z}[k] \triangleq U_2z[k]$, where $\widetilde{A}_1 \in \mathbb{R}^{(n-q)\times(n-q)}$, $E_1 \in \mathbb{R}^{(n-q)\times m}$, $\widetilde{C}_1 \in \mathbb{R}^{(m-q)\times(n-q)}$, and $\overline{z} \in \mathbb{R}^{m-q}$.

Remark 5. *D is full column rank, $rank\ CD = rank\ D$, and $rank \begin{bmatrix} sI_{n-q} - \overline{A}_{11} & -\overline{A}_{12} \\ CN & CD \end{bmatrix} = n$ $\forall s \in \mathbb{C}, Re(s) \geq 0 \rightarrow \{\widetilde{A}_1, \widetilde{C}_1\}$ is observable [72].*

State vector can be decoupled into known and unknown state subvectors. Known state subvector can be estimated as

$$\hat{\bar{x}}_1[k+1] \triangleq (\tilde{A}_1 - L[k]\tilde{C}_1)\hat{\bar{x}}_1[k] + \bar{B}_1u[k] + L^*[k]z[k], \tag{22}$$

where $\hat{\bar{x}}_1 \in \mathbb{R}^{n-q}$ is known state estimator ($\hat{\bar{x}}_1[k] \rightarrow \bar{x}_1[k]$ as $k \rightarrow \infty$), $L^*[k] \triangleq L[k]U_2 + E_1$, $L^* \in \mathbb{R}^{(n-q)\times m}$, and $L \in \mathbb{R}^{(n-q)\times(m-q)}$ is Kalman gain, calculated as

$$L[k] \triangleq \tilde{A}_1\Sigma[k]\tilde{C}_1^T\left(\tilde{C}_1\Sigma[k]\tilde{C}_1^T + \Theta[k]\right)^{-1}, \tag{23}$$

where $\Sigma \in \mathbb{R}^{(n-q)\times(n-q)}$ can be recursively calculated as

$$\Sigma[k+1] = \tilde{A}_1[\Sigma[k] - \Sigma[k]\tilde{C}_1^T(\tilde{C}_1\Sigma[k]\tilde{C}_1^T + \Theta[k])^{-1}\tilde{C}_1\Sigma[k]]\tilde{A}_1^T + D\Xi[k]D^T. \tag{24}$$

Unknown state subvector can be estimated as

$$\hat{\bar{x}}_2[k] \triangleq U_1z[k] - U_1CN\hat{\bar{x}}_1[k], \tag{25}$$

where $\hat{\bar{x}}_2 \in \mathbb{R}^q$ is unknown state estimator ($\hat{\bar{x}}_2[k] \rightarrow \bar{x}_2[k]$ as $k \rightarrow \infty$). Unknown input vector can be estimated as

$$\hat{d}[k] \triangleq U_1z[k+1] + G_{d,1}[k]\hat{\bar{x}}_1 + G_{d,2}[k]z[k] + G_{d,3}u[k], \tag{26}$$

where $\hat{d} \in \mathbb{R}^q$ is unknown input estimator, $G_{d,1} \in \mathbb{R}^{q\times(n-q)}$, $G_{d,2} \in \mathbb{R}^{q\times m}$, and $G_{d,3} \in \mathbb{R}^{q\times p}$. A controller can be further designed based on $\hat{d}$, z, and x_{des}, where $x_{des} \in \mathbb{R}^n$ is our desired state vector [73–77].

5. Test Scenario

Let us consider a traffic with ten driver types and fourteen vehicle models operating in manual, automated, and cooperative automated modes over US06 and Cycle D driving schedules with given conditions in Table 11 under malicious increases of 1 ft/s^2, 3 ft/s^2, and 5 ft/s^2 in transmitted accelerations, where

$$x[k] \triangleq \left[S^T[k]\ v_{i+1}^T[k]\ a_{i+1}^T[k]\ v_i^T[k]\ a_i^T[k]\right]^T, \tag{27}$$

$$u[k] \triangleq \left[K_p e_v^T[k]\ K_i e_p^T[k]\ K_d a_i^T[k]\ v_{ref}^T[k]\ a_{ref}^T[k]\right]^T, \tag{28}$$

$$A := \begin{bmatrix} 1 & -\Delta t & 0 & \Delta t & 0 \\ 0 & 1 & \Delta t & 0 & 0 \\ 0 & 0 & 1 & 0 & 0 \\ 0 & 0 & 0 & 1 & 0 \\ 0 & 0 & 0 & 0 & 1 \end{bmatrix}, \text{ and } B := \begin{bmatrix} 0 & 0 & 0 & 0 & 0 \\ 0 & 0 & 0 & 0 & 0 \\ 1 & 1 & 1 & 0 & 0 \\ 0 & 0 & 0 & 1 & 0 \\ 0 & 0 & 0 & 0 & 1 \end{bmatrix}. \tag{29}$$

Table 11. Input parameters.

Parameter	Value	Unit	Parameter	Value	Unit
Δt	0.1	s	$S_{des}[1]$	5	ft
ρ	0.002377 *	slug/ft^3	α	2	-
θ	0	-	β	4	-
Drivetrain Type	Front-Wheel-Drive	-	$K_{p,cr}[1]$	1	s^{-1}
μ	1 **	-	$K_{p,a}[1]$	−1	s^{-2}
l_r	$L/2$	-	$K_{d,a}[1]$	1	s^{-1}
η_b	0.95	-	$K_{p,c}[1]$	1	s^{-1}
γ_b	1.04	-	$K_{i,c}[1]$	−1	s^{-2}
τ_s	1.0, 0.6, 0.0	s	$K_{d,c}[1]$	1	-
τ_c	0.0, 0.1	s	FFS	110	ft/s
Driver Type	5 #	-	h_{min}	2	s
$x[1]$	100 #, 0 ##	ft	q	1800	veh/h
$v[1]$	0 ##	ft/s	Warm-Up Period	900	s
$a[1]$	0 ##	ft/s^2	Replications	20	-
T_{set}	1.1,0.6	s			

* for 0 ft altitude, 59° F temperature, and 14.7 lb/in^2 pressure, ** for good and dry pavement, # leader, ## follower.

6. Results

Distance gap, speed, and acceleration profiles are shown in Figures 2–4, arranged from shortest to longest time elapsed till crash occurs (showed as dashed lines)—2011 Ford F150 (7.5 s), 2004 Pontiac Grand Am (7.5 s), 2006 Honda Civic Si (7.6 s), 2009 Honda Civic (8.1 s), 2005 Mazda 6 (8.1 s), 2008 Chevy Impala (8.5 s), 2002 Chevy Silverado (8.5 s), 2004 Chevy Tahoe (9.0 s), 1998 Buick Century (9.3 s), 1998 Chevy S10 Blazer (9.3 s), intermediate semi-trailer (42.5 s), single-unit truck (42.8 s), interstate semi-trailer (44.8 s), and double semi-trailer (45.2 s). Results show that (1) proposed state and unknown input estimation model can be used to design a safe cooperative automated longitudinal control function under measurement noise, process noise, natural fault, and malicious fault; (2) vehicles over Cycle D driving schedule are more sensitive to fault magnitude than vehicles over US06 driving schedule (see Table 15), since vehicles over Cycle D driving schedule have lower average speeds and, therefore, maintain shorter time gaps than vehicles over US06 driving schedule; (3) passenger cars are more sensitive to fault magnitude than trucks, particularly at lower magnitude faults (see Table 15), since passenger cars maintain shorter time gaps than trucks; (4) errors in distance gap, speed, and acceleration are proportional to fault magnitude (see Tables 12–14); (5) errors in distance gap, speed, and acceleration are not sensitive to driving schedule; (6) distance gap is most sensitive state; (7) acceleration is least sensitive state; (8) adding 3.4 ft to estimated distance gaps, deducting 2.6 ft/s from estimated speeds, or deducting 0.8 ft/s^2 from estimated accelerations can mitigate impacts of up to malicious increase of 5 ft/s^2 in transmitted accelerations (as a hypothesis) (see Tables 12–14), (9) higher magnitude faults lead to earlier crashes (see Table 15).

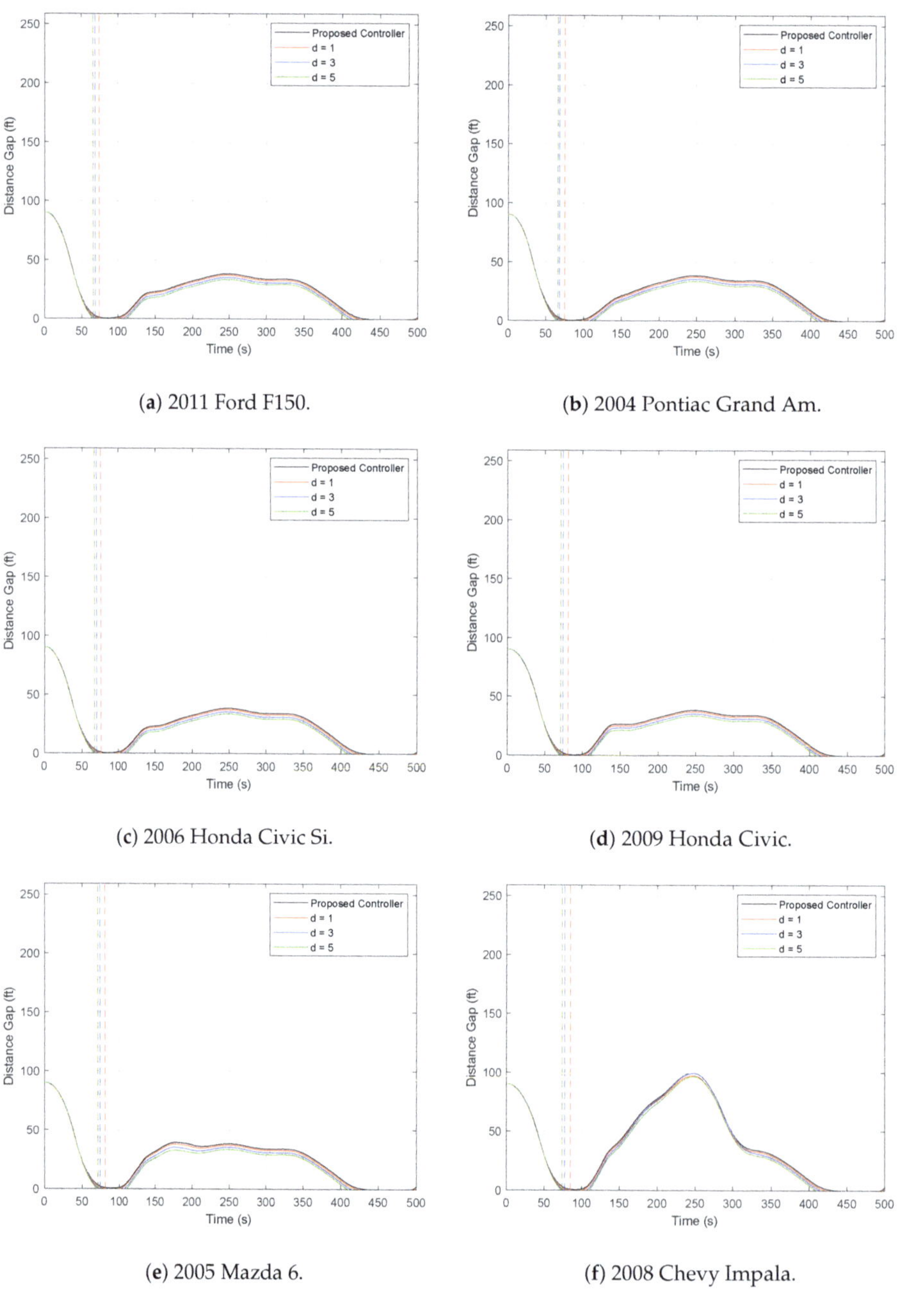

(**a**) 2011 Ford F150. (**b**) 2004 Pontiac Grand Am. (**c**) 2006 Honda Civic Si. (**d**) 2009 Honda Civic. (**e**) 2005 Mazda 6. (**f**) 2008 Chevy Impala.

Figure 2. *Cont.*

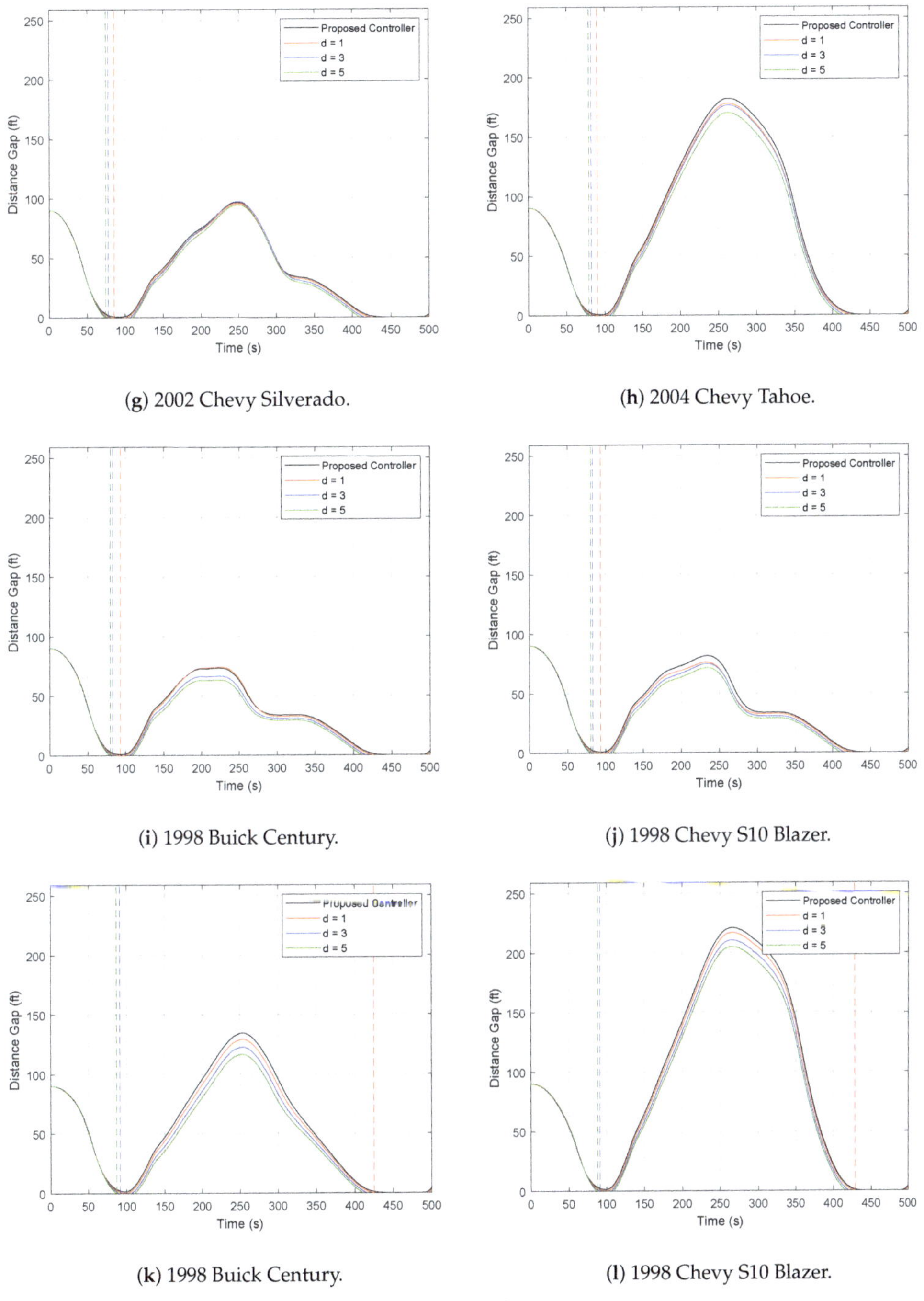

(**g**) 2002 Chevy Silverado. (**h**) 2004 Chevy Tahoe.

(**i**) 1998 Buick Century. (**j**) 1998 Chevy S10 Blazer.

(**k**) 1998 Buick Century. (**l**) 1998 Chevy S10 Blazer.

Figure 2. *Cont.*

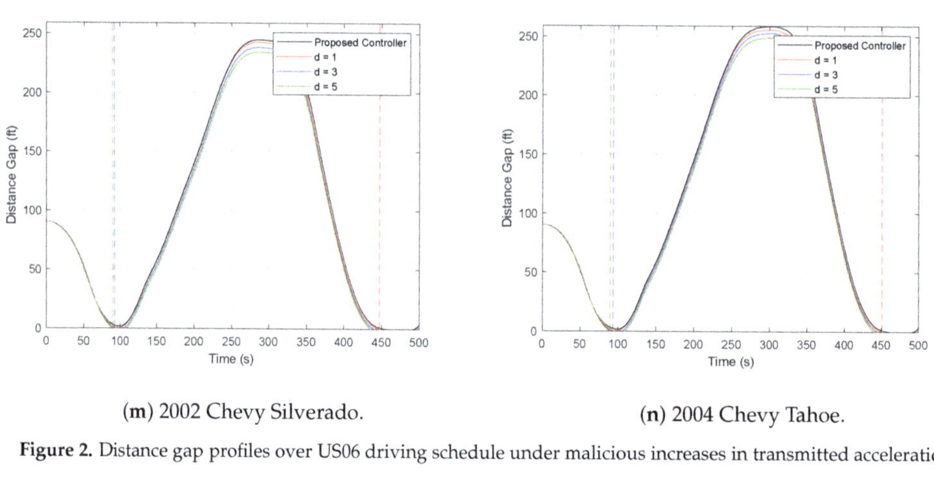

(**m**) 2002 Chevy Silverado.

(**n**) 2004 Chevy Tahoe.

Figure 2. Distance gap profiles over US06 driving schedule under malicious increases in transmitted accelerations.

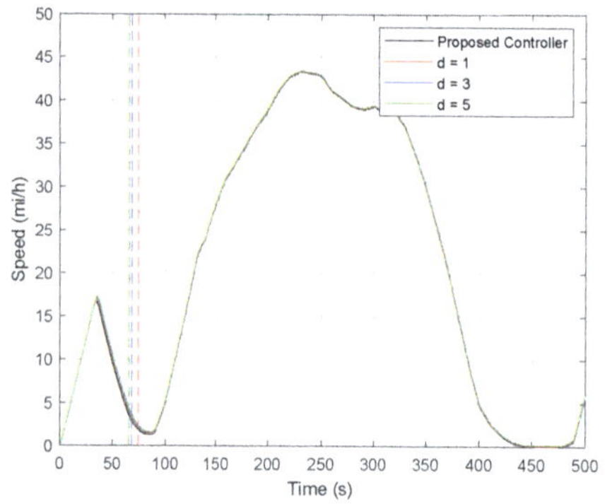

(**a**) 2011 Ford F150.

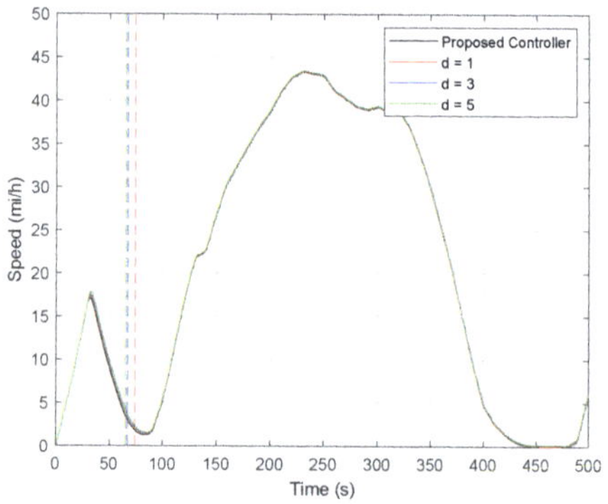

(**b**) 2004 Pontiac Grand Am.

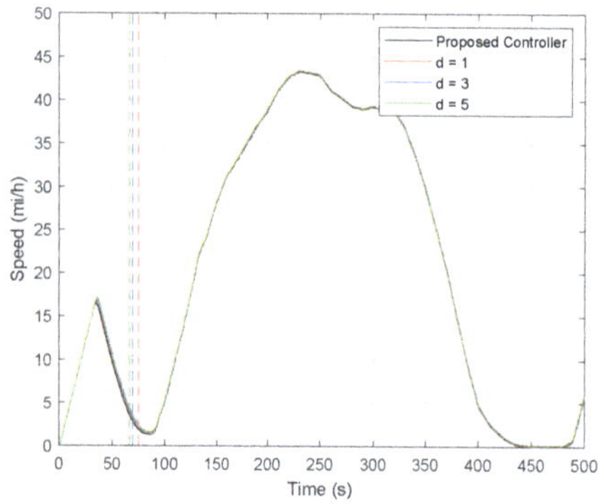

(**c**) 2006 Honda Civic Si.

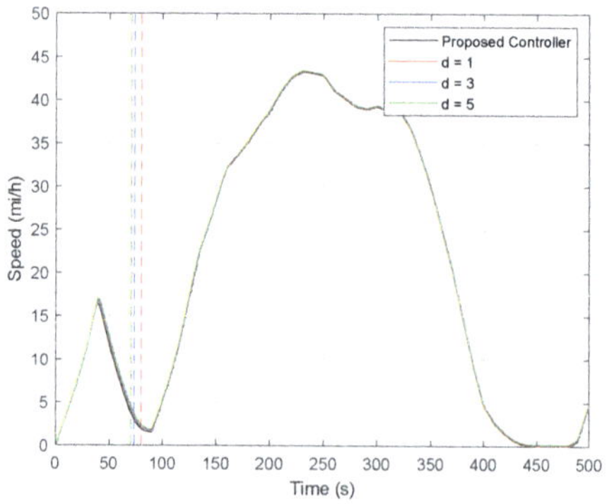

(**d**) 2009 Honda Civic.

Figure 3. *Cont.*

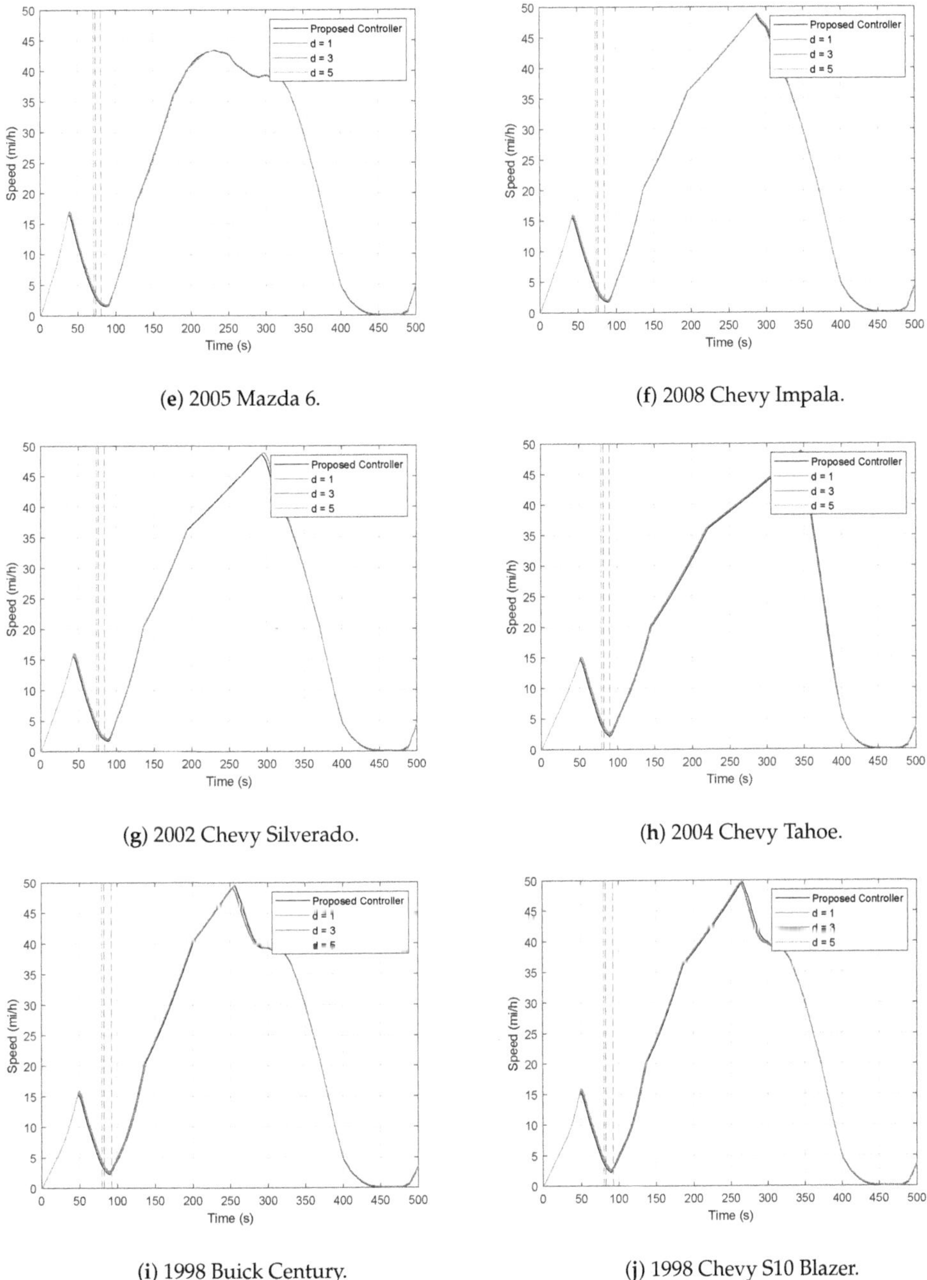

(e) 2005 Mazda 6.

(f) 2008 Chevy Impala.

(g) 2002 Chevy Silverado.

(h) 2004 Chevy Tahoe.

(i) 1998 Buick Century.

(j) 1998 Chevy S10 Blazer.

Figure 3. *Cont.*

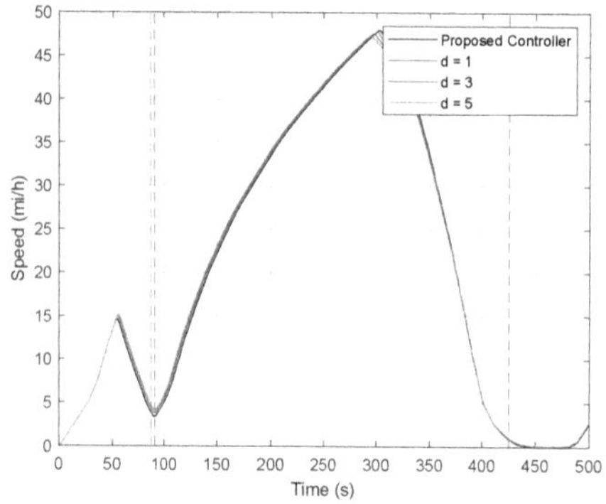

(**k**) 1998 Buick Century.

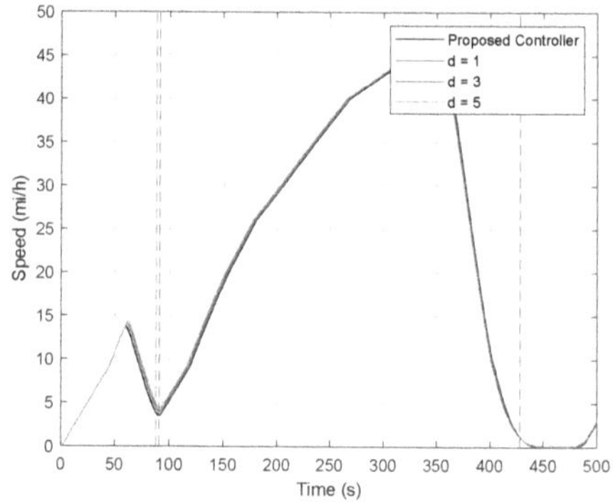

(**l**) 1998 Chevy S10 Blazer.

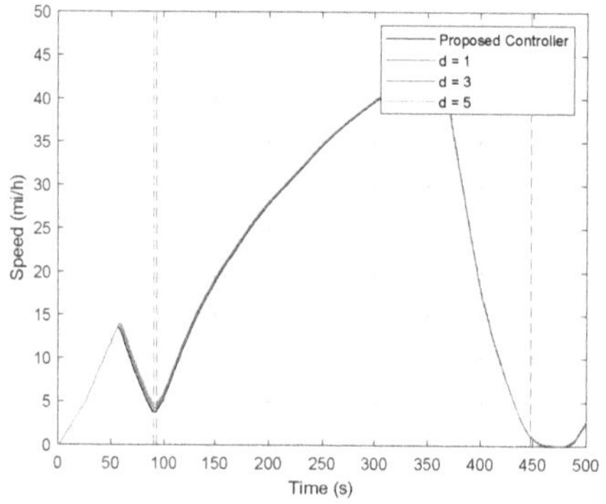

(**m**) 2002 Chevy Silverado.

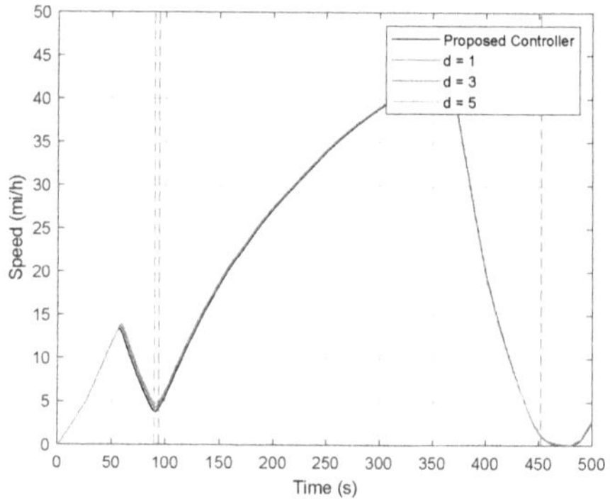

(**n**) 2004 Chevy Tahoe.

Figure 3. Speed gap profiles over US06 driving schedule under malicious increases in transmitted accelerations.

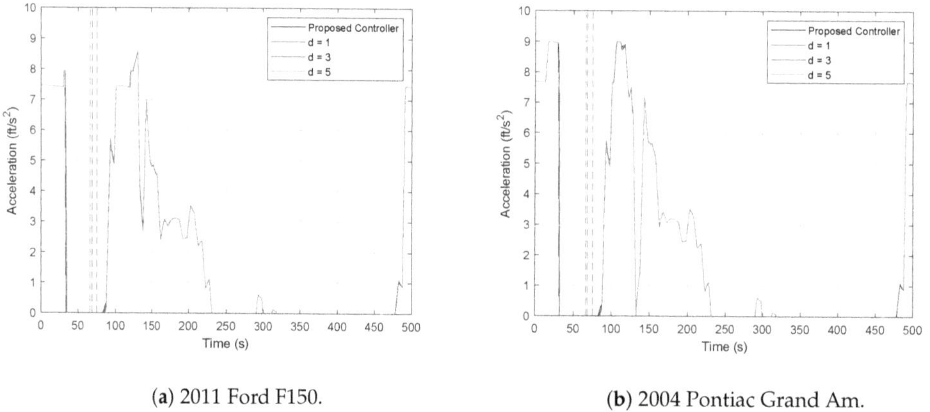

(**a**) 2011 Ford F150.

(**b**) 2004 Pontiac Grand Am.

Figure 4. *Cont.*

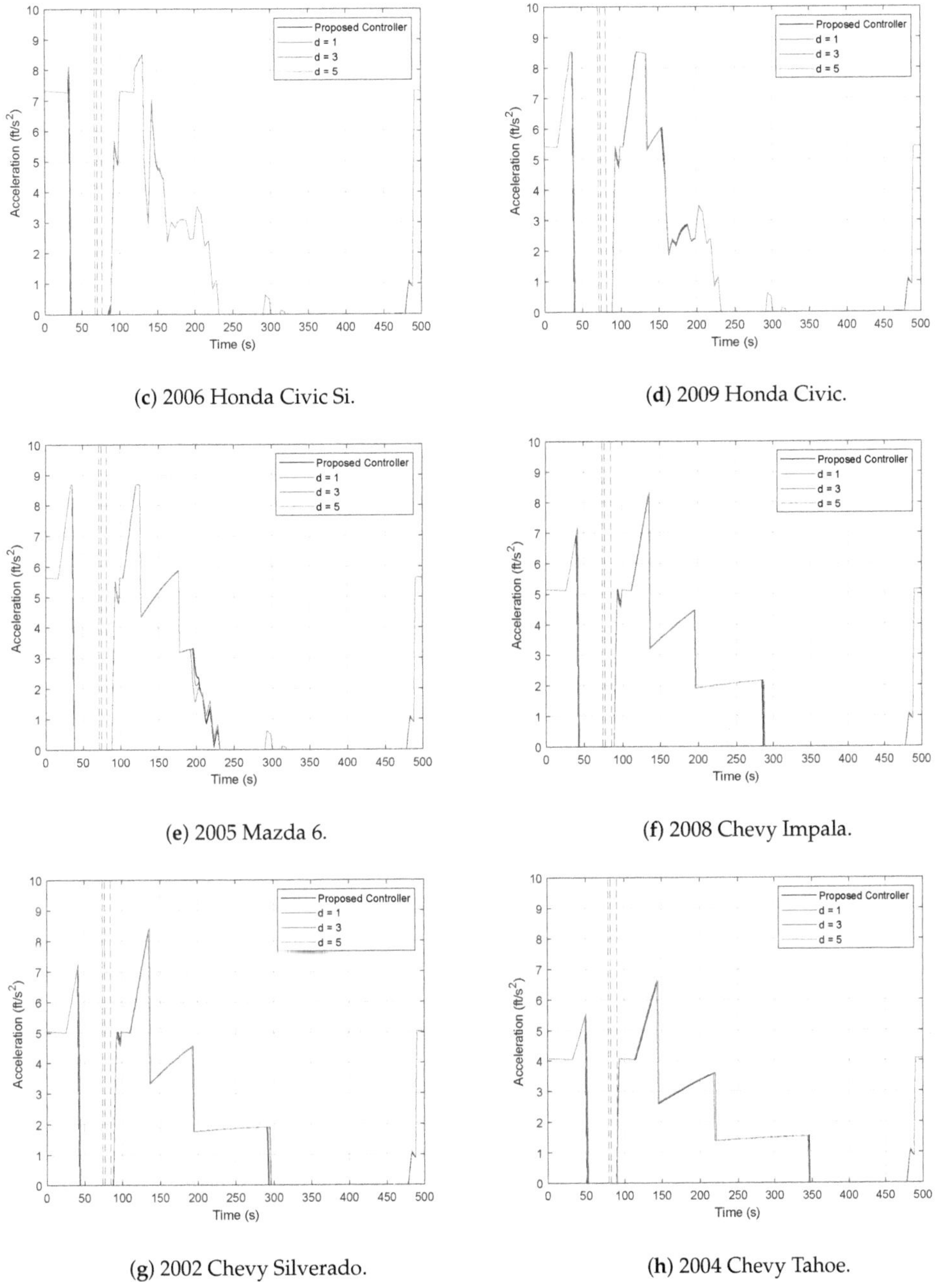

(**c**) 2006 Honda Civic Si.

(**d**) 2009 Honda Civic.

(**e**) 2005 Mazda 6.

(**f**) 2008 Chevy Impala.

(**g**) 2002 Chevy Silverado.

(**h**) 2004 Chevy Tahoe.

Figure 4. *Cont.*

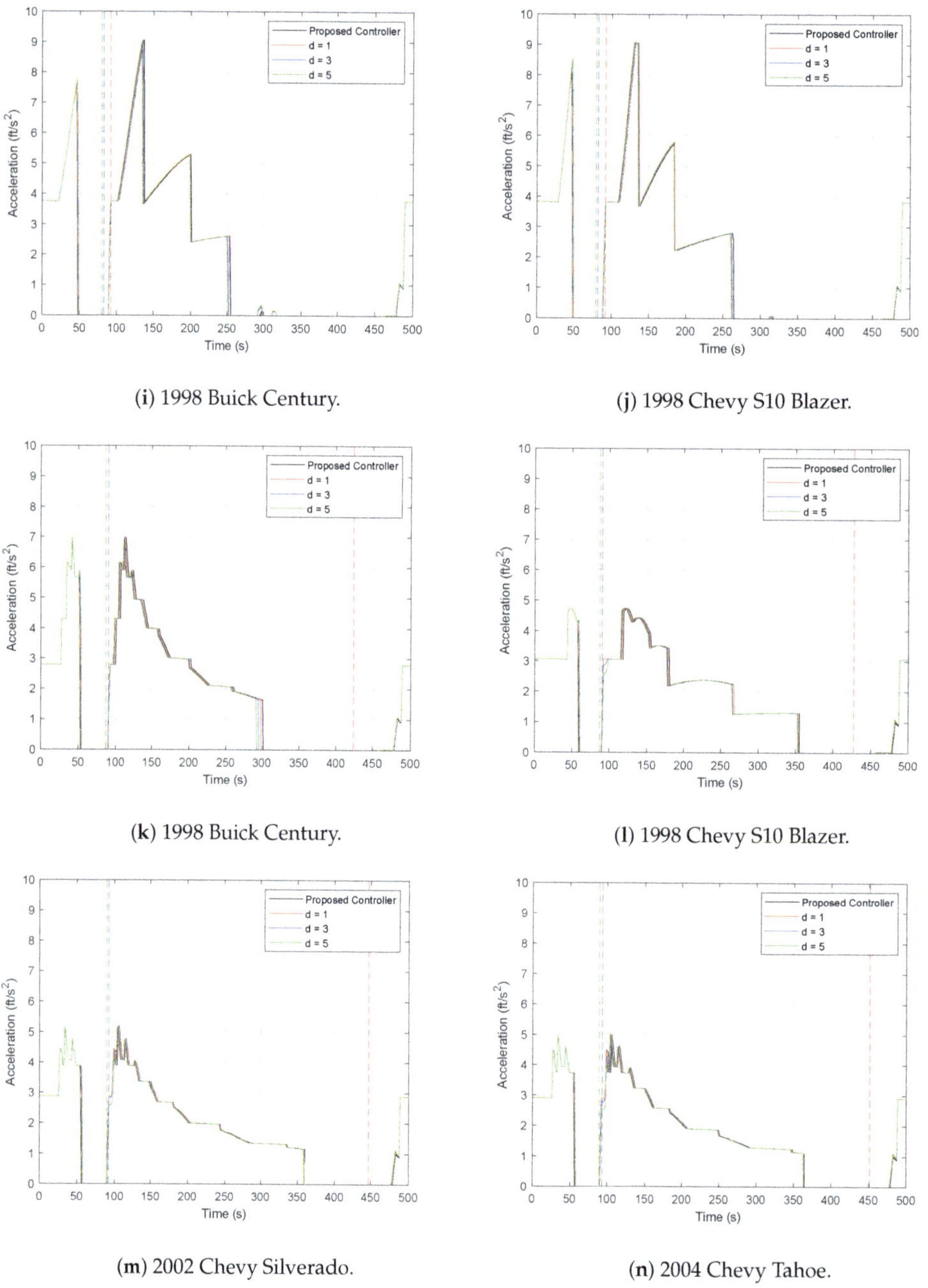

(**i**) 1998 Buick Century.

(**j**) 1998 Chevy S10 Blazer.

(**k**) 1998 Buick Century.

(**l**) 1998 Chevy S10 Blazer.

(**m**) 2002 Chevy Silverado.

(**n**) 2004 Chevy Tahoe.

Figure 4. Acceleration gap profiles over US06 driving schedule under malicious increases in transmitted accelerations.

Table 12. Distance gap errors * (ft) before crash occurs **.

	US06			Cycle D		
Malicious Fault (ft/s^2)	1	3	5	1	3	5
2011 Ford F150	0.8	2.2	3.3	0.8	2.2	3.3
2004 Pontiac Grand Am	0.8	2.2	3.4	0.8	2.2	3.3
2006 Honda Civic Si	0.8	2.2	3.3	0.8	2.2	3.3
2009 Honda Civic	0.8	2.2	3.3	0.8	2.2	3.3
2005 Mazda 6	0.8	2.2	3.3	0.8	2.2	3.3
2008 Chevy Impala	0.8	2.2	3.2	0.8	2.2	3.2
2002 Chevy Silverado	0.8	2.2	3.2	0.8	2.2	3.2
2004 Chevy Tahoe	0.8	2.1	3.1	0.8	2.1	3.1
1998 Buick Century	0.9	2.2	3.3	0.8	2.1	3.2
1998 Chevy S10 Blazer	0.9	2.2	3.3	0.8	2.1	3.1
Intermediate Semi-Trailer	0.9	2.2	3.2	0.8	2.1	3.1
Single-Unit Truck	1.0	2.1	3.1	0.8	2.0	3.0
Interstate Semi-Trailer	0.9	2.2	3.2	0.8	2.1	3.0
Double Semi-Trailer	1.0	2.2	3.2	0.8	2.0	3.1

* calculated as $S_{i+1}^{normal}[k-1] - S_{i+1}^{faulty}[k-1]$, where S^{normal} is distance gap in normal conditions, and S^{faulty} is distance gap in faulty conditions, ** in absence of our proposed state and unknown input estimation model.

Table 13. Speed errors * (ft/s) before crash occurs **.

	US06			Cycle D		
Malicious Fault (ft/s^2)	1	3	5	1	3	5
2011 Ford F150	−0.2	−1.1	−2.1	−0.3	−1.1	−2.1
2004 Pontiac Grand Am	−0.2	−1.0	−1.9	−0.2	−1.0	−2.0
2006 Honda Civic Si	−0.2	−1.0	−2.0	−0.3	−1.0	−2.0
2009 Honda Civic	−0.2	−1.1	−2.1	−0.3	−1.1	−2.1
2005 Mazda 6	−0.2	−1.0	−2.0	−0.3	−1.1	−2.0
2008 Chevy Impala	−0.2	−1.1	−2.1	−0.3	−1.1	−2.1
2002 Chevy Silverado	−0.2	−1.1	−2.2	−0.3	−1.1	−2.2
2004 Chevy Tahoe	−0.2	−1.2	−2.4	−0.3	−1.3	−2.4
1998 Buick Century	−0.2	−1.1	−2.1	−0.3	−1.1	−2.2
1998 Chevy S10 Blazer	−0.2	−1.1	−2.2	−0.3	−1.2	−2.3
Intermediate Semi-Trailer	−0.1	−1.1	−2.2	−0.3	−1.2	−2.2
Single-Unit Truck	0.1	−0.8	−2.5	−0.3	−1.3	−2.6
Interstate Semi-Trailer	0.0	−1.1	−2.2	−0.3	−1.2	−2.3
Double Semi-Trailer	0.0	−1.1	−2.2	−0.3	−1.2	−2.3

* calculated as $v_{i+1}^{normal}[k-1] - v_{i+1}^{faulty}[k-1]$, where v^{normal} is speed in normal conditions, and v^{faulty} is speed in faulty conditions, ** in absence of our proposed state and unknown input estimation model.

Table 14. Acceleration errors * (ft/s^2) before crash occurs **.

	US06			Cycle D		
Malicious Fault (ft/s^2)	1	3	5	1	3	5
2011 Ford F150	−0.1	−0.4	−0.6	−0.1	−0.4	−0.6
2004 Pontiac Grand Am	−0.1	−0.3	−0.5	−0.1	−0.3	−0.5
2006 Honda Civic Si	−0.1	−0.4	−0.6	−0.1	−0.4	−0.6
2009 Honda Civic	−0.1	−0.4	−0.6	−0.1	−0.4	−0.6
2005 Mazda 6	−0.1	−0.4	−0.6	−0.1	−0.4	−0.6
2008 Chevy Impala	−0.1	−0.4	−0.6	−0.1	−0.4	−0.6
2002 Chevy Silverado	−0.1	−0.4	−0.6	−0.1	−0.4	−0.6
2004 Chevy Tahoe	−0.1	−0.5	−0.8	−0.1	−0.5	−0.8
1998 Buick Century	−0.1	−0.4	−0.6	−0.1	−0.4	−0.6
1998 Chevy S10 Blazer	−0.1	−0.4	−0.7	−0.1	−0.4	−0.6
Intermediate Semi-Trailer	0.0	−0.4	−0.6	−0.1	−0.4	−0.5
Single-Unit Truck	−0.1	−0.5	−0.8	−0.1	−0.5	−0.8
Interstate Semi-Trailer	0.0	−0.4	−0.6	−0.1	−0.4	−0.5
Double Semi-Trailer	−0.1	−0.1	−0.6	−0.1	−0.4	−0.6

* calculated as $a_{i+1}^{normal}[k-1] - a_{i+1}^{faulty}[k-1]$, where a^{normal} is acceleration in normal conditions, and a^{faulty} is acceleration in faulty conditions, ** in absence of our proposed state and unknown input estimation model.

Table 15. Seconds elapsed till test till crash occurs *.

		US06			Cycle D	
Malicious Fault (ft/s^2)	**1**	**3**	**5**	**1**	**3**	**5**
2011 Ford F150	7.5	6.9	6.6	7.4	6.9	6.6
2004 Pontiac Grand Am	7.5	6.8	6.6	7.4	6.8	6.5
2006 Honda Civic Si	7.6	7.0	6.7	7.5	7.0	6.7
2009 Honda Civic	8.1	7.4	7.1	7.9	7.4	7.1
2005 Mazda 6	8.1	7.4	7.1	7.9	7.3	7.1
2008 Chevy Impala	8.5	7.7	7.4	8.2	7.7	7.4
2002 Chevy Silverado	8.5	7.7	7.4	8.2	7.7	7.4
2004 Chevy Tahoe	9.0	8.2	7.9	8.6	8.1	7.9
1998 Buick Century	9.3	8.3	8.0	8.7	8.2	7.9
1998 Chevy S10 Blazer	9.3	8.3	8.0	8.7	8.2	7.9
Intermediate Semi-Trailer	42.5	9.1	8.7	9.4	8.9	8.6
Single-Unit Truck	42.8	9.1	8.8	9.4	8.9	8.7
Interstate Semi-Trailer	44.8	9.3	9.0	9.6	9.1	8.8
Double Semi-Trailer	45.2	9.4	9.0	9.7	9.1	8.9

* in absence of our proposed controller.

Levels 1 and 2 automated vehicles are assumed to maintain minimum safe distance gap in a string; vehicles dedicated to automated driving systems and vehicles equipped with cooperative automated driving systems are assumed to maintain minimum safe time gap in a string; vehicles are assumed to maintain minimum safe distance gap in a platoon at each simulation time step to maximize road capacity without compromising safety or string stability. Therefore, increasing demand up to road capacity would not impact outputs (e.g., distance gap, time gap, speed, and acceleration) significantly. Demands exceeding road capacity will spill back behind entrance.

7. Discussion

Existing simulation tools may overestimate safety and road capacity improvements associated with cooperative driving automation due to not considering vehicle model and vehicle-to-vehicle communication vulnerabilities on a large scale. This research modifies a vehicle-following model for conventional vehicles, a longitudinal control function for vehicles dedicated to automated driving systems, and a longitudinal control function for vehicles equipped with cooperative automated driving systems, considering vehicle model and vehicle-to-vehicle communication vulnerabilities to maximize road capacity without compromising safety or string stability. Our proposed traffic microsimulation tool can be used to verify automated driving systems and cooperative automated driving systems in contested environments.

Drivers are assumed to drive in a single lane, and there is no lane-change maneuver, while a lane-change maneuver can temporarily affect vehicle-following behaviors. Future work can model other significant components underpinning a traffic microsimulation tool (i.e., lane-changing and gap acceptance)

- model motivation for mandatory, active, and discretionary lane-change maneuvers;
- model mandatory, active, and discretionary lane-change gap acceptance;
- model before lane-change, after lane-change, receiving, and yielding vehicle-following for each facility type (e.g., on-ramp and off-ramp);
- model lateral control for autonomous vehicles;
- model string operations (e.g., maximum platoon size, inter-platoon time gap, and cut-in and cut-out maneuvers).

Microscopic measures (e.g., distance headway and time headway) can be aggregated to macroscopic measures (e.g., density and flow) as $k \triangleq 1/s$ and $q = 3600 \times \bar{h}$, where k is density (veh/ft), and $\bar{s}$ is average distance headway (ft/veh). Future work can estimate macroscopic benefits associated with cooperative driving automation (e.g., increase in lane capacity) under various market penetration for autonomous and connected autonomous vehicles. Table 16 recommends potential improvements to our proposed longitudinal controller.

Table 16. Recommended control designs.

Cyberattack	Future Work	Description
Formulation	Fault and Delay	Most common cyberattacks can be modeled as fault (e.g., data falsification and spoofing attacks) or delay (e.g., denial-of-service attack).
Detection	Kalman Filter & Neural Network	Conventional fault-resilient longitudinal controllers are model-driven or data-driven, but not combined, potentially sacrificing accuracy or simulation speed.
Compensation	Adaptive Controller	Estimated distance gaps can be increased in proportion to cyberattack magnitude.

Author Contributions: Conceptualization, S.N., M.P. and M.N.; methodology, S.N.; software, M.P. and M.N.; validation, S.N.; formal analysis, M.P.; investigation, M.N.; resources, S.N.; data curation, S.N.; writing—original draft preparation, S.N.; writing—review and editing, M.P. and M.N.; visualization, M.N. All authors have read and agreed to the published version of the manuscript.

Funding: This work was sponsored by a contract from the Southeastern Transportation Research, Innovation, Development and Education Center (STRIDE), a Regional University Transportation Center sponsored by a grant from the U.S. Department of Transportation's University Transportation Centers Program. The contents of this report reflect the views of the authors, who are responsible for the facts and the accuracy of the information presented herein. This document is disseminated in the interest of information exchange. The report is funded, partially or entirely, by a grant from the U.S. Department of Transportation's University Transportation Centers Program. However, the U.S. Government assumes no liability for the contents or use thereof.

Conflicts of Interest: The authors declare no conflict of interest.

References

1. On-Road Automated Driving Committee. *Taxonomy and Definitions for Terms Related to Driving Automation Systems for On-Road Motor Vehicles;* Technical Report J3016; Society of Automotive Engineers: Warrendale, PA, USA, 2021.
2. National Center for Statistics and Analysis. *Critical Reasons for Crashes Investigated in the National Motor Vehicle Crash Causation Survey (Traffic Safety Facts);* Technical Report DOT HS 812 506; National Highway Traffic Safety Administration: Washington, DC, USA, 2018.
3. National Highway Traffic Safety Administration. *Federal Motor Vehicle Safety Standards; V2V Communications;* Technical Report NHTSA-2016-0126; US Department of Transportation: Washington, DC, USA, 2017.
4. On-Road Automated Driving Committee. *Taxonomy and Definitions for Terms Related to Cooperative Driving Automation for On-Road Motor Vehicles;* Technical Report J3216; Society of Automotive Engineers: Warrendale, PA, USA, 2021.
5. Olstam, J.J.; Tapani, A. *Comparison of Car-Following Models;* Volume 960 A; Swedish National Road and Transport Research Institute: Linköping, Sweden, 2004.
6. Upstream Security Ltd. *Global Automotive Cybersecurity Report;* Technical Report; Upstream Security Ltd.: Detroit, MI, USA, 2020.
7. Amoozadeh, M.; Raghuramu, A.; Chuah, C.N.; Ghosal, D.; Zhang, H.M.; Rowe, J.; Levitt, K. Security Vulnerabilities of Connected Vehicle Streams and Their Impact on Cooperative Driving. *IEEE Commun. Mag.* **2015**, *53*, 126–132. [CrossRef]
8. Sargolzaei, A.; Abbaspour, A.; Al Faruque, M.A.; Eddin, A.S.; Yen, K. Security Challenges of Networked Control Systems. In *Sustainable Interdependent Networks;* Springer: Cham, Switzerland, 2018; pp. 77–95.
9. Mouzakitis, A. Classification of Fault Diagnosis Methods for Control Systems. *Meas. Control* **2013**, *46*, 303–308. [CrossRef]
10. Abbaspour, A.; Mokhtari, S.; Sargolzaei, A.; Yen, K. A survey on active fault-tolerant control systems. *Electronics* **2020**, *9*, 1513. [CrossRef]
11. Nowakowski, C.; Shladover, S.E.; Lu, X.Y.; Thompson, D.; Kailas, A. *Cooperative Adaptive Cruise Control (CACC) for Truck Platooning: Operational Concept Alternatives;* Technical Report DTFH61-13-H-00012 Task 1.2; UC Berkeley: Berkeley, CA, USA, 2015.
12. Zambrano-Martinez, J.L.; Calafate, C.T.; Soler, D.; Cano, J.C. Towards realistic urban traffic experiments using DFROUTER: Heuristic, validation and extensions. *Sensors* **2017**, *17*, 2921. [CrossRef]

13. Krajzewicz, D.; Hertkorn, G.; Rössel, C.; Wagner, P. SUMO (Simulation of Urban MObility)–An open-source traffic simulation. In Proceedings of the 4th Middle East Symposium on Simulation and Modelling (MESM), Sharjah, United Arab Emirates, 28–30 October 2002; pp. 183–187.
14. Akçelik, R.; Besley, M. Acceleration and deceleration models. In Proceedings of the 23rd Conference of Australian Institutes of Transport Research (CAITR), Melbourne, Australia, 10–12 December 2001; p. 12.
15. Ahn, K.; Rakha, H.; Trani, A.; Van Aerde, M. Estimating vehicle fuel consumption and emissions based on instantaneous speed and acceleration levels. *J. Transp. Eng.* **2002**, *128*, 182–190. [CrossRef]
16. Fang, F.C.; Elefteriadou, L. Some guidelines for selecting microsimulation models for interchange traffic operational analysis. *J. Transp. Eng.* **2005**, *131*, 535–543. [CrossRef]
17. Kuriyama, M.; Yamamoto, S.; Miyatake, M. Theoretical study on eco-driving technique for an electric vehicle with dynamic programming. In Proceedings of the International Conference on Electrical Machines and Systems (ICEMS), Incheon, Korea, 10–13 October 2010; pp. 2026–2030.
18. Maurya, A.K.; Bokare, P.S. Study of deceleration behavior of different vehicle types. *Int. J. Traffic Transp. Eng.* **2012**, *2*, 253–270. [CrossRef]
19. Lee, J.; Park, B.B.; Malakorn, K.; So, J.J. Sustainability assessments of cooperative vehicle intersection control at an urban corridor. *Transp. Res. Part C: Emerg. Technol.* **2013**, *32*, 193–206. [CrossRef]
20. Anya, A.R.; Rouphail, N.M.; Frey, H.C.; Schroeder, B. Application of AIMSUN microsimulation model to estimate emissions on signalized arterial corridors. *Transp. Res. Rec.* **2014**, *2428*, 75–86. [CrossRef]
21. Song, G.; Yu, L.; Geng, Z. Optimization of Wiedemann and Fritzsche car-following models for emission estimation. *Transp. Res. Part D Transp. Environ.* **2015**, *34*, 318–329. [CrossRef]
22. Bokare, P.S.; Maurya, A.K. Acceleration-deceleration behaviour of various vehicle types. *Transp. Res. Procedia* **2017**, *25*, 4733–4749. [CrossRef]
23. Ramezani, H.; Shladover, S.E.; Lu, X.Y.; Altan, O.D. Micro-simulation of truck platooning with cooperative adaptive cruise control: Model development and a case study. *Transp. Res. Rec.* **2018**, *2672*, 55–65. [CrossRef]
24. Shladover, S.E.; Nowakowski, C.; O'Connell, J.; Cody, D. Cooperative adaptive cruise control: Driver selection of car-following gaps. In Proceedings of the 17th ITS World Congress, Pusan, Korea, 25–29 October 2010; p. 8.
25. van Willigen, W.H.; Schut, M.C.; Kester, L.J. Approximating safe spacing policies for adaptive cruise control strategies. In Proceedings of the Vehicular Networking Conference (VNC), Amsterdam, The Netherlands, 14–16 November 2011; pp. 9–16.
26. Horiguchi, R.; Oguchi, T. A study on car following models simulating various adaptive cruise control behaviors. *Int. J. Intell. Transp. Syst. Res.* **2014**, *12*, 127–134. [CrossRef]
27. Flores, C.; Milanés, V.; Nashashibi, F. A time gap-based spacing policy for full-range car-following. In Proceedings of the 20th International Conference on Intelligent Transportation Systems (ITSC), Yokohama, Japan, 16–19 October 2017; pp. 1–6.
28. Askari, A.; Farias, D.A.; Kurzhanskiy, A.A.; Varaiya, P. Effect of adaptive and cooperative adaptive cruise control on throughput of signalized arterials. In Proceedings of the Intelligent Vehicles Symposium (IV), Los Angeles, CA, USA, 11–14 June 2017; pp. 1287–1292.
29. Flores, C.; Milanés, V. Fractional-order-based ACC/CACC algorithm for improving string stability. *Transp. Res. Part C Emerg. Technol.* **2018**, *95*, 381–393. [CrossRef]
30. Chen, J.; Zhou, Y.; Liang, H. Effects of ACC and CACC vehicles on traffic flow based on an improved variable time headway spacing strategy. *Intell. Transp. Syst.* **2019**, *13*, 1365–1373. [CrossRef]
31. Bian, Y.; Zheng, Y.; Ren, W.; Li, S.E.; Wang, J.; Li, K. Reducing time headway for platooning of connected vehicles via V2V communication. *Transp. Res. Part C: Emerg. Technol.* **2019**, *102*, 87–105. [CrossRef]
32. PTV. Virtual Testing of Autonomous Vehicles with PTV Vissim. Available online: https://www.ptvgroup.com/en/solutions/products/ptv-vissim/areas-of-application/autonomous-vehicles-and-new-mobility/ (accessed on 16 August 2021).
33. Aimsun. Aimsun Unveils New Platform for Simulating a Driverless Future. Available online: https://www.aimsun.com/aimsun-auto-launch/ (accessed on 1 August 2020).
34. Bu, F.; Tan, H.S.; Huang, J. Design and field testing of a cooperative adaptive cruise control system. In Proceedings of the American Control Conference (ACC), Baltimore, MD, USA, 30 June–2 July 2010; pp. 4616–4621.
35. Naus, G.J.; Vugts, R.P.; Ploeg, J.; van De Molengraft, M.J.; Steinbuch, M. String-stable CACC design and experimental validation: A frequency-domain approach. *IEEE Trans. Veh. Technol.* **2010**, *59*, 4268–4279. [CrossRef]
36. Liu, H. *Using Cooperative Adaptive Cruise Control (CACC) to Form High-Performance Vehicle Streams. Microscopic Traffic Modeling;* Technical Report DTFH61-13-H-00013; UC Berkeley: Berkeley, CA, USA, 2018.
37. Ploeg, J.; Scheepers, B.T.; Van Nunen, E.; Van de Wouw, N.; Nijmeijer, H. Design and experimental evaluation of cooperative adaptive cruise control. In Proceedings of the 14th International Conference on Intelligent Transportation Systems (ITSC), Washington, DC, USA, 5–7 October 2011; pp. 260–265.
38. Shladover, S.E.; Su, D.; Lu, X.Y. Impacts of cooperative adaptive cruise control on freeway traffic flow. *Transp. Res. Rec.* **2012**, *2324*, 63–70. [CrossRef]
39. Zhao, L.; Sun, J. Simulation framework for vehicle platooning and car-following behaviors under connected-vehicle environment. *Procedia-Soc. Behav. Sci.* **2013**, *96*, 914–924. [CrossRef]

40. Segata, M.; Joerer, S.; Bloessl, B.; Sommer, C.; Dressler, F.; Cigno, R.L. Plexe: A platooning extension for Veins. In Proceedings of the Vehicular Networking Conference (VNC), Paderborn, Germany, 3–5 December 2014; pp. 53–60.
41. Milanés, V.; Shladover, S.E. Modeling cooperative and autonomous adaptive cruise control dynamic responses using experimental data. *Transp. Res. Part C Emerg. Technol.* **2014**, *48*, 285–300. [CrossRef]
42. Nikolos, I.K.; Delis, A.I.; Papageorgiou, M. Macroscopic modelling and simulation of ACC and CACC traffic. In Proceedings of the 18th International Conference on Intelligent Transportation Systems (ITSC), Gran Canaria, Spain, 15–18 September 2015; pp. 2129–2134.
43. Delis, A.I.; Nikolos, I.K.; Papageorgiou, M. Simulation of the penetration rate effects of ACC and CACC on macroscopic traffic dynamics. In Proceedings of the 19th International Conference on Intelligent Transportation Systems (ITSC), Rio de Janeiro, Brazil, 1–4 November 2016; pp. 336–341.
44. Wang, Z.; Wu, G.; Barth, M.J. Developing a distributed consensus-based cooperative adaptive cruise control system for heterogeneous vehicles with predecessor following topology. *J. Adv. Transp.* **2017**, *2017*, 1–16. [CrossRef]
45. Terruzzi, L.; Colombo, R.; Segata, M. On the effects of cooperative platooning on traffic shock waves. In Proceedings of the Vehicular Networking Conference (VNC), Turin, Italy, 27–29 November 2017; pp. 37–38.
46. Zhou, Y.; Ahn, S.; Chitturi, M.; Noyce, D.A. Rolling horizon stochastic optimal control strategy for ACC and CACC under uncertainty. *Transp. Res. Part C Emerg. Technol.* **2017**, *83*, 61–76. [CrossRef]
47. Lemessi, M. An slx-based microsimulation model for a two-lane road section. In Proceedings of the Winter Simulation Conference (Cat. No. 01CH37304), Arlington, VA, USA, 9–12 December 2001; pp. 1064–1071.
48. Rakha, H.; Ding, Y. Impact of stops on vehicle fuel consumption and emissions. *J. Transp. Eng.* **2003**, *129*, 23–32. [CrossRef]
49. Wang, X.Y.; Liu, J.S. Research of the lane utilization with microsimulation. In Proceedings of the International Conference on Machine Learning and Cybernetics (ICMLC), Guangzhou, China, 18–21 August 2005; pp. 2681–2687.
50. Ossen, S.; Hoogendoorn, S.P.; Gorte, B.G. Interdriver differences in car-following: A vehicle trajectory-based study. *Transp. Res. Rec.* **2006**, *1965*, 121–129. [CrossRef]
51. Kesting, A.; Treiber, M.; Helbing, D. General lane-changing model MOBIL for car-following models. *Transp. Res. Rec.* **2007**, *1999*, 86–94. [CrossRef]
52. Kesting, A.; Treiber, M. Calibrating car-following models by using trajectory data: Methodological study. *Transp. Res. Rec.* **2008**, *2088*, 148–156. [CrossRef]
53. Talebpour, A.; Mahmassani, H.S.; Hamdar, S.H. Multiregime sequential risk-taking model of car-following behavior: Specification, calibration, and sensitivity analysis. *Transp. Res. Rec.* **2011**, *2260*, 60–66. [CrossRef]
54. Song, G.; Yu, L.; Zhang, Y. Applicability of traffic microsimulation models in vehicle emissions estimates: Case study of VISSIM. *Transp. Res. Rec.* **2012**, *2270*, 132–141. [CrossRef]
55. Lee, J.; Park, B. Development and evaluation of a cooperative vehicle intersection control algorithm under the connected vehicles environment. *IEEE Trans. Intell. Transp. Syst.* **2012**, *13*, 81–90. [CrossRef]
56. Treiber, M.; Kesting, A. Microscopic calibration and validation of car-following models–a systematic approach. *Procedia-Soc. Behav. Sci.* **2013**, *80*, 922–939. [CrossRef]
57. Li, Z.; Elefteriadou, L.; Ranka, S. Signal control optimization for automated vehicles at isolated signalized intersections. *Transp. Res. Part C Emerg. Technol.* **2014**, *49*, 1–18. [CrossRef]
58. Tang, T.Q.; He, J.; Yang, S.C.; Shang, H.Y. A car-following model accounting for the driver's attribution. *Phys. A Stat. Mech. Its Appl.* **2014**, *413*, 583–591. [CrossRef]
59. Desiraju, D.; Chantem, T.; Heaslip, K. Minimizing the disruption of traffic flow of automated vehicles during lane changes. *IEEE Trans. Intell. Transp. Syst.* **2014**, *16*, 1249–1258. [CrossRef]
60. Liu, H.; Kan, X.D.; Shladover, S.E.; Lu, X.Y.; Ferlis, R.E. Modeling impacts of cooperative adaptive cruise control on mixed traffic flow in multi-lane freeway facilities. *Transp. Res. Part C Emerg. Technol.* **2018**, *95*, 261–279. [CrossRef]
61. Amoozadeh, M.; Deng, H.; Chuah, C.N.; Zhang, H.M.; Ghosal, D. Platoon management with cooperative adaptive cruise control enabled by VANET. *Veh. Commun.* **2015**, *2*, 110–123. [CrossRef]
62. Li, Y.; Wang, H.; Wang, W.; Xing, L.; Liu, S.; Wei, X. Evaluation of the impacts of cooperative adaptive cruise control on reducing rear-end collision risks on freeways. *Accid. Anal. Prev.* **2017**, *98*, 87–95. [CrossRef]
63. Noei, S.; Zhao, X.; Crane, C.D. Longitudinal control of vehicles in traffic microsimulation. *arXiv* **2020**, arXiv:2003.07861.
64. Mannering, F.L.; Washburn, S.S. *Principles of Highway Engineering and Traffic Analysis*; John Wiley & Sons: Hoboken, NJ, USA, 2020.
65. Ozkul, S. *Advanced Vehicle Dynamics Modeling Approach in Traffic Microsimulation with Emphasis on Commercial Truck Performance and On-Board-Diagnostics Data*; University of Florida: Gainesville, FL, USA, 2014.
66. Washburn, S.S.; Ozkul, S. *Heavy Vehicle Effects on Florida Freeways and Multilane Highways*; Technical Report BDK-77 977-15; Florida Department of Transportation: Tallahassee, FL, USA, 2013.
67. Washburn, S.; Noei, S. *Interchange Design to Accommodate Ramp Metering System*; Technical Report BDV-31 977-92; Florida Department of Transportation: Tallahassee, FL, USA, 2020.
68. Zhu, M.; Wang, X.; Tarko, A. Modeling car-following behavior on urban expressways in Shanghai: A naturalistic driving study. *Transp. Res. Part C Emerg. Technol.* **2018**, *93*, 425–445. [CrossRef]

69. Noei, S.; Santana, H.; Sargolzaei, A.; Noei, M. Reducing traffic congestion using geo-fence technology: Application for emergency car. In Proceedings of the 1st International Workshop on Emerging Multimedia Applications and Services for Smart Cities (EMASC), Orlando, FL, USA, 7 November 2014; pp. 15–20.
70. Washburn, S.S. Drivers. Available online: https://swashsim.miraheze.org/wiki/Drivers (accessed on 16 August 2021).
71. Washbun, S.S. Vehicles. Available online: https://swashsim.miraheze.org/wiki/Vehicles (accessed on 16 August 2021).
72. Khalghani, M.R.; Khushalani-Solanki, S.; Solanki, J.; Sargolzaei, A. Cyber disruption detection in linear power systems. In Proceedings of the North American Power Symposium (NAPS), Morgantown, WV, USA, 17–19 September 2017; pp. 1–6.
73. Mokhtari, S.; Abbaspour, A.; Yen, K.K.; Sargolzaei, A. A machine learning approach for anomaly detection in industrial control systems based on measurement data. *Electronics* **2021**, *10*, 407. [CrossRef]
74. Sargolzaei, A.; Crane, C.; Abbaspour, A.; Noei, S. A machine learning approach for fault detection in vehicular cyber-physical systems. In Proceedings of the 15th International Conference on Machine Learning and Applications (ICMLA), Anaheim, CA, USA, 18–20 December 2016; pp. 636–640.
75. Sargolzaei, A.; Yazdani, K.; Abbaspour, A.; Crane III, C.D.; Dixon, W.E. Detection and mitigation of false data injection attacks in networked control systems. *IEEE Trans. Ind. Inform.* **2019**, *16*, 4281–4292. [CrossRef]
76. Noei, M.; Abadeh, M.S. A genetic asexual reproduction optimization algorithm for imputing missing values. In Proceedings of the 9th International Conference on Computer and Knowledge Engineering (ICCKE), Mashhad, Iran, 24–25 October 2019; pp. 214–218.
77. Parvizimosaed, M.; Noei, M.; Yalpanian, M.; Bahrami, J. A containerized integrated fast IoT platform for low energy power management. In Proceedings of the 7th International Conference on Web Research (ICWR), Tehran, Iran, 19–20 May 2021; pp. 318–322.

MDPI

Article

Resilient Networked Control of Inverter-Based Microgrids against False Data Injections

Mohammad Reza Khalghani [1,*], Vishal Verma [2], Sarika Khushalani Solanki [2] and Jignesh M. Solanki [2]

1 Department of Electrical and Computer Engineering, Florida Polytechnic University, Lakeland, FL 33805, USA
2 Lane Department of Computer Science and Electrical Engineering, West Virginia University, Morgantown, WV 26506, USA; vv0001@mix.wvu.edu (V.V.); sarika.khushalani-solanki@mail.wvu.edu (S.K.S.); jignesh.solanki@mail.wvu.edu (J.M.S.)
* Correspondence: khalghani@ieee.org

Abstract: Inverter-based energy resource is a fast emerging technology for microgrids. Operation of micorgrids with integration of these resources, especially in an islanded operation mode, is challenging. To effectively capture microgrid dynamics and also control these resources in islanded microgrids, a heavy cyber and communication infrastructure is required. This high reliance of microgrids on cyber interfaces makes these systems prone to cyber-disruptions. Hence, the hierarchical control of microgrids, including primary, secondary, and tertiary control, needs to be developed to operate resiliently. This paper shows the vulnerability of microgrid control in the presence of False Data Injection (FDI) attack, which is one type of cyber-disruption. Then, this paper focuses on designing a resilient secondary control based on Unknown Input Observer (UIO) against FDI. The simulation results show the superior performance of the proposed controller over other standard controllers.

Keywords: inverter-based energy resources; islanded microgrids; networked control systems; resilient control design; secondary control; false data injection

Citation: Khalghani, M.R.; Verma, V.; Khushalani Solanki S.; Solanki, J.M. Resilient Networked Control of Inverter-Based Microgrids against False Data Injections. *Electronics* **2022**, *11*, 780. https://doi.org/10.3390/electronics11050780

Academic Editor: Carlos Andrés García-Vázquez

Received: 31 January 2022
Accepted: 15 February 2022
Published: 3 March 2022

1. Introduction

Due to the environmentally friendly characteristics of renewable energies and distributed energy resources, integrating these energy resources into distribution grids is growing significantly. There is an ongoing shift of system configuration from traditional distribution grids to small distributed and controllable microgrids. Microgrids can locally supply loads through distributed energy resources that include renewable energy resources and operate in grid-connected and islanded modes. Islanded mode of operation allows the microgrid to provide energy without the support of the main grid, which is critically important when the main grid cannot exchange energy with other local microgrids, such as during natural hazards and extreme weather events.

Sustaining the islanded operation of microgrids is challenging since the grid relies on a limited number of energy resources. This challenging task can be addressed by utilizing hierarchical control methods comprised of primary, secondary, and tertiary controls [1]. The primary control response is the immediate regulation of power output by the governor or electronic controller in response to changes in the grid frequency. Considering the limited capability of the primary control loop to address frequency changes, it is necessary to design the secondary control to control the grid dynamics [2]. Secondary control is a supervisory control that utilizes measurements communicated through cyber systems to capture and control fast microgrid dynamics. Hierarchical control can coordinate inverter-based energy resources to effectively track the sudden load changes and nondispatchable generators. Tertiary control focuses on optimal power flow between the main grid and microgrids, which is not within the scope of this paper.

Since the secondary control heavily relies on communications and cyber infrastructures in a closed control loop, it can be labeled as a networked control system [3]. Similar to all networked control systems, the secondary control of microgrids is vulnerable to cyber threats. Although there is a wide variety of cyberattacks, more common types are Denial-of-Service (DoS) and False Data Injection (FDI). The DoS targets the availability of machines or networks to temporarily or indefinitely distort the service to its intended users. The FDI manipulates the data exchanges throughout the network, which misleads the control center and disturbs the system operation [3,4].

Many researchers proposed a secure secondary control for microgrids. In [5], the authors review cybersecurity threats and introduce cyber attack prevention, detection, and response measures for the integration of inverter-based resources to grids. Researchers in [6] focus on the security of distributed secondary control of microgrids by proposing a Weighted Mean Subsequence Reduced algorithm at each inverter-based resource. A distributed secondary control based on a consensus algorithm is proposed to control the frequency of an islanded microgrid in [7]. Another distributed control strategy was developed in [8] where the authors designed a resilient control against cyber attacks on communication links, local controllers, and master controllers of microgrids. Also, researchers in [9] proposed a distributed control strategy based on blockchain to enhance cyber vulnerability of microgrids equipped with distributed energy resources. [10] constructs an attack detector based on the stable kernel representation of an islanded microgrid. A resilient secondary control method is proposed in [11] that mitigates Denial-of-Service (DoS) attacks from inverter-based microgrids. This proposed control method is mode-dependent and assumes that the random DoS attack follows a homogeneous Markov process. Another distributed control method is developed for islanded microgrids in [12] to secure the system from malicious attacks on the communication network, including links and nodes. The authors in [13] present secondary control for energy storage systems to control the frequency and voltage of an islanded microgrid. This control strategy uses an event-trigger scheme to lower the communication burden in the network. Most research work does not focus on inverter-based microgrids and additionally does not focus on a resilient controller that considers fast dynamics, especially under cyber disruptions.

This paper focuses on designing a resilient secondary controller against false data injection for an islanded microgrid equipped with inverter-based energy resources. The rest of the paper is organized as follows. Section 2 presents a general model for inverter-based microgrid model. The proposed secondary controller based on Unknown Input Observer (UIO) is described in Section 3. A vulnerability analysis of the islanded microgrid against FDI is elaborated in Section 4. Simulation and results are demonstrated in Section 5, and the conclusion is presented in Section 6.

2. Model of Inverter-Based Microgrids

This section introduces the inverter-based microgrid model equipped with primary control. Also, this section proposes the optimal resilient secondary control for the islanded microgrid.

2.1. The Microgrid State-Space Representation

This islanded microgrid model includes three major elements of inverters, grid topology, and loads. Inverters dynamics cover output filter, coupling inductor, power-sharing controller, and current and voltage controllers. Each inverter has a separate reference frame, e.g., axis $(d-q)_i$ for ith inverter, whose rotating frequency ω_i is tuned by its power-sharing controller. Rotating frequency ω_{com} denotes the reference frame of one of the microgrid inverters. A common reference frame with axis $(d-q)_i$ and rotating frequency ω_{com} characterize the loads and the network's dynamic equations. Through the following transformation matrix, other inverters' reference frames are translated to the common reference frame [14]:

$$[f_{DQ}] = [T_i][f_{dq}] \tag{1}$$

$$[T_i] = \begin{bmatrix} cos(\delta_i) & -sin(\delta_i) \\ sin(\delta_i) & cos(\delta_i) \end{bmatrix} \quad (2)$$

where δ_i is the angle of the reference frame of ith inverter with respect to the common reference frame. In this paper, all inverters are Voltage Source Inverters (VSI) which are usually applied to interface distributed generators (DGs) to grids.

2.1.1. VSI State-Space Model

Figure 1 shows that the DG inverter control process has three different parts, such as voltage, current, and power control loops. The power control loop regulates the magnitude and frequency for the fundamental component of the inverter's voltage according to the droop characteristic set for the active and reactive powers. Voltage and current controllers are utilized to eliminate high-frequency disturbances and obtain proper damping for the output LC filter [14,15]. Since frequency control of this islanded microgrid is within the scope of this paper, we only describe the VSI power controller (not voltage and current controllers). The detailed model can be learned from [14,16].

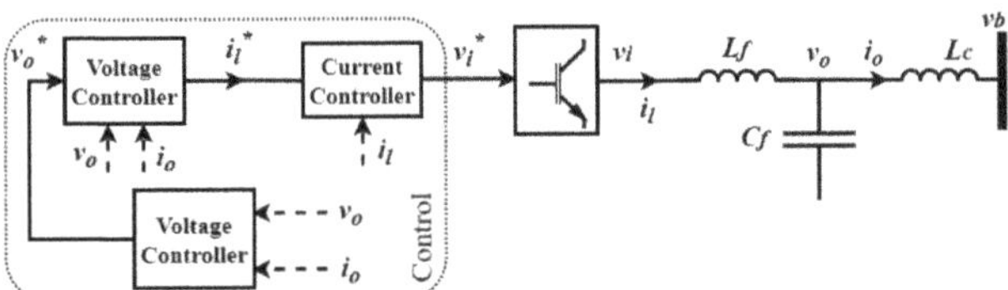

Figure 1. General diagram of DG inverter connected to Microgrids [14].

Utilizing the droop control for DG inverters, we mimic the governor behavior of synchronous generators in power systems. When there is a load rise, the microgrid frequency is reduced. Besides, when there is a voltage decrease, the reactive power is regulated proportionally. The power control diagram of VSIs is shown in Figure 2. Examining Figure 2, instantaneous active $\tilde{p}$ and reactive power $\tilde{q}$ are provided from the measured current and voltage in $(d-q)$ frame as in (3) and (4):

$$\tilde{p} = v_{od}i_{od} + v_{oq}i_{oq} \quad (3)$$

$$\tilde{q} = v_{od}i_{oq} + v_{oq}i_{od} \quad (4)$$

Also, these instantaneous power elements are passed through low-pass filters with ω_c cut-off frequency to obtain the fundamental real and reactive power as in (5) and (6):

$$P = \frac{\omega_c}{s+\omega_c}\tilde{p} \quad (5)$$

$$Q = \frac{\omega_c}{s+\omega_c}\tilde{q} \quad (6)$$

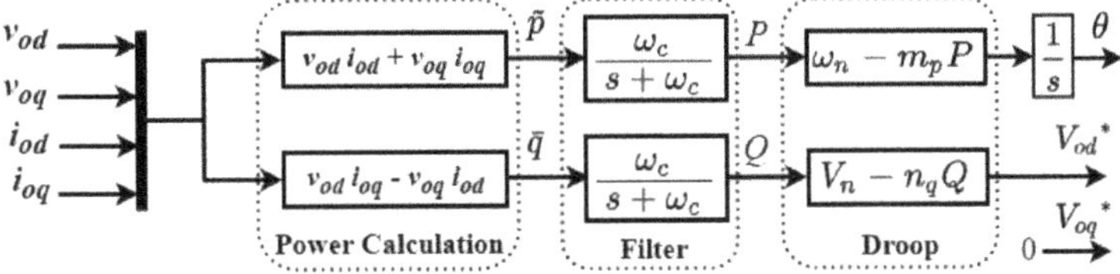

Figure 2. External power controller diagram of DG inverter [14].

The artificial droop can share active and reactive powers in the inverter frequency. As seen in (7), the frequency ω is set based on the droop coefficient m_p, and the phase θ is set by the frequency integration. ω_n indicates the nominal frequency set-point, and α indicates the inverter reference frame angle with nominal rotating frequency ω_n [14]:

$$\begin{gathered} \omega = \omega_n - m_p P \\ \dot{\theta} = \omega, \qquad \theta = \omega_n t - \int m_p P dt \\ \alpha = -\int m_p P dt, \qquad \dot{\alpha} = -m_p P \end{gathered} \tag{7}$$

Similar to active power, an artificial droop is used to share reactive power through specifying the output voltage magnitude, which is set to the d-axis of the inverter reference frame (the q-axis reference is set to zero) as in (8):

$$v_{od}^* = V_n - n_q Q, \qquad v_{oq}^* = 0 \tag{8}$$

The droop coefficients m_p and n_q are provided using the maximum and minimum limits of frequency and voltage magnitude as :

$$m_p = \frac{\omega_{max} - \omega_{min}}{P_{max}}, \qquad n_q = \frac{V_{odmax} - V_{odmin}}{Q_{max}} \tag{9}$$

As expressed earlier, one of the inverters' reference frames is picked as the common frame to build-up the complete model on a common reference frame. An angle δ is specified for each inverter as in (10) to transfer the variables of other inverters into the common reference frame:

$$\delta_i = \int (\omega_i - \omega_{com}) \qquad \Delta\dot{\delta}_i = \Delta\omega_i - \Delta\omega_{com} = \Delta\omega_{ni} - \Delta\omega_{n1} - (m_i \Delta P_i - m_1 \Delta P_1) \tag{10}$$

where ω_{ni} denotes the rated frequency set-point for each inverter [17]. As seen in (10), all inverter angle dynamics $\Delta\dot{\delta}_i$ are a function of the first inverter active power ΔP_1. After considering dynamics of voltage and current controllers and reorganizing all equations, the combined small-signal model for "s" number of DG inverters on a common reference frame is as in (11) and (12):

$$[\Delta \dot{x}_{INV}] = A_{INV}[\Delta x_{INV}] + B_{INV}[\Delta v_{bDQ}] + B_{com}[\Delta\omega_{com}] + B_n[\Delta\omega_n] \tag{11}$$

$$[\Delta i_{oDQ}] = C_{INVn}[\Delta x_{INV}] \tag{12}$$

where $[\Delta x_{INV}] = [\Delta x_{inv1} \quad \Delta x_{inv2} \quad \dots \quad \Delta x_{invs}]$ and $[\Delta\omega_n] = [\Delta\omega_{n1} \quad \Delta\omega_{n2} \quad \dots \quad \Delta\omega_{n3}]$.

2.1.2. Network Model

It is assumed that the islanded microgrid here has n lines, m nodes, s inverters, and p loads, as shown in Figure 3. The dynamic equations of line current of ith line connected between nodes j and k on the common reference frame are obtained as (13) and (14) [14]:

$$\frac{di_{lineDi}}{dt} = \frac{-r_{linei}}{L_{linei}} i_{lineDi} + \omega i_{lineQi} + \frac{1}{L_{linei}} v_{bDj} - \frac{1}{L_{linei}} v_{bDk} \tag{13}$$

$$\frac{di_{lineQi}}{dt} = \frac{-r_{linei}}{L_{linei}} i_{lineQi} - \omega i_{lineDi} + \frac{1}{L_{linei}} v_{bQj} - \frac{1}{L_{linei}} v_{bQk} \tag{14}$$

Therefore, the small-signal state-space representation of the microgrid is obtained as in (15):

$$[\Delta \dot{i}_{lineDQ}] = A_{NET}[\Delta i_{lineDQ}] + B_{1NET}[\Delta v_{bDQ}] + B_{2NET}\Delta\omega \tag{15}$$

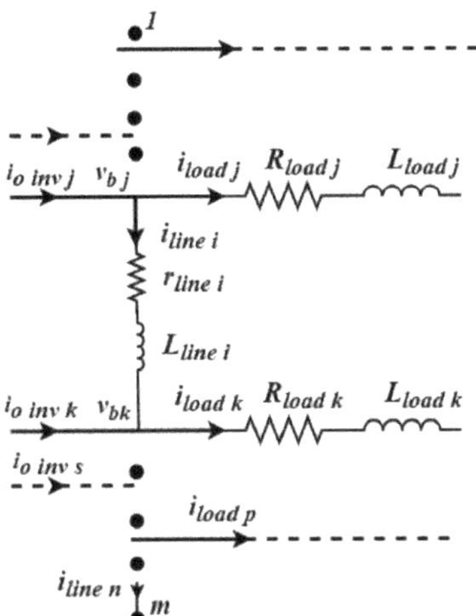

Figure 3. Network topology of inverter-based microgrid model [14].

2.1.3. Load Model

The dynamic equation for the resistive and inductive load connected at the ith node is obtained in (16):

$$\frac{di_{loadDi}}{dt} = \frac{-R_{loadi}}{L_{loadi}} i_{loadDi} + \omega i_{loadQi} + \frac{1}{L_{loadi}} v_{bDi} \tag{16}$$

$$\frac{di_{loadQi}}{dt} = \frac{-R_{loadi}}{L_{loadi}} i_{loadQi} - \omega i_{loadDi} + \frac{1}{L_{loadi}} v_{bQi} \tag{17}$$

Therefore, the small-signal state-space representation of loads is generally obtained as in (18):

$$[\Delta \dot{i}_{loadDQ}] = A_{load}[\Delta i_{loadDQ}] + B_{1LOAD}[\Delta v_{bDQ}] + B_{2LOAD}\Delta\omega \tag{18}$$

2.1.4. Complete Microgrid Model

The microgrid model is obtained by augmenting all these modules of VSI inverters, network lines, and loads together as in (19):

$$\begin{cases} \begin{bmatrix} \Delta x_{INV} \\ \Delta x_{Net} \\ \Delta x_{Load} \end{bmatrix}^{\cdot} = A \begin{bmatrix} \Delta x_{INV} \\ \Delta x_{Net} \\ \Delta x_{Load} \end{bmatrix} + B[\Delta\omega_n] + Dd(t) = Ax(t) + Bu(t) = Dd(t) \\ y = Cx(t) + v(t) \end{cases} \tag{19}$$

where A, B are microgrid characteristic matrices and [17] can be referred for more information. With this microgrid's state-space model, we can design an optimal secondary control for this system and enhance the primary control performance.

2.2. Secondary Control of the Inverter-Based Microgrid

To apply an optimal control based on Linear Quadratic Regulator (LQR), we need to ensure the pair (A, B) is controllable; otherwise, designing such controllers for this system is not possible. However, this system is not controllable since the small-signal transient, and the steady-state responses obtained from the first state $\Delta\delta_1$ is zero ($\Delta\dot{\delta}_1 = \Delta\omega_1 - \Delta\omega_{com} = 0$) [18]. Therefore, this state must be skipped by removing the corresponding row and column in A and B or using the minimum realization technique for this system. After making the reduced-order model, we can utilize the LQR controller on the microgrid model.

The microgrid eigenvalues with and without the state corresponding to $\Delta\delta_1$ are shown in Table 1.

Table 1. Eigenvalues of the microgrid (*A* matrix).

	Standard Inverter-Based Microgrid Model	**Reduced-Order Inverter-Based Microgrid Model**
Eigenvalues	−9.44e6 + j3.14e2	−9.44e6 + j3.14e2
	−9.44e6 − j3.14e2	−9.44e6 − j3.14e2
	−3.63e6 + j3.14e2	−3.63e6 + j3.14e2
	−3.63e6 − j3.14e2	−3.63e6 − j3.14e2
	−2.85e6 + j3.14e2	−2.85e6 + j3.14e2
	−2.85e6 − j3.14e2	−2.85e6 − j3.14e2
	−2.94e3 + j7.38e3	−2.94e3 + j7.38e3
	−2.94e3 − j7.38e3	−2.94e3 − j7.38e3
	−2.79e3 + j6.84e3	−2.79e3 + j6.84e3
	−2.79e3 − j6.84e3	−2.79e3 − j6.84e3
	−2.84e3 + j4.89e3	−2.84e3 + j4.89e3
	−2.84e3 − j4.89e3	−2.84e3 − j4.89e3
	−2.53e3 + j4.43e3	−2.53e3 + j4.43e3
	−2.53e3 − j4.43e3	−2.53e3 − j4.43e3
	−2.86e3 + j2.92e3	−2.86e3 + j2.92e3
	−2.86e3 − j2.92e3	−2.86e3 − j2.92e3
	−2.21e3 + j2.20e3	−2.21e3 + j2.20e3
	−2.21e3 − j2.20e3	−2.21e3 − j2.20e3
	−1.49e3 + j2.51e3	−1.49e3 + j2.51e3
	−1.49e3 − j2.51e3	−1.49e3 − j2.51e3
	−1.29e3 + j2.10e3	−1.29e3 + j2.10e3
	−1.29e3 − j2.10e3	−1.29e3 − j2.10e3
	−1.31e3 + j1.71e3	−1.31e3 + j1.71e3
	−1.31e3 − j1.71e3	−1.31e3 − j1.71e3
	−1.22e3 + j1.65e3	−1.22e3 + j1.65e3
	−1.22e3 − j1.65e3	−1.22e3 − j1.65e3
	−1.14e3 + j1.54e3	−1.14e3 + j1.54e3
	−1.14e3 − j1.54e3	−1.14e3 − j1.54e3
	−1.11e3 + j1.50e3	−1.11e3 + j1.50e3
	−1.11e3 − j1.50e3	−1.11e3 − j1.50e3
	−2.00e1 + j3.13e2	−2.00e1 + j3.13e2
	−2.00e1 − j3.13e2	−2.00e1 − j3.13e2
	−2.50e1 + j3.13e2	−2.50e1 + j3.13e2
	−2.50e1 − j3.13e2	−2.50e1 − j3.13e2
	−1.42e2 + j2.10e2	−1.42e2 + j2.10e2
	−1.42e2 − j2.10e2	−1.42e2 − j2.10e2
	−1.23e2 + j1.50e2	−1.23e2 + j1.50e2
	−1.23e2 − j1.50e2	−1.23e2 − j1.50e2
	−13.48 + j30.21	−13.48 + j30.21
	−13.48 − j30.21	−13.48 − j30.21
	−15.53 + j10.59	−15.53 + j10.59
	−15.53 − j10.59	−15.53 − j10.59
	−20.84	−20.84
	−28.25	−28.25
	−31.38	−31.38
	−31.40	−31.40
	0	Removed

3. The Proposed Control Method

Standard controllers based on observes apply the entire grid states for the estimation and control procedure. Frequency deviation Δf, which is one of the states in microgrid model, reflects frequency distortion because of intermittent behavior of renewable energy sources, loads, and any cyber anomalies in the system. Secondary controllers are often designed to guarantee the frequency stability of grids. We utilize a combined UIO and LQR controllers formed in two layers to control the microgrid . The first layer estimates the microgrid states and detects the UIs using (19). The second layer decreases the frequency discrepancy using (30)–(32).

3.1. Design of Unknown Input Observer

In UIO, the states $x(k) = \begin{bmatrix} x_1 & x_2 \end{bmatrix}^T$ are separated into two groups of states: states corresponding to known inputs, $x_1(k)$ and states corresponding to unknown inputs, $x_2(k)$ [19]. Here, $x_2(k)$ state indicates the frequency discrepancy $\Delta f(k)$ subject to the UI. Now, we can independently estimate both groups of states, including $x_1(k)$ and $x_2(k) = \Delta f$, and ultimately and identify the UI. In this case, the frequency deviation $\Delta f(k)$ does not distort the state estimation as well as the frequency regulation of the microgrid. For separating the states to two groups, a nonsingular matrix $\Psi = \begin{bmatrix} N & D \end{bmatrix}$ is determined where N is the arbitrary matrix selected such that Ψ is cannot be singular $(N \in \mathbb{R}^{n(n-q)})$ [20]. This transition matrix Ψ is multiplied to both sides of (19) to obtain its equivalent representation (20). Using this representation, we obtain another model representation $\overline{x} = \begin{bmatrix} \overline{x}_1 & \overline{x}_2 \end{bmatrix}^T = \Psi^{-1}x = \Psi^{-1}\begin{bmatrix} x_1 & x_2 \end{bmatrix}^T$ with $\overline{x}_1 \in \mathbb{R}^{n-q}$ and $\overline{x}_2 \in \mathbb{R}^{q}$, and new constant characteristic matrices specified in (21):

$$\begin{cases} \overline{x}(k+1) = \overline{A}\overline{x}(k) + \overline{B}u(k) + \overline{D}d(k) + w(k) \\ y(k) = \overline{C}\overline{x}(k) + v(k) \end{cases} \tag{20}$$

$$\begin{gathered} \overline{A} = \begin{bmatrix} \overline{A}_{11} & \overline{A}_{12} \\ \overline{A}_{21} & \overline{A}_{22} \end{bmatrix} = \Psi^{-1}A\Psi, \quad \overline{B} = \begin{bmatrix} \overline{B}_1 & \overline{B}_2 \end{bmatrix}^T = \Psi^{-1}B, \\ \overline{D} = \Psi^{-1}D, \qquad \overline{C} = C\Psi = \begin{bmatrix} CN & CD \end{bmatrix} \end{gathered} \tag{21}$$

Further, the states $\overline{x}_2(k)$ subject to the UI are eliminated to find microgrid model free from unknown input as in (22) [20]:

$$\begin{cases} \begin{bmatrix} I_{n-q} & 0 \end{bmatrix}\overline{x}(k+1) = \begin{bmatrix} \overline{A}_{11} & \overline{A}_{12} \end{bmatrix}\overline{x}(k) + \overline{B}_1 u(k) \\ y(k) = \begin{bmatrix} CN & CD \end{bmatrix}\overline{x}(k) + v(k) \end{cases} \tag{22}$$

Assuming $\overline{x}_2(k)$ can be attained from the measurement output $y(k)$, (22) can be reorganized to a linear representation. The transfer matrix $U = \begin{bmatrix} CD & \Gamma \end{bmatrix}$ is comprised of CD, which is a full-column rank matrix, and $\Gamma \in \mathbb{R}^{m\times(m-q)}$, which is an arbitrary matrix defined such that U is a nonsingular matrix. Therefore, we have $U^{-1} = \begin{bmatrix} U_1 & U_2 \end{bmatrix}^T$ with $U_1 \in \mathbb{R}^{q\times m}$ and $U_2 \in \mathbb{R}^{(m-q)\times m}$. If the measurement equations in (22) is multiplied by U^{-1}, we obtain (23) and (24):

$$U_1 y(k) = U_1 CN\overline{x}_1(k) + \overline{x}_2(k) \tag{23}$$

$$U_2 y(k) = U_2 CN\overline{x}_1(k) \tag{24}$$

Substituting (23) in (22) and merging it with (24), we obtain (25) with revised state matrix of $\widetilde{A} = \overline{A}_{11} - \overline{A}_{12}U_1 CN$, modified measurement matrix $\widetilde{C} = U_2 CN$, modified measurement vector $\overline{y}(k) = U_2 y(k)$, and $E = \overline{A}_{12}U_1$:

$$\begin{cases} \overline{x}_1(k+1) = \widetilde{A}\overline{x}_1(k) + \overline{B}_1 u(k) + Ey(k) + w_1(k) \\ \overline{y}(k) = \widetilde{C}\overline{x}_1(k) + v_1(k) \end{cases} \tag{25}$$

A Luenberger observer can be developed for this microgrid, in case the pair $(\widetilde{A}, \widetilde{C})$ is observable. The observability conditions of this microgrid is reviewed in [20]. The

Luenberger observer with $L \in \mathbb{R}^{(n-q)\times(m-q)}$ for $\overline{x}_1(k)$ is designed as in (26) to estimate $\overline{x}_1(k)$ with $\hat{\overline{x}}_1 \in \mathbb{R}^{n-q}$:

$$\hat{\overline{x}}_1(k+1|k) = (\widetilde{A} - L\widetilde{C})\hat{\overline{x}}_1(k|k-1) + \overline{B}u(k) + L^*\overline{y}(k) \tag{26}$$

where $L^* = LU_2 + E$ and the Luenberger observer coefficient matrix L is calculated from solving the discrete Riccati equation $L = (AP_1C^T)(CP_1C^T + Q_1 + R_1)$. Also, P_1 solves the algebraic Riccati equation that reduces the steady-state error covariance $P_1 = \lim_{k\to\infty} E[(x_1 - \hat{\overline{x}}_1)(x_1 - \hat{\overline{x}}_1)^T]$. Now, all estimated states can be found in (27) using (23) and (26):

$$\hat{x}(k|k) = \Psi\hat{\overline{x}} = \Psi\begin{bmatrix}\hat{\overline{x}}_1(k|k)\\ \hat{\overline{x}}_2(k|k)\end{bmatrix} \tag{27}$$

where $\hat{\overline{x}}_2(k|k) = U_1y(k) - U_1CN\hat{\overline{x}}_1(k|k)$. This indicates that the frequency discrepancy $x_2(k) = \Delta f(k)$ is estimated from measured data $y(k)$ and remaining estimated states $x_1(k)$. To identify the unknown input $d(k)$, including false data injection to the microgrid actuators, (27) is replaced in (20) to obtain (28):

$$\begin{aligned}&\begin{bmatrix}\hat{\overline{x}}_1(k+1|k)\\ U_1y(k+1) - U_1CN\hat{\overline{x}}_1(k+1|k)\end{bmatrix} = \\ &\begin{bmatrix}A_{11} & A_{12}\\ A_{21} & A_{22}\end{bmatrix}\begin{bmatrix}\hat{\overline{x}}_1(k|k)\\ U_1y(k) - U_1CN\hat{\overline{x}}_1(k|k)\end{bmatrix} + \begin{bmatrix}\overline{B}_1\\ \overline{B}_2\end{bmatrix}u(k) + \begin{bmatrix}0\\ I_q\end{bmatrix}\hat{d}(k)\end{aligned} \tag{28}$$

Simplifying (28), the UI can be identified as $\hat{d}(k) = U_1y(k+1) + U_1CN\hat{\overline{x}}_1(k+1|k) - \overline{A}_{12}\hat{\overline{x}}_1(k|k) - \overline{A}_{22}(U_1y(k) - U_1CN\hat{\overline{x}}_1(k|k)) - \overline{B}_2u(k)$. Equation (29) shows the detected UI:

$$\hat{d}(k) = F_1\overline{y}(k+1) + F_2\hat{\overline{x}}_1(k|k) + F_3y(k) + F_4u(k) \tag{29}$$

where $F_1 = U_1$, $F_2 = U_1CNLU_2CN + U_1CN\overline{A}_{12}U_1CN - U_1CN\overline{A}_{11} - \overline{A}_{21} + \overline{A}_{22}U_1CN$, $F_3 = -U_1CNLU_2 - U_1CN\overline{A}_{12}U_1 - \overline{A}_{22}U_1$, and $F_4 = -U_1CN\overline{B}_1 - \overline{B}_2$, $\overline{y}$ is a filtered measurement signal of y. These estimated states $\hat{x}(k)$ of the microgrid and the detected UIs $\hat{d}(k)$ are applied in the second control layer to compensate the frequency deviation.

3.2. Unknown Input Compensator Design

The UI compensator is developed by $\hat{u}(k) = G_1\hat{x}(k)$, which is utilized in the closed-loop control of the microgrid matrix $A - BG_1$. G_1 is an optimized state feedback coefficient designed to guarantee that the Eigenvalues of $A - BG_1$ fall within stable control region of the microgrid . To design this compensator coefficient, the pair (A, B) must be controllable that means the rank of the controlability matrix is equal to the rank of microgrid model, or $rank[B \quad AB \quad A^2B \quad A^3B \quad A^4B] = rank\ A$. In the second layer, the optimal compensator coefficient G_1 is obtained using input $u(k)$ and the estimated microgrid states $\hat{x}(k)$ in (27) to minimize the performance index J in (30). Q_2 and R_2 are weight matrices for control performance and input energy respectively, in (30). If $\Delta f_{ref} = 0$, we have the proposed control law $u^*(k) = -\hat{u}(k) - G_2\hat{d}(k)$, which is a linear combination of the detected unknown inputs $\hat{d}(k)$ with $G_2 = D$, and the compensator coefficient G_1 in (31). The first term of the proposed control law G_1 is obtained from (32) to optimize the performance index J. In Equation (32), P_2 implies the unique positive-definite solution of discrete-time algebraic Riccati equation in (30) specified as $P_2 = A^TP_2A - A^TP_2B(R_2 + B^TP_2B)^{-1}B^TP_2A + Q_2$. The second term of the proposed control law is the detected unknown inputs $\hat{d}(k)$ subtracted from the first term of the proposed control law to eliminate the UI. The diagram of the proposed control strategy is shown in Figure 4 :

$$\begin{aligned}&\text{Minimize } J = \sum_{n=1}^{\infty}(x(k)^TQ_2x(k) + u(k)^TR_2u(k))\\ &\text{Subject to. } x(k+1) = Ax(k) + Bu(k)\end{aligned} \tag{30}$$

$$u^*(k) = -(G_1\hat{x}(k) + G_2\hat{d}(k)) \tag{31}$$

$$G_1 = (R_2 + B^TP_2B)^{-1}B^TP_2A \tag{32}$$

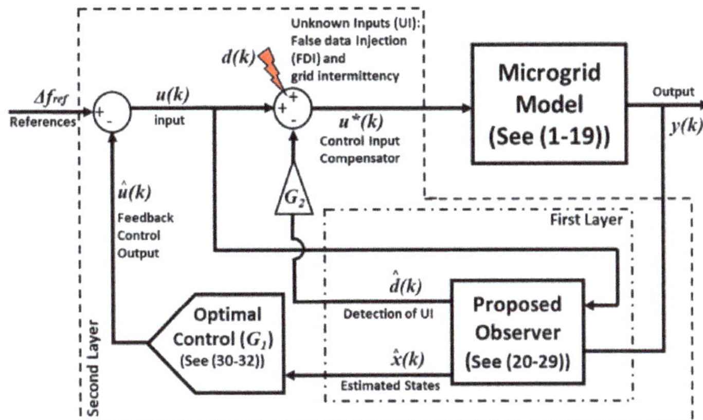

Figure 4. The proposed control strategy for islanded microgrid.

4. Vulnerability Analysis of Microgrids with Inverter-Based Resources

This section shows the vulnerability of microgrid control based on the secondary control design of the microgrid model obtained in Section 2. According to Equation (10), all inverter angle dynamics $\Delta\dot{\delta}_i$ are direct functions of the first inverter active power ΔP_1 in this model, which is introduced as (desired) reference frequency of inverters in some papers [21]. All droop gains, also called the active power sharing gains, including m_1, are selected based on the microgrid's power rating of energy resources. Typically, these active power sharing gains are chosen as the inverse proportion of the nominal power to guarantee the accuracy of active power-sharing, i.e., $\frac{P_i}{P_1} = \frac{m_i}{m_1}$. If this proportion ratio is distorted, it may destabilize the microgrid.

Disclosing and injecting the data packets of the first inverter active power ΔP_1 can enormously distort the stability and performance of the entire grid for three critical reasons:

- Stealthy FDIs always severely impact systems since this class of cyber attacks cannot be easily identified [22]. Most detecting methods fail correctly to discover the stealthy FDIs since they mainly rely on residual-based detection that may not trigger the alarm in the presence of this cyber attack type;
- Islanded microgrids have limited access to energy resources, and the poor performance of these inverter-based resources can draw the grid to an unstable control region;
- To control this microgrid model and regulate the frequency, all inverters' angles are highly dependent on the first inverter (reference) active power. In other words, any inaccuracy in the first inverter active power can degrade or destabilize the energy resources in a microgrid, as shown further.

Stealthy FDI of this information $m_{1f}\Delta P_1$ can target the droop gain of the first inverter $m_1\Delta P_1$ and change it as follows:

$$\begin{aligned}\Delta\dot{\delta}_i &= \Delta\omega_{ni} - \Delta\omega_{n1} - (m_i\Delta P_i - m_1\Delta P_1) + (m_{1f}\Delta P_1) = \\ &\Delta\omega_{ni} - \Delta\omega_{n1} - (m_i\Delta P_i - (m_1 + m_{1f})\Delta P_1)\end{aligned} \tag{33}$$

Disclosed data exchanges of the first inverter's active power (reference active power) can be injected by adversaries into the actuators of inverters' control inputs. This imposed change is not visible since these injected data are a part of the microgrid data. This data manipulation also can devastate the control mechanism of inverter-based resources in microgrids since standard secondary controls rely on all data packets of actuators. The Simulation and Results section shows that standard controllers cannot address this type of FDI to the microgrid since they need all actuators' data in their control procedure.

5. Simulation and Results

This section implements our proposed UIO-based control technique on the secondary control of an islanded inverter-based microgrid model that was introduced in [14]. We show that the effect of the false data injected to the inverter angle dynamic of this system is well compensated using our proposed control methodology. This microgrid model consists of three main modules of inverters, network (line topology), and loads. Inverters dynamics comprises power-sharing controller, output filter, coupling inductor, and current and voltage controller. Each inverter has its own reference frame whose rotation frequency is adjusted by its power sharing controller. This case study is a 220 V (per phase Root Mean Square), 60 Hz microgrid equipped with three inverters of equal rating (10 kVA), supplying two loads as shown in Figure 5. Figure 5 illustrates the microgrid topology. In this figure, the inverters supply two loads connected to the microgrid through lines 1 and 2. More information about the microgrid parameters and initial microgrid conditions are provided in Tables 2 and 3.

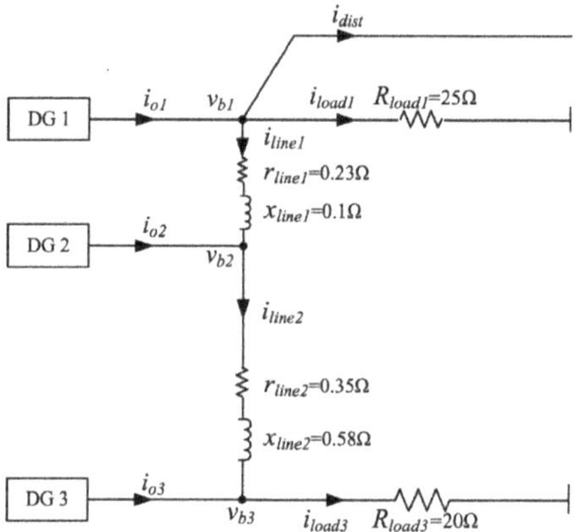

Figure 5. Case study: microgrid topology.

Table 2. Microgrid parameters.

Inverter Parameters			
Parameter	Value	Parameter	Value
f_s	8 kHz	m_p	9.4×10^{-5}
L_f	1.35 mH	n_q	1.3×10^{-3}
C_f	50 μF	K_{pv}	0.05
r_f	0.1 Ohm	K_{iv}	390
L_c	0.35 mH	K_{pc}	10.5
r_{Lc}	0.03 Ohm	K_{ic}	16×10^{e3}
ω_c	31.41	F	0.75

Table 3. Microgrid initial conditions

Initial Conditions			
Parameter	Value	Parameter	Value
V_{od}	[380.8 381.8 380.4]	V_{oq}	[0 0 0]
I_{od}	[11.4 11.4 11.4]	I_{oq}	[0.4 −1.45 1.25]
I_{id}	[11.4 11.4 11.4]	I_{lq}	[−5.5 −7.3 −4.6]
V_{bd}	[379.5 380.5 379]	V_{bq}	[−6 −6 −5]
ω_0	[314]	δ_0	[0 1.9 $\times 10^{-3}$−3 −0.0113]
I_{line1d}	[−3.8]	I_{line1q}	[0.4]
I_{line2d}	[7.6]	I_{line2q}	[−1.3]

To show the efficacy of the proposed UIO based secondary control, in particular, and secondary control, in general, three simulation cases for this model are shown as follows. In these simulation cases, the secondary control based on our proposed UIO is compared with the secondary control equipped with Linear Quadratic Gaussian (LQG), Luenberger-based control, and the microgrid without secondary control (just primary control response).

5.1. Case 1

In this case, the effect of a step load change is investigated. The step load change occurs for load3 at $t = 1$ s. The angles of inverters 2 and 3 for the microgrid with only primary control and with three secondary controllers are demonstrated in Figures 6 and 7. As Figures 6 and 7 displays, the primary control only is not enough to appropriately control microgrid. These results determine that the proposed UIO controller performs significantly better than the other two secondary controllers. Also, the inverters' frequencies presented in Figures 8 and 9 show that the proposed UIO controller compensates the inverters' frequencies better than other secondary controllers.

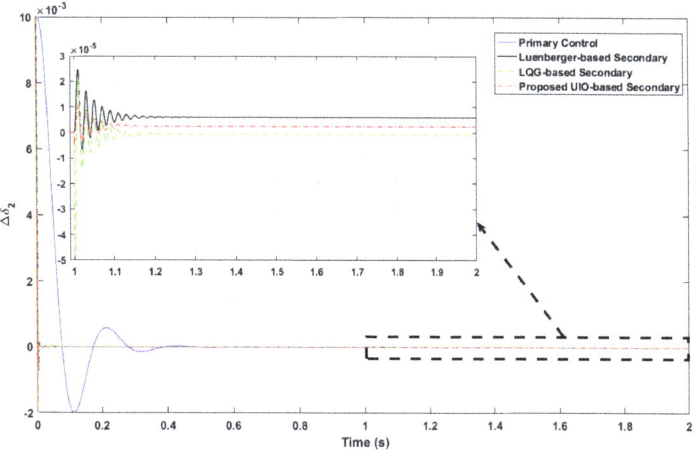

Figure 6. Second inverter angle δ_2 in scenario 1 (Step load change).

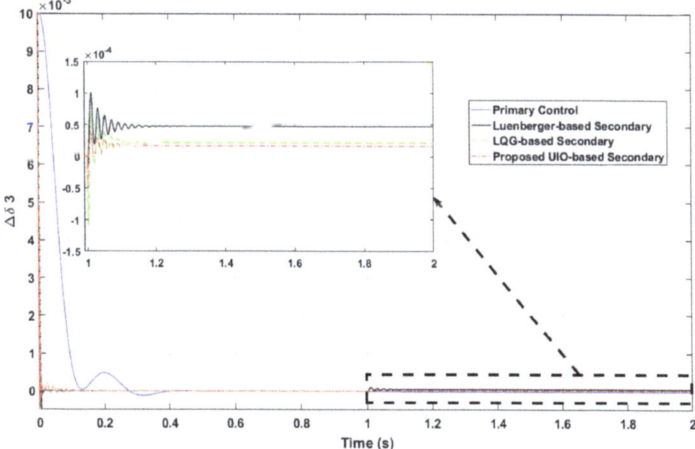

Figure 7. Third inverter angle δ_3 in scenario 1 (Step load change).

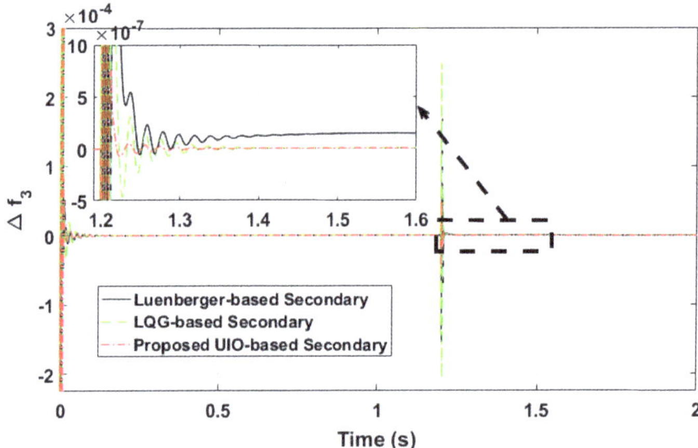

Figure 13. Third inverter frequency dynamic Δf_3 in scenario 2 (FDI into second inverter's actuator).

5.3. Case 3

In this case, the stealthy FDI effect to one inverter angle is investigated. As seen in 10, all inverter angle dynamics $\Delta\dot{\delta}_i$ are a function of the first inverter active power ΔP_1. Thus, one destructive FDI can be launched once ΔP_1 is stealthily disclosed and used in the data injection. The FDI is added to the actuator of inverter angle 2 at $t = 1.4$ s as:

$$d(t) = 0.5\Delta P_1(t) \qquad t \geq 1.4 \text{ s}$$

The angles of inverters 2 and 3 for the microgrid with only primary control and with three secondary controllers are demonstrated in Figures 14 and 15. Considering the results, the proposed UIO controller can still resiliently control the microgrid; however, this stealthy FDI can destabilize the microgrid equipped with LQG and Luenberger-based secondary control. The Luenberger-based secondary controller operates significantly more unstable compared to the LQG-based controller. Also, the inverters' frequencies shown in Figures 16 and 17 prove that the proposed UIO controller, unlike the other secondary controllers, compensates the inverters' frequencies. In fact, this simulation case verifies that the proposed UIO-based controller properly maintains the microgrid resiliency.

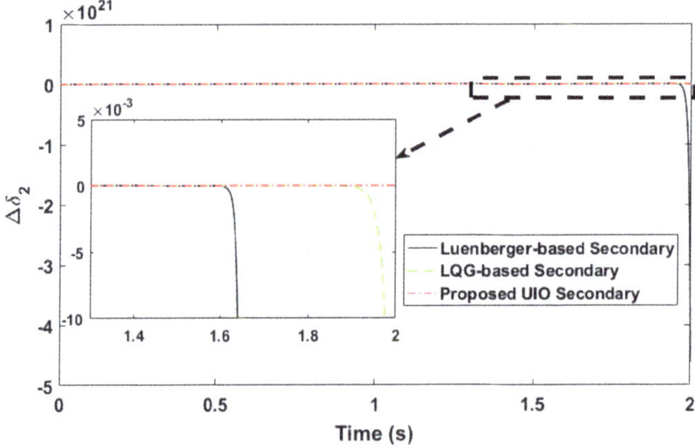

Figure 14. Second inverter angle δ_2 in scenario 3 (Stealthy FDI to second inverter's actuator).

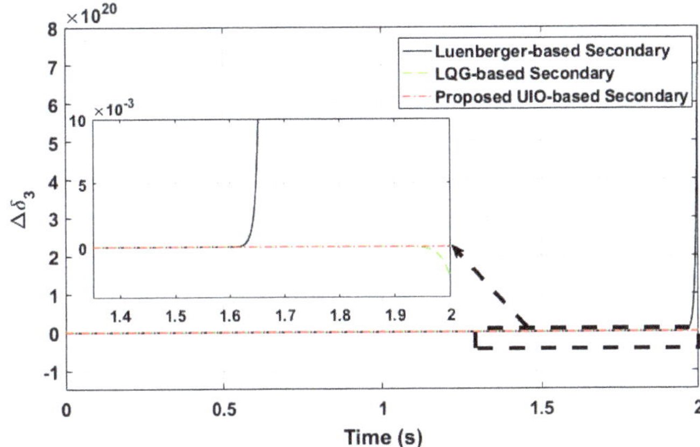

Figure 15. Third inverter angle δ_3 in scenario 3 (Stealthy FDI to second inverter's actuator).

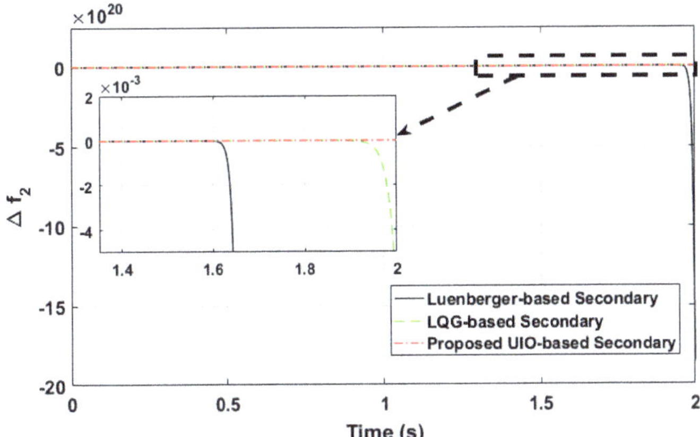

Figure 16. Second inverter frequency dynamic Δf_2 in scenario 3 (Stealthy FDI to second inverter's actuator).

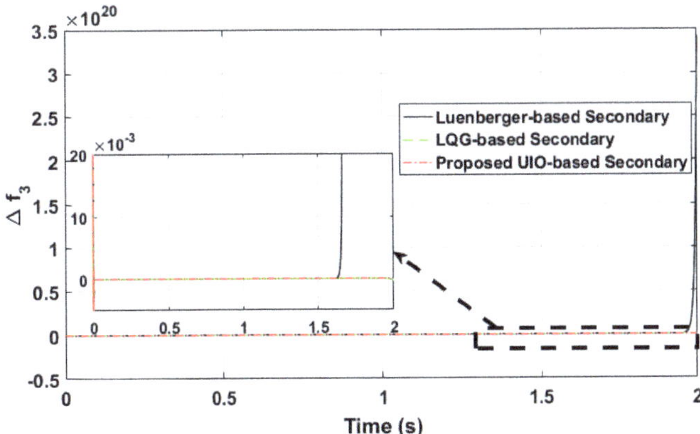

Figure 17. Third inverter frequency dynamic Δf_3 in scenario 3 (Stealthy FDI to second inverter's actuator).

6. Conclusions

This paper shows a procedure to develop a secondary control for an islanded microgrid. This islanded microgrid model includes inverter-based energy resources. To design a secure microgrid, the secondary control needs to operate resiliently against cyber anomalies, e.g., False Data Injection (FDI). In this paper, we show that the inverter-based microgrid can be tremendously vulnerable to stealthy FDI, and this type of cyber-disruption can make the microgrid unstable. We develop a secondary control based on Unknown Input Observer (UIO) and optimal compensator to regulate frequency of an islanded microgrid. The proposed secondary control can be effective against different types of FDIs.

Author Contributions: Methodology, M.R.K.; Writing, M.R.K., S.K.S. and J.M.S.; Validation, M.R.K., V.V. All authors have read and agreed to the published version of the manuscript.

Funding: Partial support of this research was provided by the Woodrow W. Everett, Jr. SCEEE Development Fund in cooperation with the Southeastern Association of Electrical Engineering Department Heads and National Science Foundation's Division Of Electrical, Communication & Cyber System Grant # 1351201. Any opinions, findings, and conclusions, or recommendations expressed in this material are those of the author(s) and do not necessarily reflect the views of the sponsoring agency.

Conflicts of Interest: The authors declare no conflict of interest.

References

1. Khalghani, M.R.; Solanki, J.; Solanki, S.K.; Khooban, M.H.; Sargolzaei, A. Resilient Frequency Control Design for Microgrids Under False Data Injection. *IEEE Trans. Ind. Electron.* **2021**, *68*, 2151–2162. [CrossRef]
2. Khalghani, M.R.; Solanki, J.; Khushalani-Solanki, S.; Sargolzaei, A. Stochastic Load Frequency Control of Microgrids Including Wind Source Based on Identification Method. In Proceedings of the 2018 IEEE International Conference on Environment and Electrical Engineering and 2018 IEEE Industrial and Commercial Power Systems Europe (EEEIC/I CPS Europe), Palermo, Italy, 12–15 June 2018; pp. 1–6. [CrossRef]
3. Victorio, M.; Sargolzaei, A.; Khalghani, M.R. A Secure Control Design for Networked Control Systems with Linear Dynamics under a Time-Delay Switch Attack. *Electronics* **2021**, *10*, 322 . [CrossRef]
4. Khalghani, M.R.; Solanki, J.; Solanki, S.K.; Sargolzaei, A. Resilient and Stochastic Load Frequency Control of Microgrids. In Proceedings of the 2019 IEEE Power Energy Society General Meeting (PESGM), Atlanta, GA, USA, 4–8 August 2019; pp. 1–5. [CrossRef]
5. Qi, J.; Hahn, A.; Lu, X.; Wang, J.; Liu, C.C. Cybersecurity For Distributed Energy Resources And Smart Inverters. *IET Cyber-Phys. Syst.: Theory Appl.* **2016**, *1*, 28–39. [CrossRef]
6. Bidram, A.; Poudel, B.; Damodaran, L.; Fierro, R.; Guerrero, J.M. Resilient and Cybersecure Distributed Control of Inverter-Based Islanded Microgrids. *IEEE Trans. Ind. Inform.* **2020**, *16*, 3881–3894. [CrossRef]
7. Nguyen, T.L.; Tran, Q.T.; Caire, R.; Gavriluta, C.; Nguyen, V.H. Agent Based Distributed Control Of Islanded Microgrid—Real-Time Cyber-Physical Implementation. In Proceedings of the 2017 IEEE PES Innovative Smart Grid Technologies Conference Europe (ISGT-Europe), Turin, Italy, 26–29 September 2017; pp. 1–6. [CrossRef]
8. Zhou, Q.; Shahidehpour, M.; Alabdulwahab, A.; Abusorrah, A. A Cyber-Attack Resilient Distributed Control Strategy In Islanded Microgrids. *IEEE Trans. Smart Grid* **2020**, *11*, 3690–3701. [CrossRef]
9. Mahmud, R.; Seo, G.S. Blockchain-Enabled Cyber-Secure Microgrid Control Using Consensus Algorithm. In Proceedings of the 2021 IEEE 22nd Workshop on Control and Modelling of Power Electronics (COMPEL), Cartagena, Colombia, 2–5 November 2021; pp. 1–7. [CrossRef]
10. Zografopoulos, I.; Konstantinou, C. Detection of Malicious Attacks In Autonomous Cyber-Physical Inverter-Based Microgrids. *IEEE Trans. Ind. Inform.* **2021**, 1. [CrossRef]
11. Liu, S.; Hu, Z.; Wang, X.; Wu, L. Stochastic Stability Analysis and Control of Secondary Frequency Regulation for Islanded Microgrids Under Random Denial of Service Attacks. *IEEE Trans. Ind. Inform.* **2019**, *15*, 4066–4075. [CrossRef]
12. Lu, L.Y.; Liu, H.J.; Zhu, H.; Chu, C.C. Intrusion Detection in Distributed Frequency Control of Isolated Microgrids. *IEEE Trans. Smart Grid* **2019**, *10*, 6502–6515. [CrossRef]
13. Nguyen, T.L.; Wang, Y.; Tran, Q.T.; Caire, R.; Xu, Y.; Besanger, Y. Agent-based Distributed Event-Triggered Secondary Control for Energy Storage System in Islanded Microgrids-Cyber-Physical Validation. In Proceedings of the 2019 IEEE International Conference on Environment and Electrical Engineering and 2019 IEEE Industrial and Commercial Power Systems Europe (EEEIC/I CPS Europe), Genova, Italy, 11–14 June 2019; pp. 1–6. [CrossRef]
14. Pogaku, N.; Prodanovic, M.; Green, T.C. Modeling, Analysis and Testing of Autonomous Operation of an Inverter-Based Microgrid. *IEEE Trans. Power Electron.* **2007**, *22*, 613–625.

15. Banadaki, A.D.; Mohammadi, F.D.; Feliachi, A. State Space Modeling of Inverter Based Microgrids Considering Distributed Secondary Voltage Control. In Proceedings of the 2017 North American Power Symposium (NAPS), Morgantown, WV, USA, 17–19 September 2017, pp. 1–6.
16. Banadaki, A.D.; Feliachi, A.; Kulathumani, V.K. Fully Distributed Secondary Voltage Control in Inverter-Based Microgrids. In Proceedings of the 2018 IEEE/PES Transmission and Distribution Conference and Exposition (T&D), Denver, CO, USA, 16–19 April 2018; pp. 1–9.
17. Keshtkar, H.; Mohammadi, F.D.; Solanki, J.; Solanki, S.K. Multi-Agent Based Control of a Microgrid Power System in Case of Cyber Intrusions. In Proceedings of the 2020 IEEE Kansas Power and Energy Conference (KPEC), Manhattan, KS, USA, 13–14 July 2020; pp. 1–6. [CrossRef]
18. Rasheduzzaman, M.; Mueller, J.A.; Kimball, J.W. Reduced-Order Small-Signal Model of Microgrid Systems. *IEEE Trans. Sustain. Energy* **2015**, *6*, 1292–1305. [CrossRef]
19. Khalghani, M.R.; Khushalani-Solanki, S.; Solanki, J.; Sargolzaei, A. Cyber Disruption Detection In Linear Power Systems. In Proceedings of the 2017 North American Power Symposium (NAPS), Morgantown, WV, USA, 17–19 September 2017; pp. 1–6.
20. Hou, M.; Muller, P.C. Design of Observers for Linear Systems with Unknown Inputs. *IEEE Trans. Autom. Control.* **1992**, *37*, 871–875. [CrossRef]
21. Guo, F.; Wen, C.; Mao, J.; Song, Y.D. Distributed Secondary Voltage and Frequency Restoration Control of Droop-Controlled Inverter-Based Microgrids. *IEEE Trans. Ind. Electron.* **2015**, *62*, 4355–4364. [CrossRef]
22. Liu, C.; Deng, R.; He, W.; Liang, H.; Du, W. Optimal Coding Schemes for Detecting False Data Injection Attacks in Power System State Estimation. *IEEE Trans. Smart Grid* **2022**, *13*, 738–749. [CrossRef]

Article

FedResilience: A Federated Learning Application to Improve Resilience of Resource-Constrained Critical Infrastructures

Ahmed Imteaj [1,2,†], **Irfan Khan** [3], **Javad Khazaei** [4] and **Mohammad Hadi Amini** [1,2,*,†]

1 Knight Foundation School of Computing and Information Sciences, Florida International University, Miami, FL 33199, USA; aimte001@fiu.edu
2 Sustainability, Optimization, and Learning for InterDependent Networks Laboratory (Solid Lab), FIU, Miami, FL 33199, USA
3 Clean and Resilient Energy Systems (CARES) Lab, Texas A&M University at Galveston, Galveston, TX 77554, USA; irfankhan@tamu.edu
4 Department of Electrical and Computer Engineering, Lehigh University, Bethlehem, PA 18015, USA; khazaei@lehigh.edu
* Correspondence: moamini@fiu.edu
† Current address: 11200 SW 8th St, Miami, FL 33199, USA.

Abstract: Critical infrastructures (e.g., energy and transportation systems) are essential lifelines for most modern sectors and have utmost significance in our daily lives. However, these important domains can fail to operate due to system failures or natural disasters. Though the major disturbances in such critical infrastructures are rare, the severity of such events calls for the development of effective resilience assessment strategies to mitigate relative losses. Traditional critical infrastructure resilience approaches consider that the available critical infrastructure agents are resource-sufficient and agree to exchange local data with the server and other agents. Such assumptions create two issues: (1) uncertainty in reaching convergence while applying learning strategies on resource-constrained critical infrastructure agents, and (2) a huge risk of privacy leakage. By understanding the pressing need to construct an effective resilience model for resource-constrained critical infrastructure, this paper aims at leveraging a distributed machine learning technique called Federated Learning (FL) to tackle an agent's resource limitations effectively and at the same time keep the agent's information private. Particularly, this paper is focused on predicting the probable outage and resource status of critical infrastructure agents without sharing any local data and carrying out the learning process even when most of the agents are incapable of accomplishing a given computational task. To that end, an FL algorithm is designed specifically for a resource-constrained critical infrastructure environment that could facilitate the training of each agent in a distributed fashion, restrict them from sharing their raw data with any other external entities (e.g., server, neighbor agents), choose proficient clients by analyzing their resources, and allow a partial amount of computation tasks to be performed by the resource-constrained agents. We considered a different number of agents with various stragglers and checked the performance of FedAvg and our proposed FedResilience algorithm with prediction tasks for a probable outage, as well as checking the agents' resource-sharing scope. Our simulation results show that if the majority of the FL agents are stragglers and we drop them from the training process, then the agents learn very slowly and the overall model performance is negatively affected. We also demonstrate that the selection of proficient agents and allowing them to complete only parts of their tasks can significantly improve the knowledge of each agent by eliminating the straggler effects, and the global model convergence is accelerated.

Keywords: power system resilience; disaster; Federated Learning; edge intelligence; resilience management systems; resource-limitations; demand response

Citation: Imteaj, A.; Khan, I.; Khazaei, J.; Amini, M.H. FedResilience: A Federated Learning Application to Improve Resilience of Resource-Constrained Critical Infrastructures. *Electronics* **2021**, *10*, 1917. https://doi.org/10.3390/electronics10161917

Academic Editor: Rui Pedro Lopes

Received: 13 July 2021
Accepted: 4 August 2021
Published: 10 August 2021

1. Introduction

In this section, we present the motivation for the development of an application to improve the resilience of critical infrastructures using a novel FL model. We discuss the prior works that have recently been published and explain how our proposed FL model can be beneficial for improving the resilience of critical infrastructures. We also show the novelty of this paper, which is followed by a brief description of the paper's organization.

1.1. Motivation

Critical infrastructures such as power systems, transportation, fuel, water, and gas are interconnected and are all parts of distinct or interdependent networks that operate cooperatively to produce and distribute essential services [1]. When a sudden interruption occurs in any of the infrastructures, the resilience technique can assist in the continuation of the operations through its ability to resist, avoid, adapt, and recover swiftly from a disastrous situation. However, it is vital to ensure the resilience of the critical infrastructures as the sectors are interdependent. Using an inter-network communication scheme, entities within the same network can exchange resources, and in an intra-network communication infrastructure, clients residing in different network domains can provide support to each other to enhance the resilience of the system [2]. For instance, most of the critical infrastructures—e.g., transportation networks, telecommunication, and finance and banking sectors—rely on a continuous and stable power supply. However, the prolonged disruption of the operations of critical infrastructures can incur a significant economic loss. One of the recent surveys found that power outages occur for a minimum of one out of four companies in every month [3]. Specifically, in large companies, power outage loss costs over a million dollars an hour and around 150 USD annually [4]. Several works have already been conducted with the aim of improving the resilience of critical infrastructures by applying machine learning (ML) [5], deep learning (DL) [6], distributed edge computing [7,8], and transfer learning [9]. All these works constructed their prediction model either by collecting data from the agents (e.g., ML, DL) or receiving an update for some data from the distributed agents (e.g., distributed ML). However, sharing such sensitive data could be privacy-intrusive. An attacker can expose or tamper with data, which may cause the failure of the whole resilience system, and a company may thus face a huge loss. Besides, in a distributed system, we may observe straggler clients that learn very slowly due to resource-limitation issues and degrade the overall performance of the prediction model [10].

The conventional resilience approaches that are constructed on the theme of the learning and forecasting of probable outages consider that all critical infrastructure agents (CIAs) have available resources and can perform an assigned computational task. However, in a real-world scenario, any agent may possess low system configurations or may run out of resources. In consequence, some agents may not be able to complete the assigned computational tasks due to the shortage of resources. Therefore, our main motivation for this research is to mitigate the outage loss of the CIAs by developing a novel FL-based prediction model that can preserve privacy and handle the straggler issues in the case of resource-constrained network agents. We propose a novel FL-based strategy that consists of a local CIA, which acts as an intelligent decision-making entity or an FL agent; e.g., a smart factory or an autonomous micro-grid can act as a CIA. A CIA can generate a model based on its available local data (i.e., power demand and resource availability) and share the model with a central fusion center that acts as an FL server. Similarly, the neighboring CIAs share their model to pursue a common goal, and the coordinator generates a global model that learns the outage information and resource sharing scope of all the CIA agents. In case any CIA agent has limited computational resources (e.g., low processing capability, bandwidth) and cannot generate a learning model, the central coordinator is enabled to select proficient clients for the training rounds, allowing partial amounts of work from the resource-constrained CIAs considering their available resources. Therefore, the resource-constrained issues of the grids related to model training would be resolved, and distributed resources could be supplied from the neighbors in case any CIA fails to continue its operation.

1.2. Literature Review

The concept of resilience can be generalized for any discipline as a system's capacity to predict and withstand forthcoming shocks, restore the system's normal state swiftly, and adapt with an improved action for handling future catastrophic events. Managing and improving the infrastructure resilience of critical infrastructures has recently attracted the attention of several researchers and, in consequence, several studies related to the modeling and upgrading of systems and networks resilience have been proposed [11–13]. The authors in [14] focused on reducing the peak load of the critical infrastructure of power systems by considering multi-agent-based power generation, network grids, and relative demand response status. In their proposed approach, the agents could share their local information only with a central fusion center and were unable to interact with neighboring agents to exchange local resources. Besides, a comprehensive study on modeling the resilience of large-scale critical infrastructure was presented in [15]. However, centralized resilience systems become overly complex when a large amount of data is stored, processed, analyzed, and shared from a central fusion center [16,17]. The drawbacks of centralized resilience systems (e.g., scalability, computational power, storage) can be handled with distributed systems and learning resilience schemes [18]. The authors of [19] presented a detailed analysis of multi-agent systems (MAS), leveraging distributed intelligence among the network agents through peer-to-peer communication and sharing demand and load status to achieve a common goal. Besides, the authors of [20] proposed an adaptive synchronization approach for heterogeneous MAS against actuator fault by developing a multi-objective optimization technique to measure the installation capacity of network agents considering power-resilience against disasters [21,22]. Moreover, several works have adapted the strategy of utilizing infrastructure resources to improve the resilience of relative operations [23,24]. Further, some recent works developed resilience management systems for power systems [8,25–28], transport [29–32], urban areas [33–35], healthcare [36–38], and production systems [39–41] by adding intelligence to the CIAs so that the agents could make autonomous decisions by analyzing the demand–response state. In summary, all the prior works proposed the improvement of resilience either by passing local sensitive data of infrastructure agents to a central fusion center or by sharing such local sensitive data with neighboring agents. However, sharing the sensitive data that reside in the CIAs leads to the risk of privacy violation and can also interrupt the infrastructure operation through data falsification. To prevent that, a recently invented distributed ML technique called Federated Learning (FL)was proposed that can generate a smart model by utilizing edge resources and keeping an agent's information private. As the FL process is completely dependent on the agent's local model update, one of the challenges that the FL process presents is the straggler issues that arise due to the heterogeneity of the systems. System heterogeneity can be referred to as the heterogeneous nature of the agents in terms of their computational power, memory, battery life, or bandwidth. If we apply the FL process considering Internet of Things (IoT) devices, then there is a high chance of observing straggler agents during a training process [42]. This is because IoT devices are resource-constrained and vulnerable [43]. If we consider the state-of-the-art FedAvg algorithm [44], then it simply drops the straggler agents from the training process. However, dropping the stragglers can degrade the model performance, and also some agents may have valuable data. Instead of this approach, we need a strategy that can effectively handle the stragglers by counting every contribution, irrespective of its size. The authors of [45] proposed the FedProx algorithm, which can enable partial amounts of work to be collected from the agents; however, they randomly selected agents for the training round. According to the authors of [46], FedMax outperforms FedProx in terms of communication rounds by applying a strategy of limiting activation-divergence across multiple devices.

To tackle the above-mentioned issues in the context of the resilient operation of critical infrastructures and analyzing the existing works of FL, this paper is the first to propose a novel FL-based strategy that can predict the probable outages and resource-sharing capabilities of the network agents with the aim of improving resilience. Our FedResilience algorithm can select proficient agents by examining their resources and handle the stragglers by assigning feasible local computational tasks based on their capabilities. Our proposed technique relies on sharing local models of the infrastructure agents instead of sharing sensitive data and, finally, exploits the collaboratively learned knowledge on the probable outages and resource availability status of the whole FL network to enhance the resilience operations.

1.3. Contribution

The main contributions of this paper are given below:

- To the best of our knowledge, this is the first FL application that can improve the resilience of critical infrastructures through early prediction;
- We present a pathway of collaborative learning for CIAs that enables on-device learning without sharing any data and by exchanging only model information;
- We choose only the proficient agents for the FL training process and enable partial works to be collected from the resource-constrained agents to resolve the straggler issues;
- To demonstrate the effectiveness of Federated Learning in improving resilience by the early prediction of the outages and resource-sharing scope of the agents, we evaluate the prediction performance considering a varying number of stragglers and compare the model with the popular FedAvg [44] algorithm.

1.4. Organization

The rest of this paper is organized as follows. Section 2 presents the overview of FL and explains how our developed FL model can be effective in improving the resilience operations of critical infrastructures in detail. Section 3 presents our experiment results and is followed by Section 4, which concludes the paper.

2. Proposed System Description

FL is a distributed machine learning technique that allows the on-device training of network clients with their local data instead of sharing raw data with the server. Each client generates a local model by optimizing its local objective function that is shared with the FL server. After receiving local models from all participating FL clients, the FL server performs aggregation on the received models and updates a global model which is initialized as well as shared with all network clients at the initial stage of FL training. After that, the updated global model is disseminated to all FL clients, and each FL client tunes their local model by learning from the global model. The FL client–server interaction process is continued until the global model achieves a desired accuracy; hence, the model reaches a target convergence. The overall FL process is presented in Figure 1.

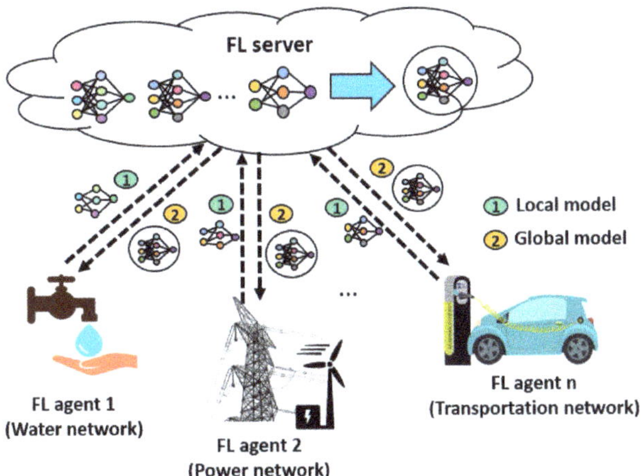

Figure 1. FL process considering critical infrastructure agents (CIAs).

In critical infrastructure networks, we may observe heterogeneous agents with varying system configurations and data volumes. Therefore, it is not viable to assign a uniform number of tasks to all the agents that participate in the FL process. The authors in [10] conducted a comprehensive survey on leveraging FL for IoT devices, where they discussed the possible challenges faced while applying FL on resource-constrained agents. Due to the varying and limited resource statuses, while one agent could perform a given computational task efficiently, another might turn into a straggler. An agent may become a straggler if the assigned computational task is overwhelming compared to its available resources. If the majority of the participating agents in the FL process turn into stragglers, then target convergence may never be obtained. Besides this, the IoT-enabled infrastructure agents are generally more prone to attacks that may cause divergent local model updates [47]. To improve the resilience, it is crucial to predict the infrastructure's outage, and in distributed systems, the main hindrance in the agent's learning process is stragglers. Therefore, it is essential to monitor the resource status and ensure that all agents are effectively operating by avoiding straggler issues. The typical FedAvg algorithm [44] assumes that all FL agents are resource-proficient and capable of accomplishing any given computational tasks. However, if a real-world FL-based IoT scenario is considered, then the majority of the agents may possess very few resources. Therefore, it is not effective to randomly select a fraction of agents for the training process. If the agents' resource availability statuses are tracked and the weak agents are filtered out from the training process, then it may possible to move one step closer towards resolving straggler issues. To infuse resource-awareness functionalities into the FL process, the task publisher (i.e., FL server) needs to acknowledge each network agent's minimum requirements for accomplishing a published task. After that, all interested agents share their resource information (e.g., memory, processing ability, bandwidth, and battery-life) with the task publisher. By examining the interested agents' resources, the task publisher prepares a list of proficient agents and randomly selects a subset of agents for that task.

2.1. Handling Systems and Statistical Heterogeneity of Critical Infrastructure Agents (CIAs)

In this segment, we discuss how a strategy of allowing partial works from the FL agents can be adopted through a generalization of the FedAvg algorithm [44]. In Section 2, we explain how the comparatively proficient agents can be selected for an FL process. However, it is possible that, among the selected agents, some agents would not be able to accomplish their entire task. Particularly, this can occur when all the interested and available FL agents have constrained resources and there are no other options without considering a subset of those agents for the training phase. Now, if the conventional FedAvg algorithm is applied [44], which instructs the center to assign uniform local computational tasks to all the selected agents, then straggler effects may be observed that can slow down the model convergence, or we may never be able to reach the target convergence. Instead, if we allow the selected agents to perform computational tasks based on their resources, then we would not require the straggler agents to be dropped, and every agent could contribute towards constructing a global model. In Figure 2, the high-level view of allowing partial works from the FL agents is presented. From the figure, it can be observed that the water network agent and transportation agent have limited resources while the power network has sufficient resource availability. Considering the resource status, the water network and transportation agents are performing 30% and 48%, respectively, of the overall computational tasks, while the power network is performing the entire task. Let us assume that the task publisher defines a local epoch of 100 that needs to be performed by all chosen FL agents. However, some of the agents are not capable of performing 100 local epochs on their data to generate a local model. In such a case, if an agent is capable of performing 30% of the overall computational task (i.e., 30 local epochs on their own data) rather than the whole task, then the agent would be allowed to perform that amount of the computational task and send back the model to the server. This proposed strategy solves two issues: first, the FL server does not need to wait a long time for a straggler agent, and second, every individual contribution from the agents can be counted.

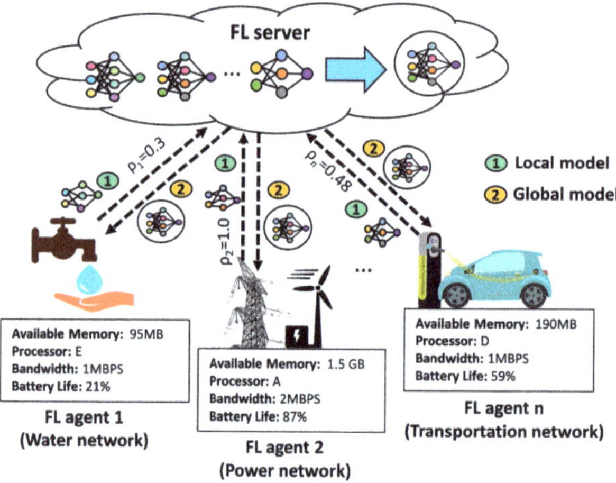

Figure 2. Allowing partial amounts of work from the FL agents.

To reduce communication overheads, a popular strategy in federated optimization is that for each iteration period, each agent tries to achieve a local objective function that is used as a replacement of a global objective function. In each training round, a subset of agents is chosen, and each agent uses its resources to optimize the local objective function. After that, the agents share their model with the FL server, which performs aggregation and updates the global model. Allowing a flexible amount of work helps to solve the inexact nature of local objectives and assists in tuning the number of communications vs. local computations. While too many local epochs can overfit the model, a smaller number of local epochs increases communication overheads as well as the convergence time [47]. Therefore, it is required to set local epochs through proper tuning to ensure robust convergence. The concept of an inexact solution can be stated as follows:

Definition 1 (ϱ-inexact solution)**.** *Let us consider a function* $\mathcal{R}(\omega;\omega_0) = \mathcal{F}(\omega) + \frac{\xi}{2}\|\omega-\omega_0\|^2$*, and* $\varrho \in [0,1]$*, it can be said that* ω^* *is a* ϱ *-inexact solution of* $\min_\theta \mathcal{R}(\omega;\omega_0)$ *if* $\|\nabla\mathcal{R}(\omega^*;\omega_0)\| \leq \varrho\|\nabla\mathcal{R}(\omega_0;\omega_0)\|$*, where* $\nabla\mathcal{R}(\omega;\omega_0) = \nabla\mathcal{F}(\omega) + \xi(\omega-\omega_0)$.

To leverage the proper tuning of local computations by handling system heterogeneity, we use the concept of the inexact solution [45], which allows us to collect variable numbers of local epochs from the participated agents according to their resource availability. The ϱ_a^t -inexactness for a CIA a at training round t can be defined as follows:

Definition 2 (ϱ_a^t-inexact solution)**.** *Let us consider a function* $\mathcal{R}_a(w; w_t) = \mathcal{F}_a(\omega) + \frac{\xi}{2}\|\omega - \omega_t\|^2$, *and* $\varrho \in [0,1]$, *we call* ω^* *is a* ϱ_a^t *-inexact solution of* $\min_\omega \mathcal{R}_a(\omega; \omega_t)$ *if* $\|\nabla \mathcal{R}_a(\omega^*; \omega_t)\| \leq \varrho_a^t \|\nabla \mathcal{R}_a(\omega_t; \omega_t)\|$, *where* $\nabla \mathcal{R}_a(\omega; w_t) = \nabla \mathcal{F}_a(\omega) + \xi(\omega - \omega_t)$.

Here, the convenience of ϱ-inexactness is that it allows variable local computations to be accomplished by the selected CIAs in each training round. As the system heterogeneity causes heterogeneous progress from the agents while solving local objective functions, it is vital to enable adaptive ϱ considering agents' resource availability. We can consider a scenario from our real-life perspectives. Suppose we have a few power agents that agree to participate in an FL process and utilize their edge resources. Each agent may have some outage information about some past events and also can possess resource information about its neighboring agents. Now, if an agent wants to gather knowledge about outage events that were never seen by that agent and store resource information from the agents that are not its neighbors, then it needs to adopt a method so that it can obtain the collective knowledge of the whole network. We can infuse the collective knowledge to each agent through the power of FL. In case, if a power agent does not have sufficient resources to complete an assigned computational task, we allow that agent to perform partial works. In this way, we do not ignore any agent's local knowledge. As a consequence, each agent is more capable of predicting an outage event and can locate an agent that needs a power supply.

2.2. Proposed FedResilience Algorithm

The proposed FedResilience algorithm is presented in Algorithm 1. The goal of this algorithm is to predict the outages and resource-sharing scope of CIAs without sharing any agent's local data, utilizing the computational resources of the CIAs. Applying the FL strategy for critical infrastructures mainly involves two entities: the critical infrastructure server (CIS) and available CIAs within the networks. At the beginning of the FL process, the server initializes a global model that is disseminated to all available CIAs within the networks specifying task requirements (line **1–2**). Each interested CIA shares its current resource status with the CIS (line **3**). In each training round, the CIS examines the resource information (i.e., processing power, memory, bandwidth, battery-charge status, and data volume) of the interested CIAs by calling the **CheckResource()** function (line **4–5**). The **CheckResource()** function receives a CIA's information upon calling, stores the information in a list, and compares the resource availability status with the task requirements (line **13–15**). If the CIA's available resources satisfy the minimum task requirements, then that CIA's information is stored in another list and sent back from where the **CheckResource()** function is called (line **16–18**). Upon receiving the resource information from all the interested CIAs, the CIS sorts the eligible CIAs based on their resource status, selects a fraction from those CIAs, and randomly chooses a subset of proficient CIAs for the training phase (line **6–8**). After that, the CIS calls the selected agents to perform on-device training using the **AgentLocalUpdate()** function and shares the latest global model (line **9–10**). It is assumed that the total number of data samples within the network is n, which are distributed among the CIAs with a set of indexes $\mathcal{D}_a$ on CIA a, where $\mathcal{N}_a = |\mathcal{D}_a|$. Each CIA's local data in a communication round t are referred to by $\mathcal{N}_t$. During FL training, each selected CIA utilizes its local solver to determine the inexact minimizer ϱ_a^t to solve the local objective function (line **19–20**). Further, each CIA splits its local samples into batches, performs SGD to achieve an optimal local solution, and shares the model with the CIS (line **21–25**). The CIS aggregates the local models to generate an updated global model, and the same iteration period is continued until the global model reaches convergence (line **11–12**).

Algorithm 1: FedResilience: An FL-based approach to predict the outages and resource-sharing scopes of critical infrastructures. The $\mathcal{A}$ eligible clients are indexed by a; B = local minibatch size, $\mathcal{F}$ = client fraction, E = local epoch, and η = learning rate.

CIS executes: initialize global model w_0
Disseminate task requirements to all CIAs
Collect resource status of interested CIAs
for *each round* $t = 1, 2, \ldots$ **do**
  $\mathcal{R}_t$ = **CheckResource** $(\mathcal{P}_t, \mathcal{M}_t, \mathcal{B}_t, \mathcal{C}_t, \mathcal{V}_t)$ for all interested CIAs
  Sort CIAs based on $\mathcal{R}$ and store in a list $\mathcal{L}$
  $\mathcal{E} \leftarrow$ Top $\mathcal{L} \cdot \mathcal{F}$ CIAs
  $\mathcal{A}_t \leftarrow$ (random set of $\mathcal{E}$ CIAs)
  for *each client* $a \in \mathcal{A}_t$ *in parallel* **do**
    $w^a_{t+1} \leftarrow$ **AgentLocalUpdate** (a, w_t)
  for *each CIA* $a \in \mathcal{A}_t$ **do**
    $w_{t+1} \leftarrow w_{t+1} + \frac{N_t}{N} w^a_{t+1}$

CheckResource $(\mathcal{P}_a, \mathcal{M}_a, \mathcal{B}_a, \mathcal{C}_a, \mathcal{V}_a)$**:**
Store $(\mathcal{B}_a, \mathcal{M}_a, \mathcal{E}_a, \mathcal{V}_a)$ into a list $\mathcal{Q}_a$
Compare $\mathcal{Q}_a$ with $\mathcal{L}_{Req}$
if $\mathcal{Q}_a$ *satisfies* L_{Req} **then**
  Add $\mathcal{Q}_a$ to list $\mathcal{R}$
Return $\mathcal{R}$
AgentLocalUpdate (a, w) : // Run on CIA a
  Each CIA a finds a w^{t+1}_a which is a ϱ^t_a -inexact minimizer of: $w^{t+1}_a = F_a(w) + \frac{\xi}{2}\|w - w^t\|^2$ and measures maximum feasible round of local epochs E
  $\mathcal{B} \leftarrow$ (split $\mathcal{D}_a$ into batch size B)
  for *each CIA's local epoch e from 1 to E* **do**
    for *batch* $b \in B$ **do**
      $w \leftarrow w - \eta \nabla \ell(w; b)$
  return w to server

3. Experimental Results

To evaluate the performance of the proposed FedResilience method, various distributed mobile robots are considered as critical infrastructure agents that possess heterogeneous resources in terms of processing power, battery life, memory, and data volume. To simulate the straggler effects and the effectiveness of the proposed FedResilience algorithm, the Electro-Maps dataset [48] is used to predict the power outages and resource-sharing scopes of the agents. The dataset is preprocessed considering temperature, number of weeks, hour, holiday, and population, and an additional column of resource availability is generated from the information regarding the population, holiday, and temperature. Using the information, we target the prediction of the outages and resource-sharing scopes of the critical infrastructure agents. A similar transmission rate is set for all the distributed agents for the simplicity of the FL implementation process. To simulate the effectiveness of allowing partial works from the distributed agents, different numbers of weak distributed agents are deliberately considered to create straggler effects; i.e., some of the agents fail to generate local models due to their constrained resources. It is assumed that there remains a global cycle that is followed by each agent, and each selected agent measures the amount of the local computational task it can perform in training round i as a function of its available resources and clock cycle. The code is publicly available and has been uploaded to a GitHub repository (https://github.com/Imteaj10/FedResilience, accessed on 31 July 2021). In a conventional FL approach, a global epoch E is defined for all the participating agents to perform a particular task, and if any of the agents fail to generate a local model on time, the model simply drops that agent from the training process (no partial tasks are allowed). However, dropping slow clients from the training process may prolong the model convergence, or the model may even never reach the target convergence. To handle such issues, we adapt a generalization of the FedAvg algorithm that enables each agent to perform part of a computational task by considering the agent's resource limitations. To present the motivation behind this research, we applied the FedAvg algorithm [44] for predicting the outages and resource-sharing scopes of CIAs and presented the straggler effects. We considered a varying number of CIAs and assumed that a majority of those agents would be stragglers. At first, we considered three agents (where two were stragglers) and computed the training loss and testing accuracy during the prediction of a probable outage by applying the state-of-the-art FedAvg [44] algorithm. In Figure 3, we can see that the training loss started to decrease in the initial few communication rounds and remained almost unchanged for further communication rounds due to the dropping of the majority of clients. In contrast, in Figure 4, it is clear that the improvement of testing accuracy was quite steady and each agent learned very slowly.

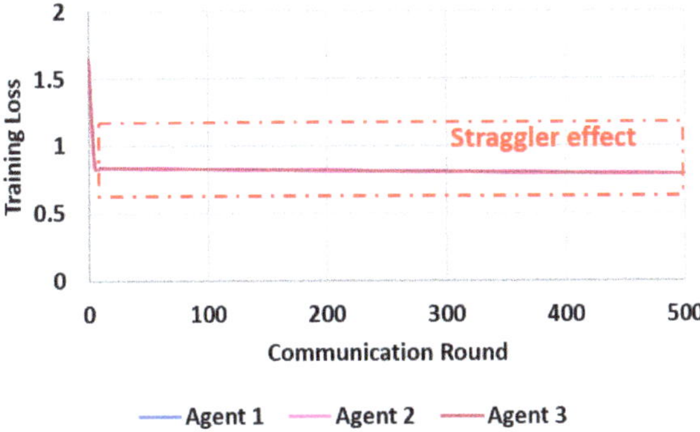

Figure 3. Straggler effects on participating FL agents' (three agents and two stragglers) model loss for the prediction of critical infrastructure outage.

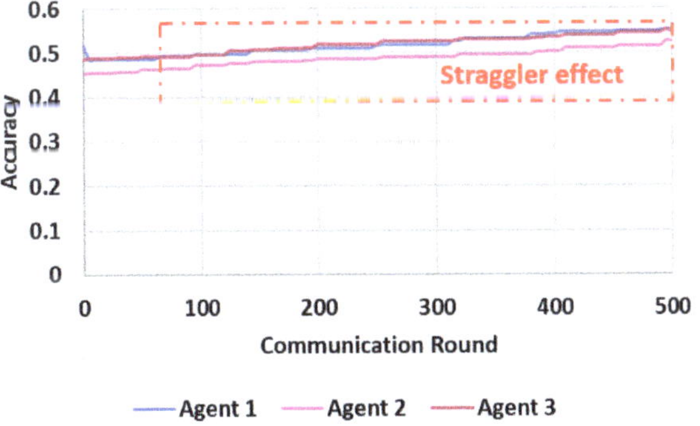

Figure 4. Straggler effects on participating FL agents' (three agents and two stragglers) model accuracy for the prediction of critical infrastructure outage.

After that, we simulated the straggler effects by increasing the number of agents (three stragglers out of five agents) and computing the training loss to predict a probable outage by applying the state-of-the-art FedAvg [44] algorithm (see Figure 5). We can see a small decrease in training loss for communication round 500. In contrast, in Figure 6, it is observable that some agents had very low accuracy while other agents had comparatively high accuracy. However, none of the agents achieved satisfactory improvements in their accuracy.

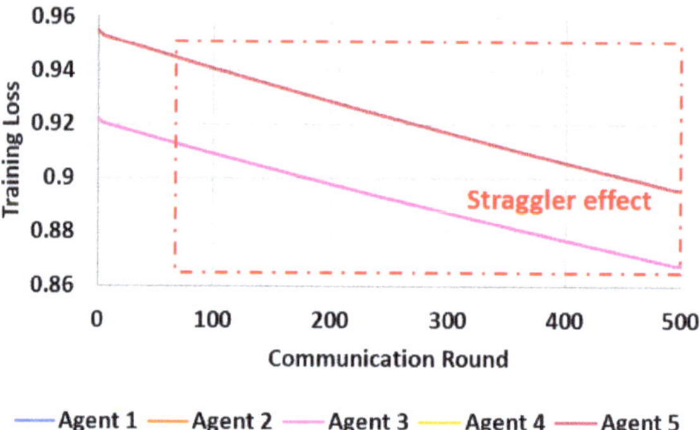

Figure 5. Straggler effects on participating FL agents' (five agents and three stragglers) model loss for the prediction of critical infrastructure outage.

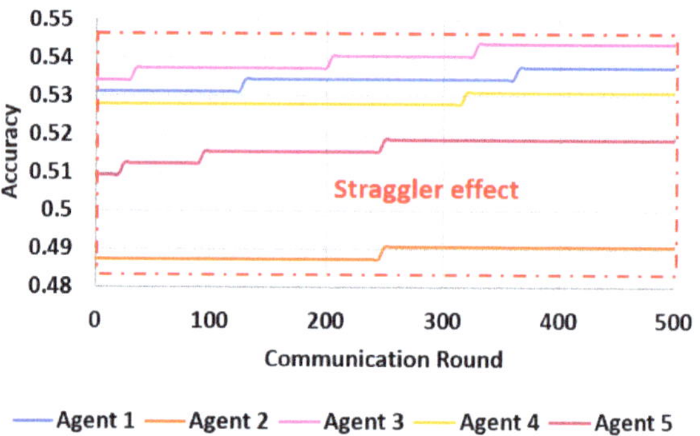

Figure 6. Straggler effects on participating FL agents' (five agents and three stragglers) model accuracy for the prediction of critical infrastructure outage.

We also simulated the straggler effects for eight agents (where six-of them were stragglers) and generated the training loss for a predicted outage by applying the state-of-the-art FedAvg [44] algorithm (see Figure 7). We can see that both of the non-straggler agents had a very slow learning process in spite of a higher communication round. In Figure 8, we can see that the clients barely learned from each other and consequently were not able to improve their model quality significantly.

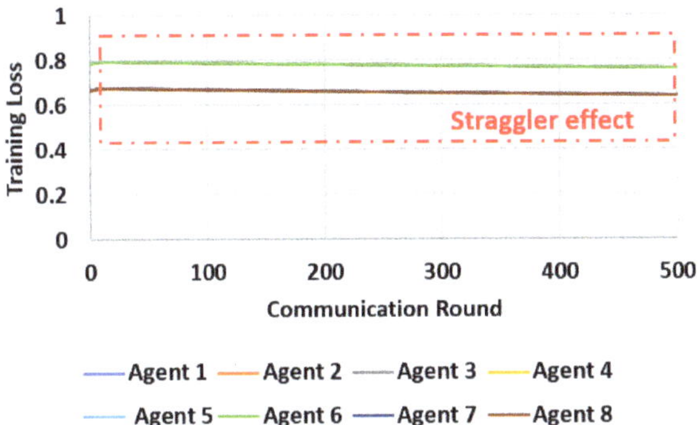

Figure 7. Straggler effects on participating FL agents' (eight agents and six stragglers) model loss for the prediction of critical infrastructure outage.

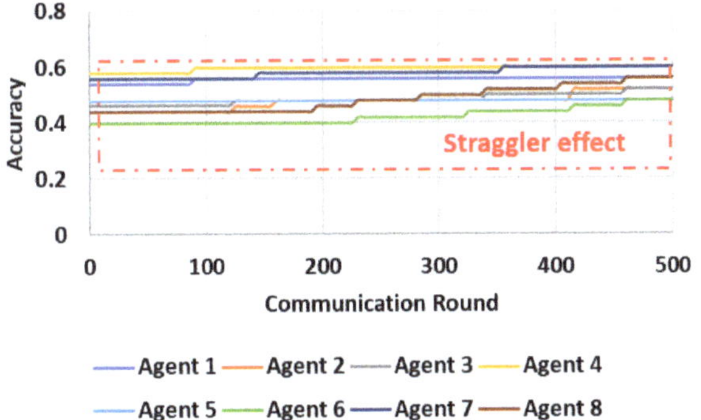

Figure 8. Straggler effects on participating FL agents' (eight agents and six stragglers) model accuracy for the prediction of critical infrastructure outage.

Next, we simulated the straggler effects during the prediction of the agents' resource-sharing scope by applying the state-of-the-art FedAvg [44] algorithm. Similar to the outage prediction, we considered three agents (where two were stragglers) and generated the training loss and testing accuracy to predict the resource-sharing scope of the agents. In Figure 9, we can see that though the training loss was comparatively lower than the outage prediction loss for the three agents–two stragglers scenario, a similar training loss was observed for all agents (i.e., the agents did not learn through collaboration). On the other hand, we can see from Figure 10, that agent 2 had a comparatively lower accuracy than other agents, but it slightly improved its accuracy through the FL process. However, after 200 communication rounds, all agents' testing accuracies improved very slowly.

Besides, we simulated the straggler effects during the prediction of the agents' resource-sharing scope by increasing the number of agents (five agents, where three of them were stragglers). We generated the training loss and testing accuracy by applying the state-of-the-art FedAvg [44] algorithm. In Figure 11, we can see that though the training loss dropped significantly in the few initial communication rounds, almost constant training loss was observed for all agents. Moreover, in Figure 12, all the agents failed to obtain a marginal improvement in their accuracy.

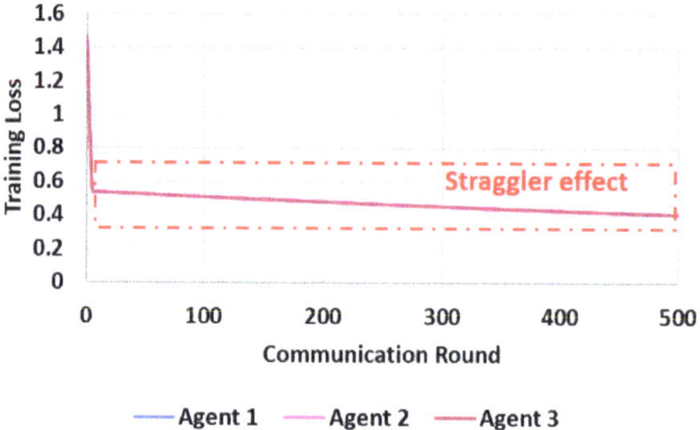

Figure 9. Straggler effects on participating FL agents' (three agents and two stragglers) model loss for the prediction of agents' resource-sharing capability.

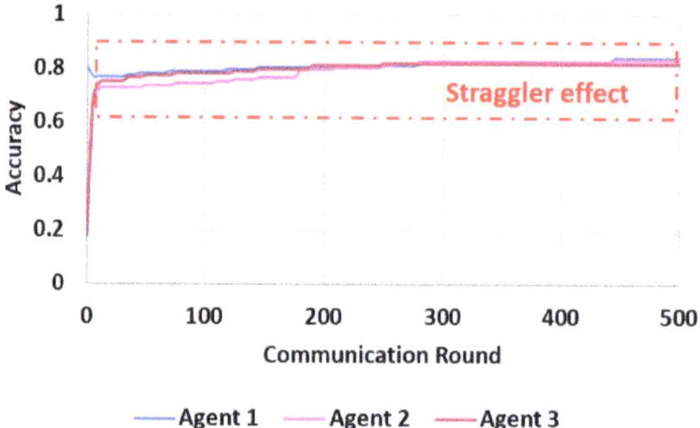

Figure 10. Straggler effects on participating FL agents' (three agents and two stragglers) model accuracy for the prediction of agents' resource-sharing capability.

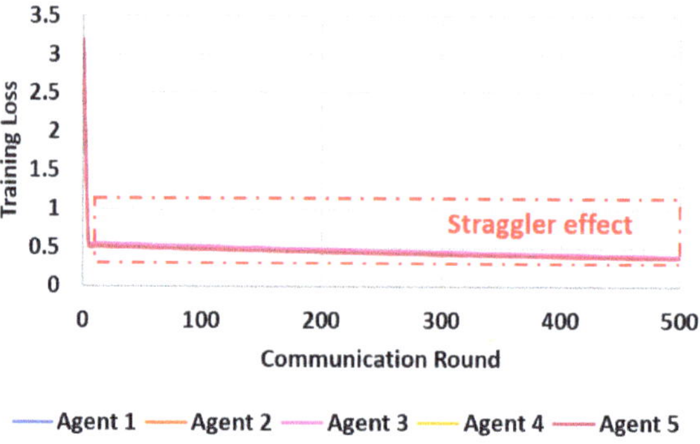

Figure 11. Straggler effects on participating FL agents' (five agents and three stragglers) model loss for the prediction of agents' resource-sharing capability.

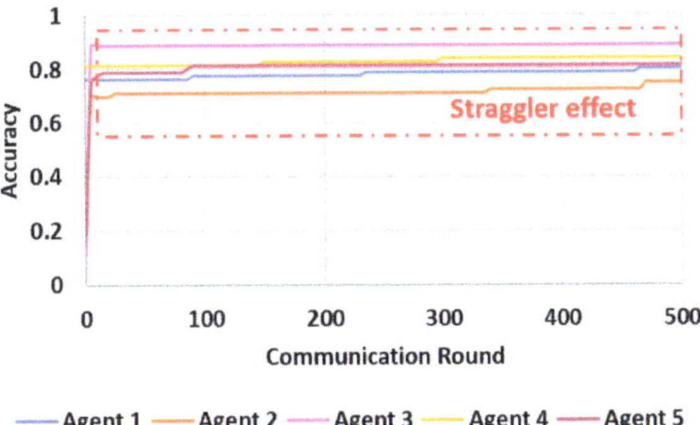

Figure 12. Straggler effects on participating FL agents' (three agents and two stragglers) model accuracy for the prediction of agents' resource-sharing capability.

Similarly, we simulated the training loss and accuracy by considering eight agents (where six were stragglers) and applied the FedAvg algorithm [44] during the prediction of the agents' resource-sharing scope. From Figures 13 and 14, we can see that the training loss and accuracy improved as we increased the number of agents; however, both accuracies improved little with the increment of communication rounds due to the straggler effects. In a summary, for all the considered cases, the agents struggled to minimize loss and remained very steady in terms of improving accuracy due to the straggler effect.

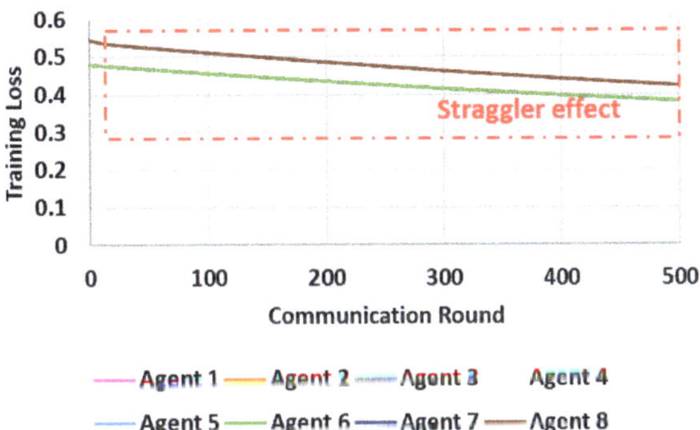

Figure 13. Straggler effects on participating FL agents' (eight agents and six stragglers) model loss for the prediction of agents' resource-sharing capability.

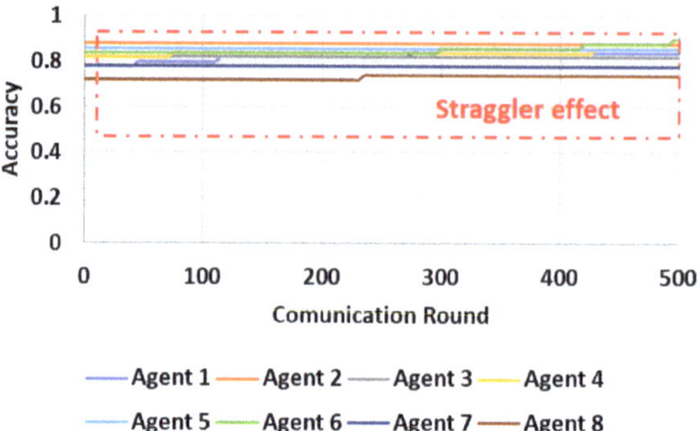

Figure 14. Straggler effects on participating FL agents' (eight agents and six stragglers) model accuracy for the prediction of agents' resource-sharing capability.

To eliminate the straggler effects during the prediction of the power outage and resource-sharing information, we proposed partial works to be allowed from the straggler agents; i.e., we assigned computational tasks based on the agents' available resources. To evaluate the performance of FedResilience, we considered the same number of agents (three, five, and eight agents) for the training rounds and observed their learning process. At first, we considered three agents (where two agents were stragglers) and checked their loss (Figure 15) and accuracy (Figure 16) during the prediction of power outages. Though the training loss increased due to the deviation of the local model updates as the stragglers performed low computational tasks, the agents started to reduce their training loss by learning from the global model and from their own data. On the contrary, the accuracy of the agents started to increase after 320 communication rounds because of the low number of resource-sufficient agents (Figure 16). We also simulated the loss (Figure 17) and accuracy (Figure 18) during the prediction of resource-sharing scope by considering the same number of agents and achieved better performance than the FedAvg [44] algorithm.

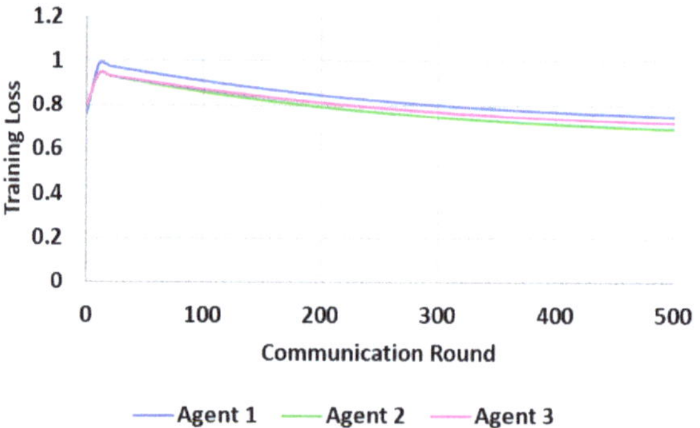

Figure 15. FedResilience's impact on participating FL agents' (three agents and two stragglers) model loss for the prediction of CIA outages.

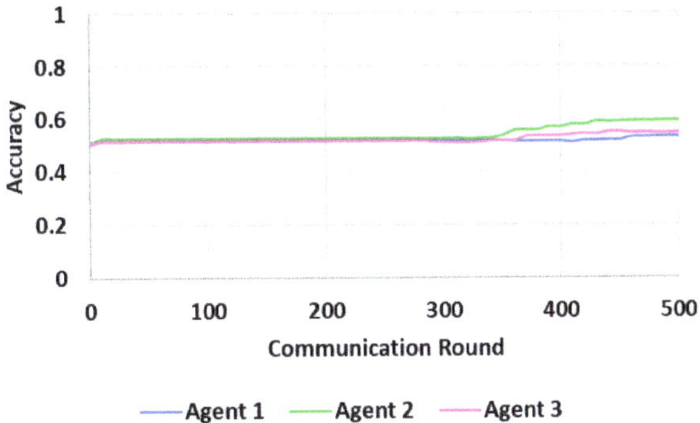

Figure 16. FedResilience's impact on participating FL agents' (three agents and two stragglers) model accuracy for the prediction of CIA outages.

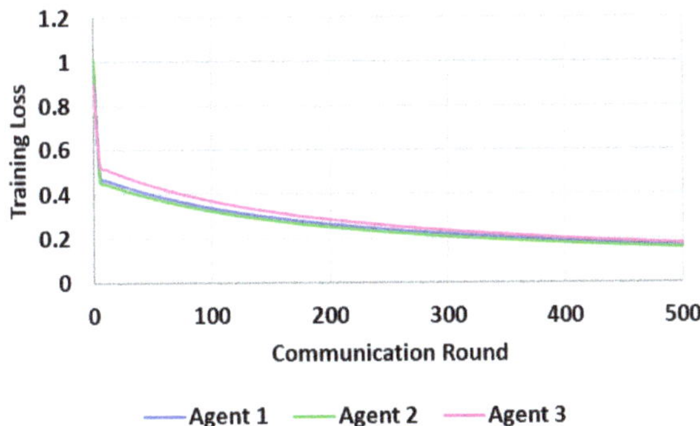

Figure 17. FedResilience's impact on participating FL agents' (three agents and two stragglers) model loss for the prediction of CIAs' resource-sharing capability.

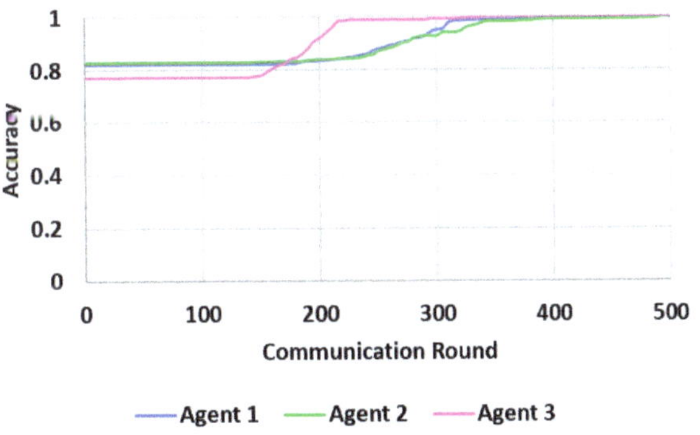

Figure 18. FedResilience's impact on participating FL agents' (three agents and two stragglers) model accuracy for the prediction of CIAs' resource-sharing capability.

After that, we considered five agents (where two agents were stragglers) and checked their loss (Figure 19) and accuracy (Figure 20) during the prediction of power outages. For these simulations, we observed similar patterns to those in Figures 15 and 16, but obtained better performance due to the higher number of active clients. The accuracy of the agents started to increase after 100 communication rounds because of the comparatively higher number of resource-sufficient agents (Figure 20).

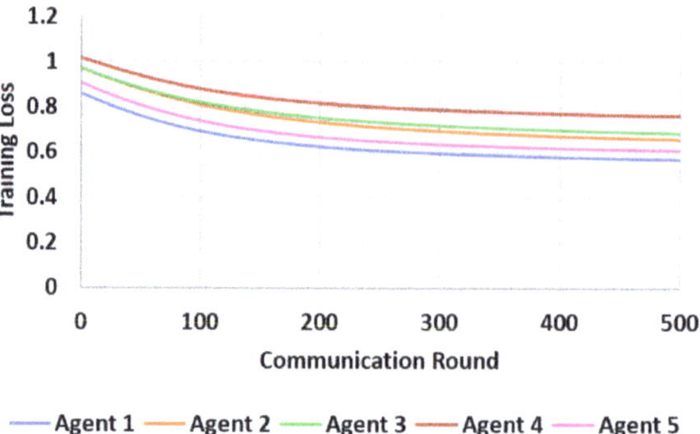

Figure 19. FedResilience's impact on participating FL agents' (five agents and three stragglers) model loss for the prediction of CIA outages.

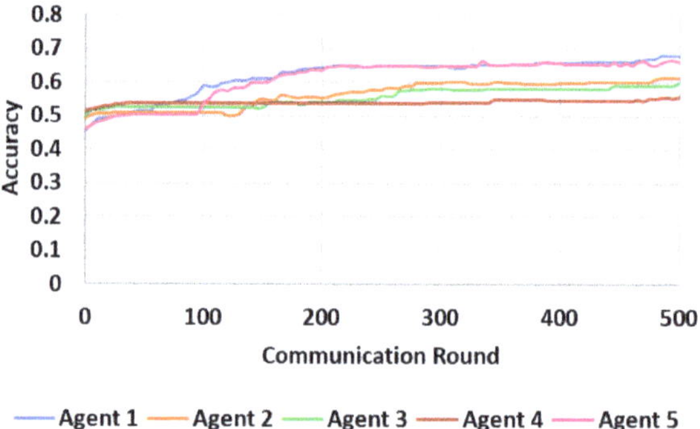

Figure 20. FedResilience impact on participating FL agents' (five agents and three stragglers) model accuracy for the prediction of CIA outages.

We also simulated the loss (Figure 21) and accuracy (Figure 22) during the prediction of resource-sharing scope by considering five agents and achieved better performance than the FedAvg [44] algorithm.

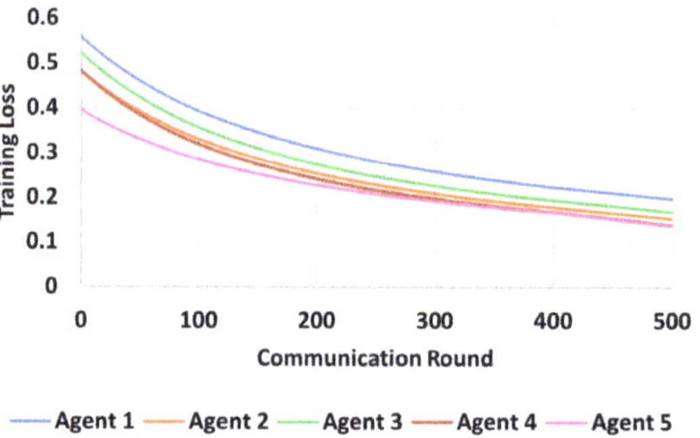

Figure 21. FedResilience's impact on participating FL agents' (five agents and three stragglers) model loss for the prediction of CIAs' resource-sharing capability.

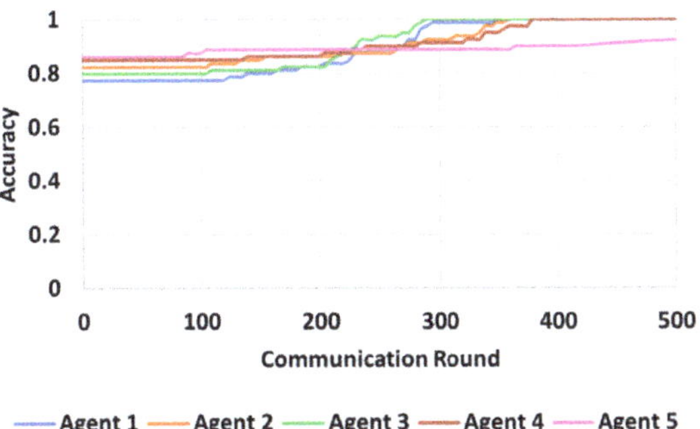

Figure 22. FedResilience's impact on participating FL agents' (five agents and three stragglers) model accuracy for the prediction of CIAs' resource-sharing capability.

Further, we simulated the performance of eight agents (where six agents were stragglers) and checked their loss (Figure 23) and accuracy (Figure 24) when predicting power outages. Here, it was clear that the agents improved their knowledge base (i.e., the training loss decreased and a significant accuracy improvement is observed) due to the improved quality of the global model.

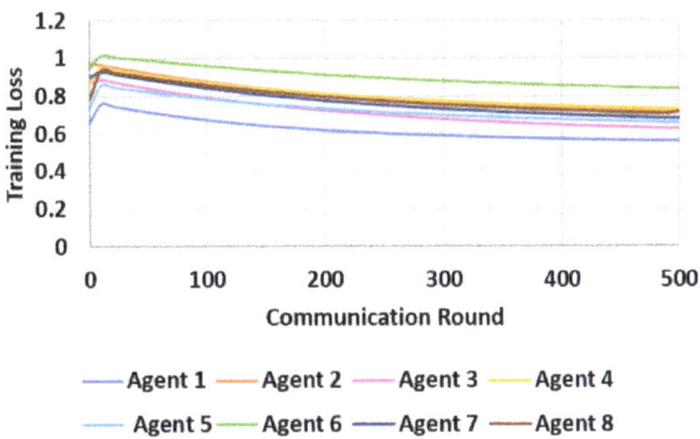

Figure 23. FedResilience's impact on participating FL agents' (eight agents and six stragglers) model loss for the prediction of CIA outages.

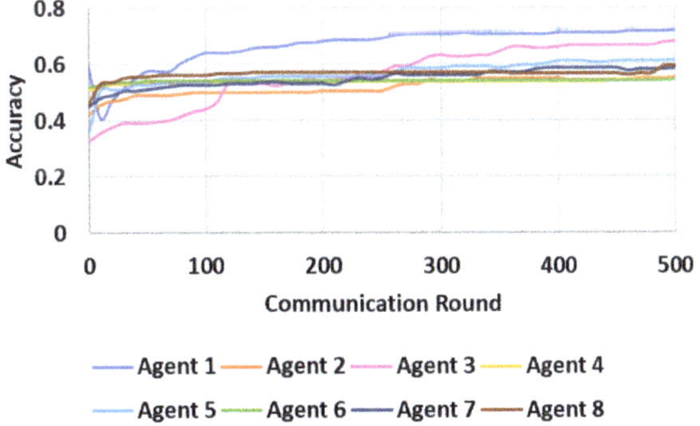

Figure 24. FedResilience's impact on participating FL agents' (eight agents and six stragglers) model accuracy for the prediction of CIA outages.

In a similar fashion, we simulated the loss (Figure 25) and accuracy (Figure 26) during the prediction of resource-sharing scope by considering eight agents and achieved a remarkable performance improvement compared to the FedAvg [44] algorithm. The training loss became close to 0.1 (Figure 25), and some of the agents achieved higher accuracy within 250 − 300 communication rounds (Figure 26).

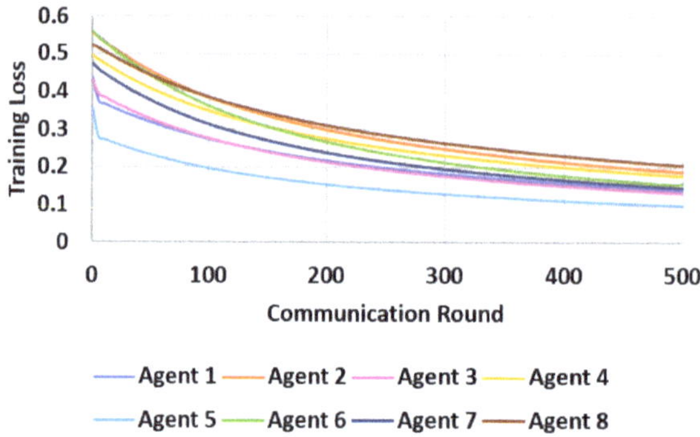

Figure 25. FedResilience's impact on participating FL agents' (eight agents and six stragglers) model loss for the prediction of CIAs' resource-sharing capability.

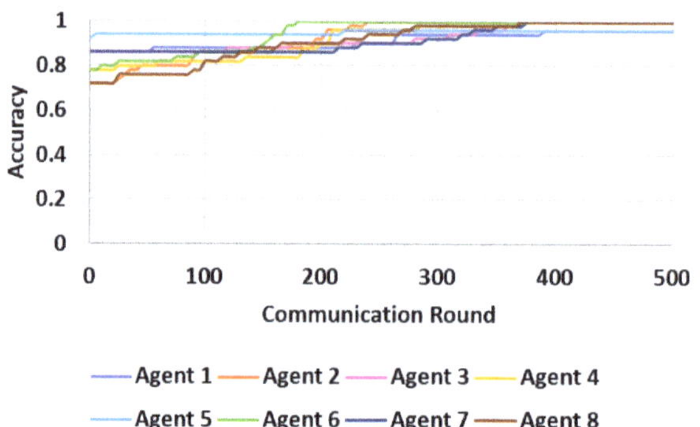

Figure 26. FedResilience's impact on participating FL agents' (eight agents and six stragglers) model accuracy for the prediction of CIAs' resource-sharing capability.

From the simulation results, it is observable that FedResilience has better performance than the conventional FedAvg model. As we increase the number of agents and count partial works from each of them, then agents can learn quickly, and the global model accuracy also increases. When we considered eight agents and six stragglers and applied the FedAvg algorithm, then the global model only contained the knowledge of the two active agents. As a result, the agent could not upgrade its knowledge base and showed a steady learning curve. However, when we counted the partial computational tasks by those stragglers, then the accumulation of those partial works generated an upgraded global model. As the quality of the global model improved and each agent tuned their local model by learning from the latest global model, the agents' learning process was accelerated. Our simulation results demonstrate two trends: first, the FedAvg algorithm is not suitable to predict outages or the resource-sharing information of resource-constrained CIAs as the algorithm cannot handle the straggler effects, which eventually slows down the agents' learning process; second, the FedResilience algorithm can handle straggler effects and is suitable even when we have a large number of stragglers within the network. In Figure 27, we can see that the FedResilience algorithm outperforms the FedAvg algorithm [44], achieving higher global model accuracy (cumulative updates of all the participating agents' local models) while predicting the outages and resource-sharing information of CIAs even with a large number of stragglers.

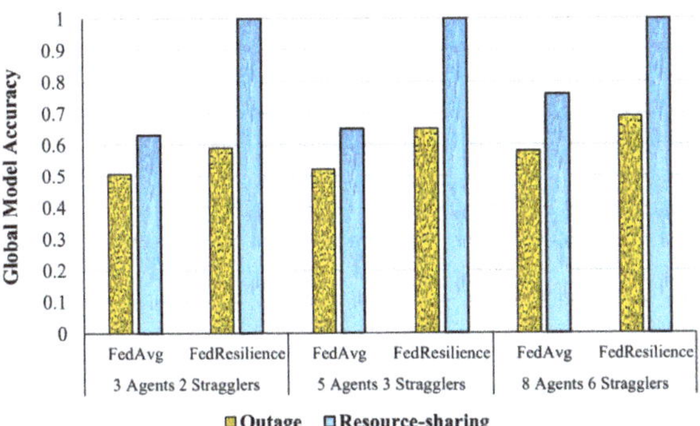

Figure 27. Comparison of global model accuracy of FedAvg and proposed FedResilience in the presence of stragglers.

In Figure 28, a linear approximation of the real system and the performance of the proposed FedResilience algorithm for a disaster event is presented. It can be seen that the real system performance index decreases after time t_d and reaches a minimal index at time t_m. In the beginning, the performance index is stable due to a preventive outage; however, as soon as the preventive outage is finished, the curve starts to move downwards and reaches a minimal performance index ($P_{\min}$). The low-performance index remains until a certain time interval, and after that, the system starts to recover. In contrast, when the FedResilience algorithm is applied during an outage, the performance index does not move down at minimal performance index ($P_{\min}$); instead, using the power of edge intelligence, the system can recover swiftly and a remarkable performance index can be achieved.

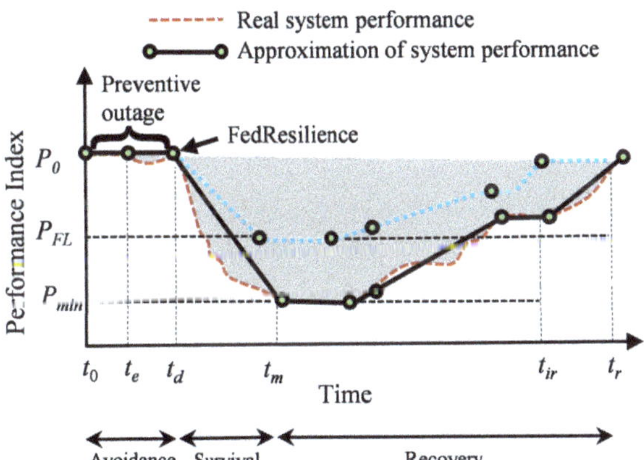

Figure 28. Eight-point linear approximation of the performance of the FedResilience algorithm during a disaster event.

4. Conclusions

This paper proposes a strategy to improve the resilience operations of critical infrastructures even when the network agents have limited resources. To evaluate our approach, the impact of straggler agents on the overall learning process is presented by considering resource-constrained distributed agents. After that, the effectiveness of our proposed FedResilience algorithm is evaluated, demonstrating the acceleration of the distributed agents' learning process despite heterogeneous system resources and model updates. By choosing proficient agents, performing on-device training, transferring knowledge, and allowing partial works, a robust and consistent FL model is achieved with higher global model accuracy compared to the state-of-the-art FedAvg algorithm; the model can also accelerate the learning process of unreliable IoT-enabled heterogeneous environments. The proposed concept can be applied to any resource-constrained heterogeneous IoT environment that is disrupted by straggler effects and struggles to reach convergence due to slow learning.

Author Contributions: Conceptualization, A.I. and M.H.A.; methodology, A.I.; software, A.I.; validation, A.I.; formal analysis, A.I.; investigation, A.I., I.K., J.K., M.H.A.; resources, M.H.A.; data curation, A.I.; writing—original draft preparation, A.I.; writing—review and editing, A.I., I.K., J.K., M.H.A.; visualization, A.I.; supervision, M.H.A.; project administration, M.H.A.; funding acquisition, M.H.A. All authors have read and agreed to the published version of the manuscript.

Funding: This research received no external funding.

Acknowledgments: A.I. and M.H.A. acknowledge the resources and support of Sustainability, Optimization, and Learning for InterDependent networks laboratory (solid lab) members at the Florida International University (www.solidlab.network) (accessed on 20 June 2021).

Conflicts of Interest: The authors declare no conflict of interest.

Abbreviations

The following abbreviations are used in this manuscript:

FL	Federated Learning
CIA	Critical infrastructure agent

References

1. Guidotti, R.; Chmielewski, H.; Unnikrishnan, V.; Gardoni, P.; McAllister, T.; van de Lindt, J. Modeling the resilience of critical infrastructure: The role of network dependencies. *Sustain. Resilient Infrastruct.* **2016**, *1*, 153–168. [CrossRef]
2. Amini, M.H.; Imteaj, A.; Pardalos, P.M. Interdependent networks: A data science perspective. *Patterns* **2020**, *1*, 100003. [CrossRef]
3. S & C Electric Company, F.S. S & C's 2018 State of Commercial & Industrial Power Reliability Report. 2013. Available online: https://www.sandc.com/globalassets/sac-electric/documents/sharepoint/documents---all-documents/technical-paper-100-t120.pdf (accessed on 23 April 2018).
4. Hussain, A. A Day Without Power: Outage Costs for Businesses. 2013. Available online: https://www.bloomenergy.com/blog/a-day-without-power-outage-costs-businesses (accessed on 8 October 2019).
5. Salehi, V.; Veitch, B.; Musharraf, M. Measuring and improving adaptive capacity in resilient systems by means of an integrated DEA-Machine learning approach. *Appl. Ergon.* **2020**, *82*, 102975. [CrossRef]
6. Dick, K.; Russell, L.; Souley Dosso, Y.; Kwamena, F.; Green, J.R. Deep learning for critical infrastructure resilience. *J. Infrastruct. Syst.* **2019**, *25*, 05019003. [CrossRef]
7. Alqahtani, A.; Abhishek, R.; Tipper, D.; Medhi, D. Disaster recovery power and communications for smart critical infrastructures. In Proceedings of the 2018 IEEE International Conference on Communications (ICC), Kansas City, MO, USA, 20–24 May 2018; pp. 1–6.
8. Amini, M.H.; Imteaj, A.; Mohammadi, J. Distributed Machine Learning for Resilient Operation of Electric Systems. In Proceedings of the 2020 International Conference on Smart Energy Systems and Technologies (SEST), Istanbul, Turkey, 7–9 September 2020; pp. 1–6.
9. Ilgen, S.; Sengers, F.; Wardekker, A. City-to-city learning for urban resilience: The case of water squares in Rotterdam and Mexico City. *Water* **2019**, *11*, 983. [CrossRef]
10. Imteaj, A.; Thakker, U.; Wang, S.; Li, J.; Amini, M.H. A Survey on Federated Learning for Resource-Constrained IoT Devices. *IEEE Internet Things J.* **2021**. [CrossRef]
11. Cai, H.; Lam, N.S.; Qiang, Y.; Zou, L.; Correll, R.M.; Mihunov, V. A synthesis of disaster resilience measurement methods and indices. *Int. J. Disaster Risk Reduct.* **2018**, *31*, 844–855. [CrossRef]
12. Pursiainen, C. Critical infrastructure resilience: A Nordic model in the making? *Int. J. Disaster Risk Reduct.* **2018**, *27*, 632–641. [CrossRef]
13. Alemzadeh, S.; Talebiyan, H.; Talebi, S.; Dueñas-Osorio, L.; Mesbahi, M. Resource Allocation for Infrastructure Resilience using Artificial Neural Networks. In Proceedings of the 2020 IEEE 32nd International Conference on Tools with Artificial Intelligence (ICTAI), Baltimore, MD, USA, 9–11 November 2020; pp. 617–624.
14. Amini, M.H.; Nabi, B.; Haghifam, M.R. Load management using multi-agent systems in smart distribution network. In Proceedings of the 2013 IEEE Power & Energy Society General Meeting, Vancouver, BC, Canada, 21–25 July 2013; pp. 1–5.
15. Wang, J.; Zuo, W.; Rhode-Barbarigos, L.; Lu, X.; Wang, J.; Lin, Y. Literature review on modeling and simulation of energy infrastructures from a resilience perspective. *Reliab. Eng. Syst. Saf.* **2019**, *183*, 360–373. [CrossRef]

16. Lee, E.H.; Choi, Y.H.; Kim, J.H. Real-time integrated operation for urban streams with centralized and decentralized reservoirs to improve system resilience. *Water* **2019**, *11*, 69. [CrossRef]
17. Wang, Y.; Rousis, A.O.; Strbac, G. On microgrids and resilience: A comprehensive review on modeling and operational strategies. *Renew. Sustain. Energy Rev.* **2020**, *134*, 110313. [CrossRef]
18. Arghandeh, R.; Brown, M.; Del Rosso, A.; Ghatikar, G.; Stewart, E.; Vojdani, A.; von Meier, A. The local team: Leveraging distributed resources to improve resilience. *IEEE Power Energy Mag.* **2014**, *12*, 76–83. [CrossRef]
19. Sujil, A.; Verma, J.; Kumar, R. Multi agent system: Concepts, platforms and applications in power systems. *Artif. Intell. Rev.* **2018**, *49*, 153–182. [CrossRef]
20. Chen, C.; Xie, K.; Lewis, F.L.; Xie, S.; Davoudi, A. Fully distributed resilience for adaptive exponential synchronization of heterogeneous multiagent systems against actuator faults. *IEEE Trans. Autom. Control* **2018**, *64*, 3347–3354. [CrossRef]
21. Uemichi, A.; Yagi, M.; Oikawa, R.; Yamasaki, Y.; Kaneko, S. Multi-objective optimization to determine installation capacity of distributed power generation equipment considering energy-resilience against disasters. *Energy Procedia* **2019**, *158*, 6538–6543. [CrossRef]
22. Hossain, E.; Roy, S.; Mohammad, N.; Nawar, N.; Dipta, D.R. Metrics and enhancement strategies for grid resilience and reliability during natural disasters. *Appl. Energy* **2021**, *290*, 116709. [CrossRef]
23. Hussain, A.; Bui, V.H.; Kim, H.M. Microgrids as a resilience resource and strategies used by microgrids for enhancing resilience. *Appl. Energy* **2019**, *240*, 56–72. [CrossRef]
24. Gao, H.; Chen, Y.; Mei, S.; Huang, S.; Xu, Y. Resilience-oriented pre-hurricane resource allocation in distribution systems considering electric buses. *Proc. IEEE* **2017**, *105*, 1214–1233. [CrossRef]
25. Imteaj, A.; Amini, M.H.; Mohammadi, J. Leveraging decentralized artificial intelligence to enhance resilience of energy networks. In Proceedings of the 2020 IEEE Power & Energy Society General Meeting (PESGM), Montreal, QC, Canada, 2–6 August 2020; pp. 1–5.
26. Mishra, D.K.; Ghadi, M.J.; Azizivahed, A.; Li, L.; Zhang, J. A review on resilience studies in active distribution systems. *Renew. Sustain. Energy Rev.* **2021**, *135*, 110201. [CrossRef]
27. Ghiasi, M.; Dehghani, M.; Niknam, T.; Baghaee, H.R.; Padmanaban, S.; Gharehpetian, G.B.; Aliev, H. Resiliency/cost-based optimal design of distribution network to maintain power system stability against physical attacks: A practical study case. *IEEE Access* **2021**, *9*, 43862–43875. [CrossRef]
28. Bhusal, N.; Abdelmalak, M.; Kamruzzaman, M.; Benidris, M. Power system resilience: Current practices, challenges, and future directions. *IEEE Access* **2020**, *8*, 18064–18086. [CrossRef]
29. Colon, C.; Hallegatte, S.; Rozenberg, J. Criticality analysis of a country's transport network via an agent-based supply chain model. *Nat. Sustain.* **2021**, *4*, 209–215. [CrossRef]
30. Bellini, E.; Bellini, P.; Cenni, D.; Nesi, P.; Pantaleo, G.; Paoli, I.; Paolucci, M. An IoE and Big Multimedia Data Approach for Urban Transport System Resilience Management in Smart Cities. *Sensors* **2021**, *21*, 435. [CrossRef]
31. Argyroudis, S.A.; Mitoulis, S.A.; Winter, M.G.; Kaynia, A.M. Fragility of transport assets exposed to multiple hazards: State-of-the-art review toward infrastructural resilience. *Reliab. Eng. Syst. Saf.* **2019**, *191*, 106567. [CrossRef]
32. Pan, S.; Yan, H.; He, J.; He, Z. Vulnerability and resilience of transportation systems: A recent literature review. In *Physica A: Statistical Mechanics and Its Applications*; Elsevier: Amsterdam, The Netherlands, 2021; p. 126235.
33. Nik, V.M.; Moazami, A. Using collective intelligence to enhance demand flexibility and climate resilience in urban areas. *Appl. Energy* **2021**, *281*, 116106. [CrossRef]
34. Elmqvist, T.; Andersson, E.; Frantzeskaki, N.; McPhearson, T.; Olsson, P.; Gaffney, O.; Takeuchi, K.; Folke, C. Sustainability and resilience for transformation in the urban century. *Nat. Sustain.* **2019**, *2*, 267–273. [CrossRef]
35. Cariolet, J.M.; Vuillet, M.; Diab, Y. Mapping urban resilience to disasters—A review. *Sustain. Cities Soc.* **2019**, *51*, 101746. [CrossRef]
36. Rieckert, A.; Schuit, E.; Bleijenberg, N.; Ten Cate, D.; de Lange, W.; de Man-van Ginkel, J.M.; Mathijssen, E.; Smit, L.C.; Stalpers, D.; Schoonhoven, L.; et al. How can we build and maintain the resilience of our health care professionals during COVID-19? Recommendations based on a scoping review. *BMJ Open* **2021**, *11*, e043718. [CrossRef]
37. Hines, S.E.; Chin, K.H.; Glick, D.R.; Wickwire, E.M. Trends in moral injury, distress, and resilience factors among healthcare workers at the beginning of the COVID-19 pandemic. *Int. J. Environ. Res. Public Health* **2021**, *18*, 488. [CrossRef]
38. Setiawati, Y.; Wahyuhadi, J.; Joestandari, F.; Maramis, M.M.; Atika, A. Anxiety and resilience of healthcare workers during COVID-19 pandemic in Indonesia. *J. Multidiscip. Healthc.* **2021**, *14*, 1. [CrossRef] [PubMed]
39. Jung, J.; Maeda, M.; Chang, A.; Bhandari, M.; Ashapure, A.; Landivar-Bowles, J. The potential of remote sensing and artificial intelligence as tools to improve the resilience of agriculture production systems. *Curr. Opin. Biotechnol.* **2021**, *70*, 15–22. [CrossRef] [PubMed]
40. Kusiak, A. Open manufacturing: A design-for-resilience approach. *Int. J. Prod. Res.* **2020**, *58*, 4647–4658. [CrossRef]
41. Boyacı-Gündüz, C.P.; Ibrahim, S.A.; Wei, O.C.; Galanakis, C.M. Transformation of the Food Sector: Security and Resilience during the COVID-19 Pandemic. *Foods* **2021**, *10*, 497.
42. Khan, L.U.; Saad, W.; Han, Z.; Hossain, E.; Hong, C.S. Federated learning for internet of things: Recent advances, taxonomy, and open challenges. In *IEEE Communications Surveys & Tutorials*; IEEE: Piscataway, NJ, USA, 2021.

43. Mothukuri, V.; Khare, P.; Parizi, R.M.; Pouriyeh, S.; Dehghantanha, A.; Srivastava, G. Federated Learning-based Anomaly Detection for IoT Security Attacks. *IEEE Internet Things J.* **2021**. [CrossRef]
44. McMahan, B.; Moore, E.; Ramage, D.; Hampson, S.; Agüera Arcas y, B. Communication-efficient learning of deep networks from decentralized data. In Proceedings of the 20th International Conference on Artificial Intelligence and Statistics (AISTATS), Ft. Lauderdale, FL, USA, 20–22 April 2017; pp. 1273–1282.
45. Li, T.; Sahu, A.K.; Zaheer, M.; Sanjabi, M.; Talwalkar, A.; Smith, V. Federated optimization in heterogeneous networks. *arXiv* **2018**, arXiv:1812.06127.
46. Chen, W.; Bhardwaj, K.; Marculescu, R. Fedmax: Mitigating activation divergence for accurate and communication-efficient federated learning. *arXiv* **2020**, arXiv:2004.03657.
47. Imteaj, A.; Amini, M.H. FedAR: Activity and Resource-Aware Federated Learning Model for Distributed Mobile Robots. In Proceedings of the 19th IEEE International Conference on Machine Learning and Applications (ICMLA), Miami, FL, USA, 14–17 December 2020.
48. Hegde, R.; Mukherjee, K.; Gupta, S.T. Electro-Maps. 2020. Available online: https://github.com/sabm0hmayahai/Electro-Maps (accessed on 10 May 2021).

 *electronics* 

Article

A Machine Learning Approach for Anomaly Detection in Industrial Control Systems Based on Measurement Data

Sohrab Mokhtari [1,*], Alireza Abbaspour [2], Kang K. Yen [1] and Arman Sargolzaei [3]

1 Electrical and Computer Engineering Department, Florida International University, Miami, FL 33174, USA; yenk@fiu.edu
2 Functional Safety Engineer, Tusimple Co., San Diego, CA 92093, USA; Aabba014@fiu.edu
3 Mechanical Engineering Department, Tennessee Technological University, Cookeville, TN 38505, USA; a.sargolzaei@gmail.com
* Correspondence: somokhta@fiu.edu; Tel.: +1-305-680-4338

Abstract: Attack detection problems in industrial control systems (ICSs) are commonly known as a network traffic monitoring scheme for detecting abnormal activities. However, a network-based intrusion detection system can be deceived by attackers that imitate the system's normal activity. In this work, we proposed a novel solution to this problem based on measurement data in the supervisory control and data acquisition (SCADA) system. The proposed approach is called measurement intrusion detection system (MIDS), which enables the system to detect any abnormal activity in the system even if the attacker tries to conceal it in the system's control layer. A supervised machine learning model is generated to classify normal and abnormal activities in an ICS to evaluate the MIDS performance. A hardware-in-the-loop (HIL) testbed is developed to simulate the power generation units and exploit the attack dataset. In the proposed approach, we applied several machine learning models on the dataset, which show remarkable performances in detecting the dataset's anomalies, especially stealthy attacks. The results show that the random forest is performing better than other classifier algorithms in detecting anomalies based on measured data in the testbed.

Keywords: machine learning; industrial control systems; anomaly detection; fault detection; intrusion detection system

Citation: Mokhtari, S.; Abbaspour, A.; Yen, K.K.; Sargolzaei, A. A Machine Learning Approach for Anomaly Detection in Industrial Control Systems Based on Measurement Data. *Electronics* **2021**, *10*, 407. https://doi.org/10.3390/electronics10040407

Academic Editor: Rashid Mehmood
Received: 1 January 2021
Accepted: 1 February 2021
Published: 8 February 2021

1. Introduction

The industrial control system (ICS) consists of devices, networks, and controllers to automate industrial processes. ICS contains several types of control systems, such as supervisory control and data acquisition (SCADA) systems, and distributed control systems (DCSs). ICSs are widely used in different critical infrastructures such as smart grids, power distribution, transportation systems, water treatment plants, and manufacturing [1,2]. In the power plants, ICS's key role is evident, and a multitude of automated systems are operating in a SCADA framework. The automated systems' entanglements could endanger the entire system's performance, where a small fault or malfunction would lead to a cascade failure. Thus, fault detection in ICSs, especially in critical infrastructures such as large-scale power plants, has attracted much attention in recent years [3–6].

Generally, communication between ICS components is based on an information technology stack (ITS) and remote connectivity. The reliance on communication networks to transmit measurements could increase the possibility of intentional attacks against physical plants. Conventionally, network traffic is secured by mechanisms such as authentication, data encryption, and message integrity techniques. However, these methods cannot completely protect the entire levels of an ICS network against a wide range of malicious activities. Figure 1 illustrates the five distinct levels of an ICS architecture. These conventional mechanisms try to secure the network traffic transmitting between ICS levels and do not investigate the compatibility of the physical plant measurements. This makes the system vulnerable against malicious activities such as insider sabotages, spoofing, and stealthy attacks [7]. One solution to tackle this problem is the intrusion detection system (IDS). The two main IDS strategies are signature-based and anomaly-based, which differ in their detection mechanisms [8]. The signature-based strategy trains the system to find specific anomalies while the anomaly-based strategy searches for any deviation from a pre-known normal activity. Generally, IDS investigates the network traffic in an ICS and tries to detect abnormal activities in the transmitting data packets. This strategy, known as the network intrusion detection system (NIDS), monitors the incoming data packets and prevents suspicious data from intruding into the system. Many studies leveraged machine learning algorithms to train a NIDS model, which is responsible for detecting attacks in the network traffic [9–11]. Although the NIDS effectively qualifies and quantifies attacks by analyzing the amount and types of attacks in the network flow, its performance against encrypted data packets, faked IP packets, and regular false positive alerts is not guaranteed.

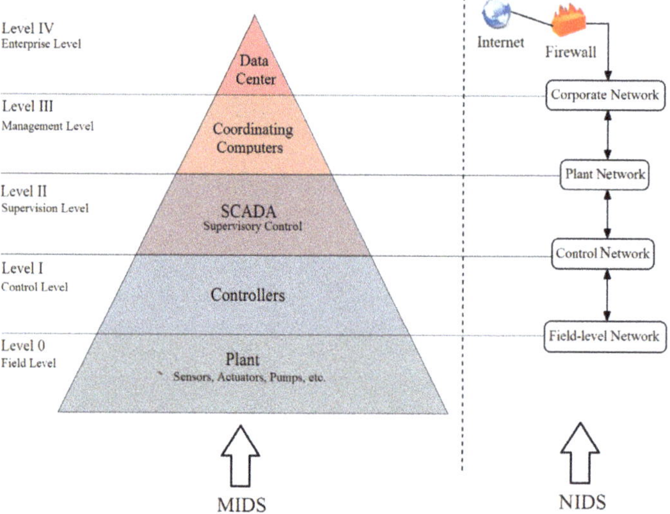

Figure 1. ICS network architecture.

On the other hand, the measurement intrusion detection system (MIDS) instead of monitoring the network traffic, investigates suspicious activities in the system's measurement data. As shown in Figure 1, the MIDS does not interact with the connections between the ICS levels, but directly inspect the measurement data in the system. This fault detection approach can find any deviation from normal performance caused by malicious activities such as changing the sensors setpoints or injecting fake data measurements into the ICS network levels. Since, in an ICS, the SCADA system (Level II) collects the data from the entire system, the MIDS can be embedded in this system's level. In comparison to the NIDS method, a few studies tried to use machine learning algorithms for training a fault detection model. Choi et al. [12] presented an IDS based on voltage measurement data to detect in-vehicle controller area network (CAN) intrusions using inimitable characteristics of electrical signals. Their approach is well designed to detect bus-off attacks [13] and performs very well to secure an in-vehicle CAN. However, relying on only one type of variable to detect suspicious activities in the system caused a high rate of false positives in the IDS. In [14], Pan et al. introduced an IDS strategy leveraging features of signature-based and specification-based detection methods which protects an electrical power transmission line from attacks. They used data from relay, network security logs, and energy management system (EMS) logs. Their method could accurately distinguish malicious activities from normal control operations. However, their proposed algorithm requires a large number of captured data scenarios, which is difficult to acquire. In another study, Ozay et al. [15] proposed an attack detection model employing state vector estimation (SVE) to detect false data injection at the physical layer of a smart grid. They showed that the model performs accurately on various IEEE test systems in detection of abnormal behaviors; however, it cannot detect the stealthy malicious activities properly.

Basically, due to the difficulties in generating a labeled dataset, which indicates different types of attacks in an ICS, most studies apply a normal activity dataset for training machine learning models. Therefore, the MIDS could only compare a set of normal data with the incoming data and detect any deviation from the normal activity. This strategy would fail while a stealthy attack that imitates a normal behavior intrudes into the system. A solution to tackle this problem is generating a labeled dataset that includes different types of attacks to train a machine learning model that is capable of detecting malicious stealthy activities in the system. But, building a labeled dataset consisting attack scenarios means the system should tolerate a set of controlled attack injected to the system, that could lead to a system failure and irreparable damages. Nevertheless, it is possible to simulate the critical infrastructures of an ICS using a hardware-in-the-loop (HIL) to prevent damaging the system. This approach could be sensible while the ICS is a vital infrastructure, and the system's security is dramatically significant [16]. The main goal of this paper is to investigate the performance of the MIDS by training machine learning algorithms leveraging a labeled dataset. To this end, we develop an experimental setup in which we can evaluate the effectiveness of fault detection by monitoring the measurement data in an ICS. For this, we employ a power generation testbed whose sensors' values are measured over several days. Different scenarios of attacks are injected into the system to generate the labeled dataset. The dataset was generated in 2020, available at [17]. Overall, this work has made the following contributions to the attack detection domain:

(1) Introducing a novel approach which can be integrated to NIDS as a second layer of defense mechanism for intrusion detection using measurement data and improving the security of the ICS system.

(2) Applying the HIL-based augmented ICS (HAI) testbed dataset [18] for the first time for training a supervised machine learning model to detect intrusions in an ICS. Unlike the previous works [12–15], our proposed design is able to detect the stealthy attacks without imposing any threat to the actual system using the advantages of labeled data obtained from the HIL testbed.

(3) Using the measurement data in the all levels of ICS (Figure 1), which can help to detect not only the sabotages in the communication links between the levels but also the insider sabotages in each level. This particular feature would help to improve the security of the system without any conflict with NIDS.

In addition, we compared different machine learning techniques to find the best learning model for the detection of stealthy attacks in the ICSs. According to the results, the random forest algorithm [19] has the best performance for the proposed dataset.

Problem Description and Motivation

ICS, including SCADA networks, consists of several parts such as controllers monitored by operators through the human–machine interface (HMI). In critical ICS infrastructures such as power plants, the communication network between parts of the system can be extended over large geographical regions, which perform under virtual private networks (VPN) or the Internet. Although connecting the communication network to the Internet or using remote connections help to have an off-site operation and management of ICS over a vast geographical distance, it puts the system at risk of malicious attacks [20]. The NIDS is widely employed to detect any kind of abnormal activity in the system's network flow to defeat these types of attacks. Nevertheless, while the NIDS could address the problem of cyber attacks at a sensible rate of accuracy, these systems are inefficient in detecting insider attacks or any other sabotage inside the system. Furthermore, the NIDS is incapable of detecting encrypted or any other faked data packets, especially stealthy attacks. To address these problems, one solution is investigating the behavior of measurement data instead of monitoring network traffic.

The MIDS is not only capable of detecting any deviation from the normal activity of the ICS, but it is also effective in the detection of stealthy attacks [21]. The presence of SCADA systems, especially in large-scale ICSs, helps deploy the MIDS without any additional devices; however, the main obstacle in practicing the MIDS is preparing a comprehensive dataset to train a machine learning model. Basically, building an attack dataset for measured data means injecting a malfunction to the system and possibly the whole system's failure. This could have dramatically high costs when the system is a critical large-scale ICS. Moreover, for each system's environment, the dataset should be built separately, and it could be extremely expensive. Fortunately, in recent years, development in processing units and computation power has helped to overcome this problem by introducing the HIL technique. This technique stimulates the critical parts of an ICS and injects attacks into the system without any threat to the existing system. Tackling the problem of building a dataset, including real-time attacks, provides a remarkable opportunity for studying the MIDS.

In this study, to investigate the MIDS's efficiency in ICS fault detection, an electrical power generation testbed is employed, which is wholly explained in Section 3. The investigation procedure includes pre-processing the data, fitting supervised learning models, and evaluating each model's classification accuracy. The standard methods of assessing models' effectiveness are the confusion matrix, the area under the curve (AUC), and the receiver operating characteristics (ROC) curve.

The remainder of the paper is organized as follows. In Section 2, the methodology for building a machine learning model is described. Section 3 includes a description of the dataset. The results and discussions of implementing the model on the dataset are proposed in Section 4. Section 5 presents the conclusion and future work directions.

2. Methodology

The developed attack detection procedure is described in this section. First, the most significant attacks in the ICSs are introduced; then, the approach to detect these attacks is explained. As shown in Figure 2, after collecting the data, the most relevant features are selected, and a trained ML model based on the corresponding features classifies the output data.

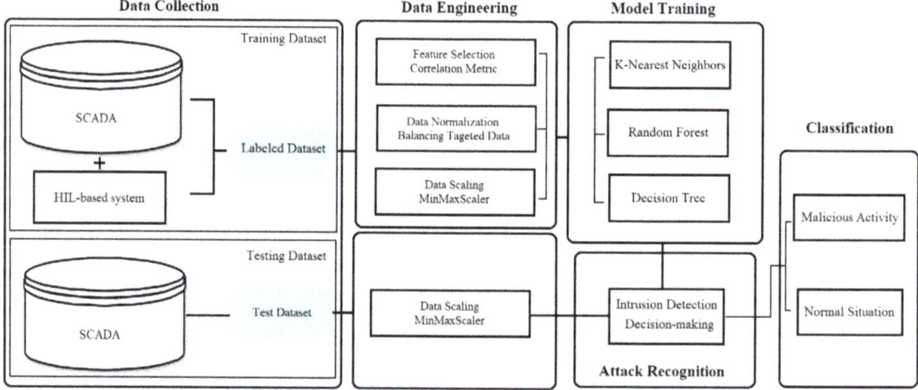

Figure 2. The framework of MIDS in ICS for the HAI dataset.

2.1. Attack Description

Anomaly detection consists of various domains, such as intrusion detection, fault detection, and event detection in sensor networks. Any deviation from a normal performance could be considered an anomaly in an ICS. It could happen due to several reasons, including a malfunction in a system's component, insider sabotage, or an intentional cyberattack. In this paper, the concept of anomaly detection based on the MIDS refers to fault detection and intrusion detection. When a malfunction or insider sabotage occurs, the MIDS tries to detect faults in the system. In addition, when an attacker attempts to intrude in the system, it is known as intrusion detection.

Anomaly detection in ICSs using the measured data captured by the SCADA system has the privilege of detecting any deviation from a normal activity even if the intrusion is not recognized in the network layer by the NIDS. Evidently, due to a deviation from the system's normal behavior, detecting a malfunction or a simple attack that directly manipulates the system's measurement data would not be a challenging task for the MIDS. On the other hand, the main concern about the MIDS effectiveness is its performance in the detection of stealthy attacks. These kinds of attacks occur when an attacker manipulates sensor measurements or control signals persistency by penetrating control networks without being detected until the system crashes. Normally, attackers attempt to imitate the system's normal behavior to stay undetected. In this paper, not only the MIDS performance in the detection of malfunctions is evaluated, but also a set of stealthy attacks are injected into the system to investigate the MIDS effectiveness in the detection of this type of attack.

2.2. Data Analyzing

The problem of imbalanced datasets in IDS modeling is a critical issue. In machine learning modeling, particularly in classification problems, having access to a balanced dataset in the training stage has a significant impact on the model's performance. In the MIDS, this problem comes from a large number of normal conditions compared to abnormal activities in the system logs.

To handle the problem of imbalanced datasets, a multitude of techniques are introduced, such as the threshold method, one-class learning, or cost-sensitive learning [22]. In fact, all balancing methods are based on oversampling or undersampling approaches. Briefly, the undersampling method tries to decrease the number of instances from the majority class; on the other hand, the oversampling method attempts to increase the number of samples of the minority class. While undersampling has the risk of losing important data, oversampling puts the model at the stake of overfitting.

One solution to tackle this problem is the Synthetic Minority Over-sampling Technique (SMOTE) method [23]. The SMOTE is a random oversampling approach that generates new instances using existing data from the rare classes. For this, any point from the minority class that smoothly moves an existing sample around its neighbors will be added to the dataset until the dataset reaches a balanced condition. Therefore, this method by generating new samples (which are not exactly the same as the existing samples) makes it possible to avoid the risk of overfitting problems [24]. In this paper, due to the imbalance in the labeled data, the SMOTE method is employed to normalize the dataset targeted data. Moreover, the balanced dataset helps the normal and abnormal data be split equally during the procedure of building the train and test datasets. In this paper, the Stratified Shuffle Split (SSS) method is applied to divide the train and test data. The test data include 0.3 of the entire dataset, and the number of re-shuffling and splitting iterations is considered as 5.

2.3. Feature Engineering

In the MIDS, features are basically the measured data collected by the SCADA system from embedded sensors. In large-scale ICSs, the quantity of sensors is normally a large number. This mentions two facts. First, in the proposed problem, the feature selection method plays a significant role in the performance of the model; second, due to a large amount of measured data, the model algorithm should be capable of fast prediction to be usable in real-time applications. The goal of feature selection is to find the most effective features that lead to training more accurate models and less computation time. Feature selection techniques can be classified as filter, wrapper, embedded, and hybrid methods [25]. In the filter method, correlation criteria are employed widely in machine learning problems. Correlation is a measure of the linear relationship between two or more parameters. In feature selection, the most correlated features with the target would be chosen to build the model. Moreover, those features should not show a high correlation with each other to avoid using redundant data. Pearson correlation technique is one of the most useful criteria in feature selection, which can be described as

$$Corr(i) = \frac{cov(a_i, b)}{\sqrt{var(a_i) * var(b)}} \tag{1}$$

where a_i is the ith feature, b is the target label, and $cov()$ and $var()$ represent the covariance and the variance functions, respectively. $Corr(i)$ also indicates the Pearson correlation technique, which shows the correlation between the ith feature and the corresponding target.

To select the features with a high correlation with the target, we need to set a threshold value for choosing the features with a higher correlation. Suppose that the selected features are correlated to each other. In that case, we can drop the one with the lowest correlation to the target. In addition, the features that show a high correlation together can be unified. To do so, the correlation of the features, two by two, are calculated, and the most correlated features are nominated for removal.

Moreover, in the pre-processing step, the input data should be scaled. This could result in a sustainable learning process. In this paper, the MinMaxScaler is employed to scale the features values. Equation (2) describes this function, where a_i^m is the ith feature from mth experiment, a_{min} and a_{max} are the minimum and the maximum values of the feature among the experiments, respectively. In addition, $a^m_{i(scaled)}$ indicates the scaled value for the ith feature of mth experiment.

$$a^m_{i(scaled)} = \frac{(a_i^m - a_{min})}{(a_{max} - a_{min})} \tag{2}$$

2.4. Machine Learning Models

Supervised anomaly detection in ICS generally uses normal activity data to build a predictive model of normal class as well as anomaly class. Then, any unforeseen data are compared with the generated model to detect its class. Several algorithms are applied in this study to train a machine learning model for detecting anomalies by the MIDS. Having access to a labeled dataset allows for applying supervised learning strategies by considering two classes of attack and normal activities. In this study, the most accurate supervised learning algorithms are chosen that are k-nearest neighbors (KNN), decision tree classifier (DTC), and random forest (RF).

- The KNN algorithm uses data to classify unforeseen data points by measuring the distances from the neighbor points. This classification method classifies new data by the plurality vote of its k neighbors which are assigned to the most similar class.
- The decision tree classifier uses a tree-like model of decisions and their possible outcomes. Normally, a decision tree classifier is used for discrete categorical targets, which, in this paper, the target is a binary variable that includes attack and normal situations.
- The random forest algorithm is a combination of tree classifiers. This classifier tries to maximize the variance by injecting randomness in data selection and to minimize the bias by increasing the tree depth to a maximum level.

2.5. Model Evaluation Metrics

The performance of algorithms in detecting anomalies in ICSs based on supervised learning is investigated by the following metrics.

- Confusion matrix: This measure is used to evaluate a classifier's performance considering a pre-known set of labeled data. For each classifier, a confusion matrix would be generated. In addition, sensitivity, specificity, precision, and F1-score metrics are calculated regarding this matrix. The sensitivity or recall metric shows the likelihood of predicting true positive, while the specificity measures the true negative rate. In addition, the precision metric represents the accuracy of the positively predicted classes, which are actually positive. The F1-Score shows the balance between sensitivity and precision. Finally, the accuracy of the model is measured by evaluating the trueness of the results. Figure 3 explains a confusion matrix and its associated metrics.
- Receiver operator characteristic (ROC) curve and area under the curve (AUC). The ROC is a graph that illustrates the performance of the classification algorithm at all classification thresholds and includes two parameters: true-positive and false-positive rates. The ROC compares the classifiers' performance among the whole range of class distributions and error costs. To compare the ROC curves, the area under the ROC curve is calculated, called the area under the curve (AUC) metric. More values of AUC implies more accuracy in the model prediction [26].

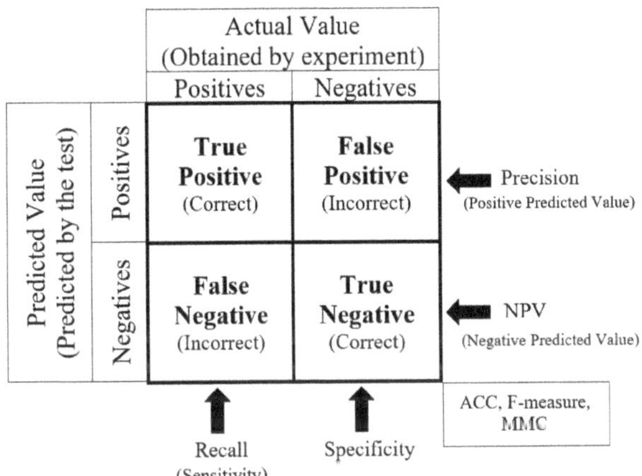

Figure 3. Explanation of confusion matrix. ACC and MMC are accuracy and Matthew's correlation coefficient, respectively [27,28].

3. Experimental Setup

3.1. ICS Testbed

Performance of anomaly detection in ICSs based on sensors measurement data is evaluated by implementing machine learning models on a dataset from a power generation system [18]. As shown in Figure 4, the testbed system has four primary processes, including a turbine process, a water-treatment process, a boiler process, and a HIL simulator. In the procedure of building the attack dataset, to protect the system from harmful damages of attacks, the HIL simulates the thermal power and the pumped-storage hydropower generators.

The boiler process, including four controllers (level controller, pressure controller, temperature controller, flow-rate controller), is responsible for heating the pumped water from the main water tank. The turbine process consists of a motor speed controller that rotates a turbine. The water-treatment process has a level controller that manages the level control pump (LCP) and the level control valve (LCV) and is in charge of transferring water from the upper to the lower reservoir and vice versa. The HIL simulator includes two generators and a power grid model that feeds an electrical load.

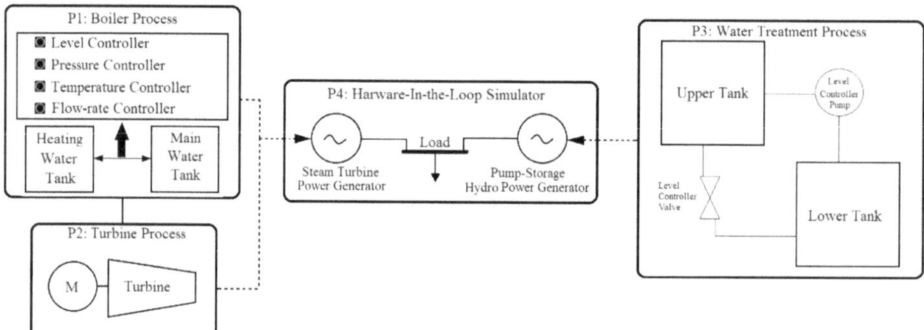

Figure 4. HIL-based augmented ICS.

3.2. Dataset

The dataset used in this paper is from a HIL-based augmented ICS security (HAI) available at [17]. The testbed dataset is built by collecting measurements of 59 sensors every second through four days. During these four days, 28 attacks are injected into the system. These attacks are a combination of 14 process control loop (PCL) primitive attacks which are affecting four points in the system: setpoints, process variables, control output, and control parameters. The attacks are stealthy type and cannot be detected easily by the conventional NIDS.

The next section is devoted to the implementation of supervised machine learning algorithms on the proposed dataset. The following section illustrates the MIDS performance on fault detection, especially stealthy attacks.

4. Results and Discussion

The proposed MIDS method based on a machine learning approach is tested on the HAI dataset, and its performance in anomaly detection is evaluated. The machine learning algorithms are trained and tested by employing Python. The procedure of generating the model is shown in Figure 2. In this paper, several classification algorithms were examined, and the most accurate ones were selected to be implemented on the MIDS model.

Due to the large number of measuring points in the dataset, the most important features are selected by employing a correlation metric. The feature selection process contains two steps. First, the most correlated features are identified and unified. In this step, from 59 features, 41 are selected. Then, among the residue features, the ones showing the most correlation with the target values are chosen, leading to 17 remaining features. Figure 5 shows the correlation matrices during the feature selection process. It should be mentioned that Figure 5b is the correlation matrix after removing the most correlated features together, and Figure 5c is the correlation matrix after removing the least correlated feature with the target.

Mostly, in the intrusion detection problems, the training datasets are suffering from imbalance targeted data. This is because of the much lower duration of attack activities compared to the normal conditions. In the testbed dataset, less than 4% of the whole data are associated with abnormal activities. This imbalance of data could affect the performance evaluation of the trained models. The SMOTE method is employed to tackle this problem. This method helps to balance the dataset without a high risk of overfitting. Figure 6 shows the target distribution in the dataset before and after normalizing.

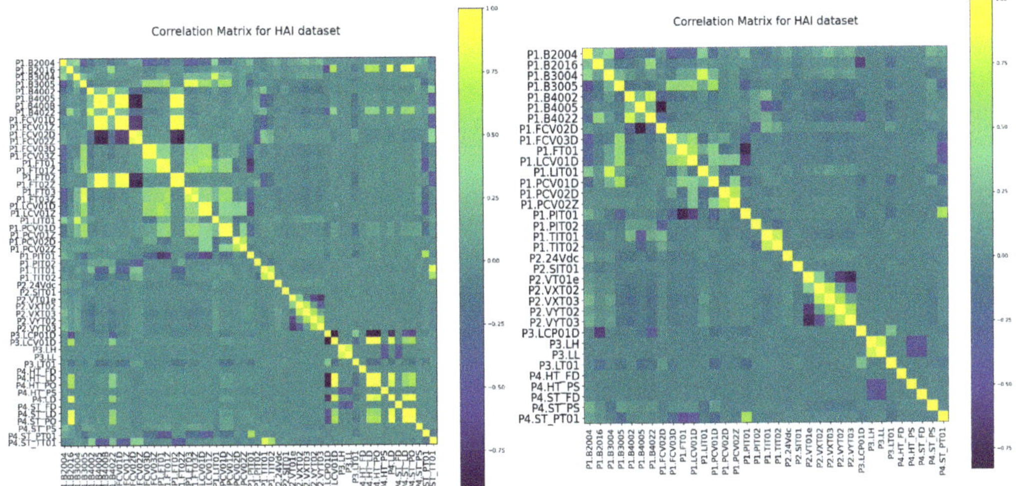

(**a**) Features correlation before feature selection.

(**b**) Features correlation after feature selection. Features with a correlation of more than 0.85 are united.

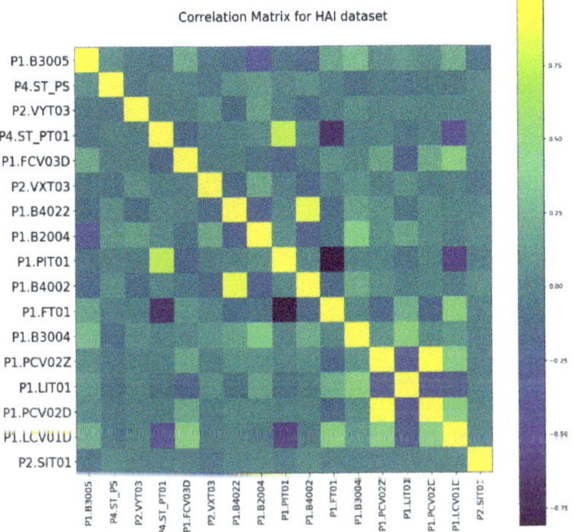

(**c**) Features correlation after feature selection. Features with a correlation of more than 0.95 with the target are united.

Figure 5. Feature selection using a correlation metric.

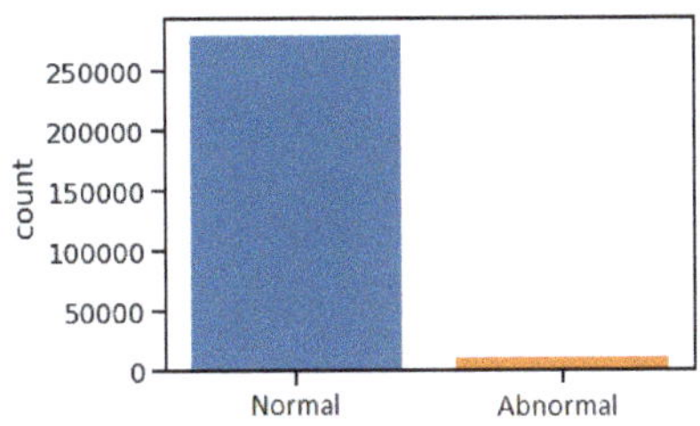

(**a**) Before performing normalization.

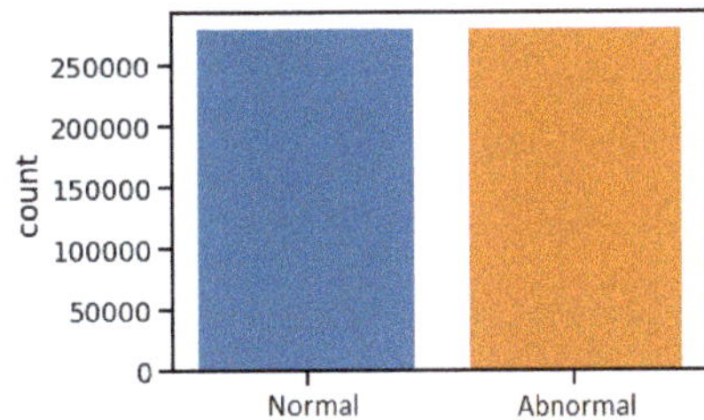

(**b**) After performing normalization (SMOTE method).

Figure 6. Normal and abnormal conditions' distribution.

It should be considered that the sensors' measurements in an ICS have a wide range of values. The unscaled data could cause significant problems during the model training procedure, and lead to an unsustainable learning process. Therefore, the MinMaxScaler function explained in Section 3 is used to scale the measurement data in an appropriate range.

After the pre-processing step, the dataset is ready to train a machine learning model. In this study, the supervised classification models that are implemented on the dataset are k-nearest neighbor (KNN), random forest (RF), and decision tree classifier (DTC). These algorithms are chosen based on their effectiveness on this particular problem. Two major factors are considered in the evaluation of the algorithms' performances. First, the accuracy in classifying the targeted output. Second, the required time for fitting and predicting processes. For the first factor, the confusion matrix is computed (Figure 7), and regarding this matrix, other metrics such as accuracy, precision, F1-score, specificity, and sensitivity are calculated. More information related to the confusion matrix concept and its metrics is available at [29]. For the second factor, the computation time for fitting the training dataset and predicting the test dataset is captured. Table 1 shows the confusion matrices along with the computation times for the selected algorithms.

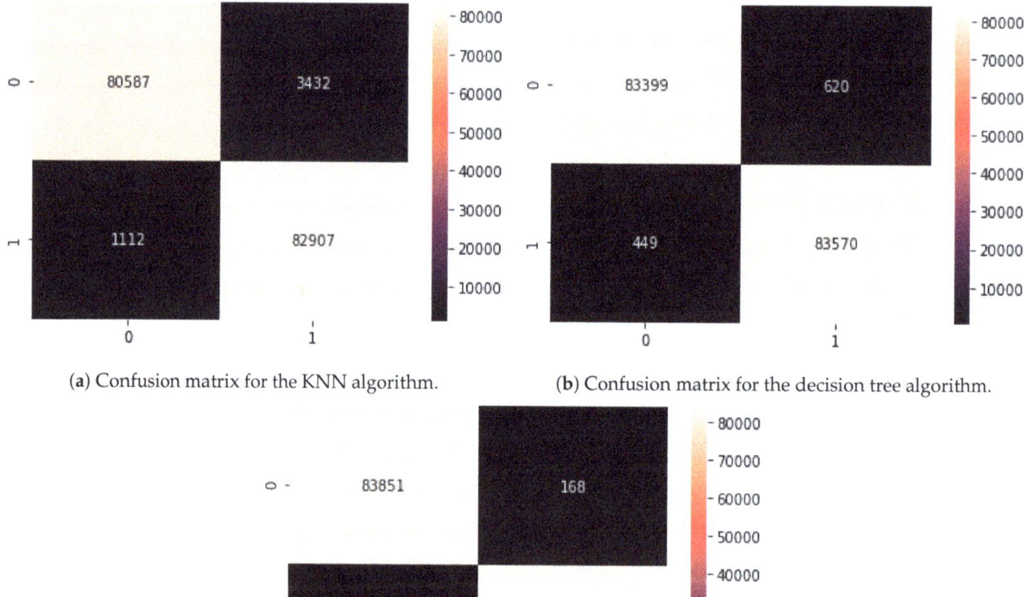

(**a**) Confusion matrix for the KNN algorithm.

(**b**) Confusion matrix for the decision tree algorithm.

(**c**) Confusion matrix for the random forest algorithm.

Figure 7. Confusion matrices.

Table 1. Models' performance comparison.

	KNN	DTC	RF
Precision	0.9732	0.9937	0.9976
Recall	0.9729	0.9937	0.9976
F1-score	0.9729	0.9937	0.9976
Accuracy	0.9729	0.9937	0.9976
AUC	0.9729	0.9937	0.9976
Fitting time [s]	173	5.8	2.21
Prediction time [s]	104	0.0283	0.0505

The random forest algorithm has the best performance in detecting the anomalies in the dataset. This algorithm consumes the least time for generating the model, while the decision tree classifier has the lowest prediction computing time. This mentions that a trade-off between accuracy and prediction time should be considered for a real-life problem. Moreover, the KNN algorithm shows a lower accuracy in predicting the anomalies and requires a longer time for fitting and predicting processes. The result implies that the MIDS can be a reliable solution for the anomaly detection problem. Applying measurement data from the SCADA system to detect attacks could be considered a protection layer in ICSs. While the NIDS can protect the network traffic from malicious intrusions, the MIDS could improve the system's reliability as the second layer of protection, especially against stealthy attacks.

Usually, stealthy attacks that intrude into the field level (level 0) of an ICS attempt to conceal the state changes by imitating a normal behavior and deceiving the protection systems. Usually, building a dataset that includes stealthy attacks is a complicated process. For the first time, by leveraging HIL systems, a real-life dataset containing stealthy attacks on sensor data measurements is provided. This dataset consists of actual data during intrusion attacks that manipulate the control parameters.

The ROC curve illustrated in Figure 8 indicates that the random forest algorithm has a remarkable performance in detecting anomalies. This figure compares the three applied algorithms' accuracy using the AUC of the ROC curves. The diagonal dashed line from the bottom left to the top right corner of Figure 8a represents a non-discriminatory test points where sensitivity = 1-specificity. As shown in Figure 8b, the random forest algorithm performs far better than the KNN and the DTC algorithms by an AUC of 1.

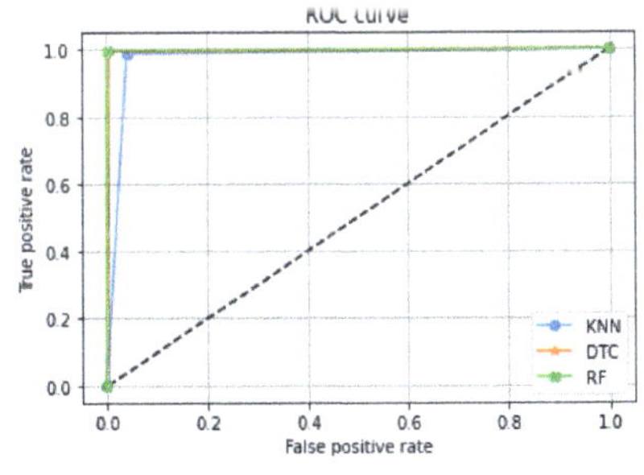

(**a**) ROC curves.

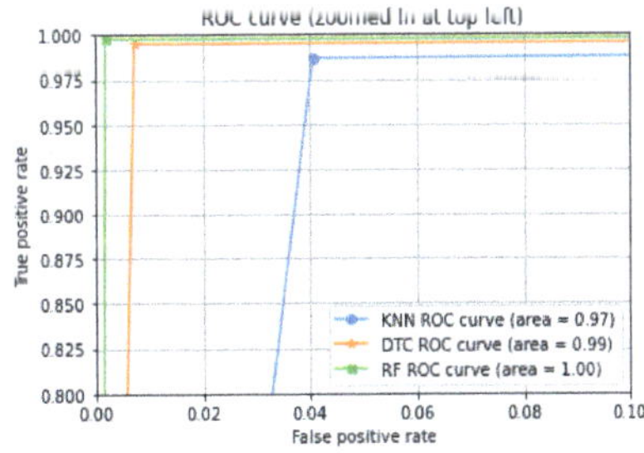

(**b**) ROC curves from a closer view.

Figure 8. ROC curves for KNN, random forest, and decision tree classifiers.

5. Conclusions

In this paper, the classification performance of the measurement intrusion detection system is investigated and a remarkable outcome is concluded, especially on stealthy attacks. The MIDS is working based on the measured data that the SCADA system collects from the ICS sensors. IDSs are mostly investigating network traffic to find malicious activities in the system, and stealthy attacks are hard to be detected by these strategies. Because the MIDS is investigating the measurement data, it could detect deceptive behaviors in the system better than the NIDS. The HAI dataset, including actual data from a power generation system, is applied to evaluate the MIDS performance in fault detection. The results show a very successful classification employing the random forest algorithm in the fault detection process with an accuracy of 99.76%.

Nevertheless, although the MIDS can greatly detect anomalies, especially stealthy attacks, it cannot prevent malicious intrusions in the layer of network traffic. Indeed, the MIDS could detect anomalies when they successfully deceive the NIDS by imitating a normal behavior in the system. Therefore, the MIDS cannot be a substitution of the NIDS; however, it can be embedded as the second layer of protection in the critical infrastructure of ICSs. By combining these two protection layers, if any malfunction in the system, including insider's sabotage, systems failure, stealthy attack, or network intrusion happens, the IDS could be sufficient in detecting it successfully.

Moreover, this study leverages the supervised learning approach to build a machine learning model. The unsupervised learning methods are also a topic of interest for our future works; however, their efficiencies in comparison with supervised learning models should be investigated. In the future, we would like to investigate unsupervised learning algorithms due to the fact that they don't require labeled data for the model training step, which leads to detecting new anomalies in the system.

Author Contributions: S.M. prepared an initial draft and overall structure of this research paper for anomaly detection in industrial control based on measurement data and invited other authors to contribute and expand the research scope. A.A. and A.S. were involved in data analysing and model generation of MIDS. K.K.Y. supervised the research and co-wrote the paper. All authors have read and agreed to the published version of the manuscript.

Funding: Partial support of this research was provided by Tennessee Technological University and the National Science Foundation under Grant No. CNS-1919855. Any opinions, findings, and conclusions, or recommendations expressed in this material are those of the author(s) and do not necessarily reflect the views of the sponsoring agency.

Data Availability Statement: The data that used in this study are openly available on kaggle website at https://www.kaggle.com/icsdataset/hai-security-dataset (accessed on 20 December 2020), entitled HAI Security Dataset V. 4, created by Hyeok-Ki Shin, Woomyo Lee, Jeong-Han Yun and HyoungChun Kim in the Affiliated Institute of ETRI, Daejeon, South Korea, licence CC BY-SA 4.0. avialable at https://creativecommons.org/licenses/by/4.0/ (accessed on 7 February 2021).

Conflicts of Interest: The authors declare no conflict of interest.

References

1. Paridari, K.; O'Mahony, N.; Mady, A.E.D.; Chabukswar, R.; Boubekeur, M.; Sandberg, H. A framework for attack-resilient industrial control systems: Attack detection and controller reconfiguration. *Proc. IEEE* **2017**, *106*, 113–128. [CrossRef]
2. Arafat, M.; Iqbal, S.; Hadi, M. Utilizing an Analytical Hierarchy Process with Stochastic Return on Investment to Justify Connected Vehicle-Based Deployment Decisions. *Transp. Res. Rec.* **2020**, *2674*, 462–472. [CrossRef]
3. Abbaspour, A.; Mokhtari, S.; Sargolzaei, A.; Yen, K.K. A Survey on Active Fault-Tolerant Control Systems. *Electronics* **2020**, *9*, 1513. [CrossRef]
4. Mokhtari, S.; Yen, K.K. A Novel Bilateral Fuzzy Adaptive Unscented Kalman Filter and its Implementation to Nonlinear Systems with Additive Noise. In Proceedings of the 2020 IEEE Industry Applications Society Annual Meeting, Detroit, MI, USA, 10–16 October 2020; pp. 1–6.
5. Fawzy, N.; Habib, H.F.; Mohammed, O.; Brahma, S. Protection of Microgrids with Distributed Generation based on Multiagent System. In Proceedings of the 2020 IEEE International Conference on Environment and Electrical Engineering and 2020 IEEE Industrial and Commercial Power Systems Europe (EEEIC/I& CPS Europe), Madrid, Spain, 9–12 June 2020; pp. 1–5.
6. Habib, H.F.; Fawzy, N.; Esfahani, M.M.; Mohammed, O.A. Enhancement of protection scheme for distribution system using the communication network. In Proceedings of the 2019 IEEE Industry Applications Society Annual Meeting, Baltimore, MD, USA, 29 September–3 October 2019; pp. 1–7.
7. Slay, J.; Miller, M. Lessons learned from the maroochy water breach. In Proceedings of the International Conference on Critical Infrastructure Protection, Hanover, NH, USA, 19–21 March 2007; Springer: Berlin/Heidelberg, Germany, 2007; pp. 73–82.

8. Wang, Y.; Meng, W.; Li, W.; Li, J.; Liu, W.X.; Xiang, Y. A fog-based privacy-preserving approach for distributed signature-based intrusion detection. *J. Parallel Distrib. Comput.* **2018**, *122*, 26–35. [CrossRef]
9. Aloqaily, M.; Otoum, S.; Al Ridhawi, I.; Jararweh, Y. An intrusion detection system for connected vehicles in smart cities. *Ad Hoc Netw.* **2019**, *90*, 101842. [CrossRef]
10. Vinayakumar, R.; Alazab, M.; Soman, K.; Poornachandran, P.; Al-Nemrat, A.; Venkatraman, S. Deep learning approach for intelligent intrusion detection system. *IEEE Access* **2019**, *7*, 41525–41550. [CrossRef]
11. Manzoor, I.; Kumar, N. A feature reduced intrusion detection system using ANN classifier. *Expert Syst. Appl.* **2017**, *88*, 249–257.
12. Choi, W.; Joo, K.; Jo, H.J.; Park, M.C.; Lee, D.H. Voltageids: Low-level communication characteristics for automotive intrusion detection system. *IEEE Trans. Inf. Forensics Secur.* **2018**, *13*, 2114–2129. [CrossRef]
13. Cho, K.T.; Shin, K.G. Error handling of in-vehicle networks makes them vulnerable. In Proceedings of the 2016 ACM SIGSAC Conference on Computer and Communications Security, Vienna, Austria, 24–28 October 2016; pp. 1044–1055.
14. Pan, S.; Morris, T.; Adhikari, U. Developing a hybrid intrusion detection system using data mining for power systems. *IEEE Trans. Smart Grid* **2015**, *6*, 3104–3113. [CrossRef]
15. Ozay, M.; Esnaola, I.; Vural, F.T.Y.; Kulkarni, S.R.; Poor, H.V. Machine learning methods for attack detection in the smart grid. *IEEE Trans. Neural Netw. Learn. Syst.* **2015**, *27*, 1773–1786. [CrossRef] [PubMed]
16. Habib, H.F.; Fawzy, N.; Brahma, S. Hardware in the Loop of a Protection Scheme for Microgrid using RTDS with IEC 61850 Communication Protocol. In Proceedings of the 2020 IEEE Industry Applications Society Annual Meeting, Detroit, MI, USA, 13–15 October 2020; pp. 1–6.
17. Choi, S. HIL-Based Augmented ICS (HAI) Security Dataset. 2020. Available online: https://github.com/icsdataset/hai (accessed on 20 December 2020).
18. Shin, H.K.; Lee, W.; Yun, J.H.; Kim, H. HAI 1.0: HIL-based Augmented ICS Security Dataset. In Proceedings of the 13th USENIX Workshop on Cyber Security Experimentation and Test (CSET 20), Boston, MA, USA, 10 August 2020.
19. Biau, G.; Scornet, E. A random forest guided tour. *Test* **2016**, *25*, 197–227. [CrossRef]
20. Van der Knijff, R.M. Control systems/SCADA forensics, what's the difference? *Digit. Investig.* **2014**, *11*, 160–174. [CrossRef]
21. Urbina, D.I.; Giraldo, J.A.; Cardenas, A.A.; Tippenhauer, N.O.; Valente, J.; Faisal, M.; Ruths, J.; Candell, R.; Sandberg, H. Limiting the impact of stealthy attacks on industrial control systems. In Proceedings of the 2016 ACM SIGSAC Conference on Computer and Communications Security, Vienna, Austria, 24–28 October 2016; pp. 1092–1105.
22. Kotsiantis, S.; Kanellopoulos, D.; Pintelas, P. Handling imbalanced datasets: A review. *GESTS Int. Trans. Comput. Sci. Eng.* **2006**, *30*, 25–36.
23. Chawla, N.V.; Bowyer, K.W.; Hall, L.O.; Kegelmeyer, W.P. SMOTE: Synthetic minority over-sampling technique. *J. Artif. Intell. Res.* **2002**, *16*, 321–357. [CrossRef]
24. Zolanvari, M.; Teixeira, M.A.; Jain, R. Effect of imbalanced datasets on security of industrial IoT using machine learning. In Proceedings of the 2018 IEEE International Conference on Intelligence and Security Informatics (ISI), Miami, FL, USA, 9–11 November 2018; pp. 112–117.
25. Chandrashekar, G.; Sahin, F. A survey on feature selection methods. *Comput. Electr. Eng.* **2014**, *40*, 16–28. [CrossRef]
26. Marzban, C. The ROC curve and the area under it as performance measures. *Weather Forecast.* **2004**, *19*, 1106–1114. [CrossRef]
27. Matthews, B.W. Comparison of the predicted and observed secondary structure of T4 phage lysozyme. *Biochim. Biophys. Acta BBA Protein Struct.* **1975**, *405*, 442–451. [CrossRef]
28. Xia, B.; Zhang, H.; Li, Q.; Li, T. PETs: A stable and accurate predictor of protein-protein interacting sites based on extremely-randomized trees. *IEEE Trans. Nanobiosci.* **2015**, *14*, 882–893. [CrossRef] [PubMed]
29. Ting, K.M. Confusion Matrix. In *Encyclopedia of Machine Learning and Data Mining*; Sammut, C., Webb, G.I., Eds.; Springer: Boston, MA, USA, 2017; p. 260. [CrossRef]

 *electronics*

Article

P1OVD: Patch-Based 1-Day Out-of-Bounds Vulnerabilities Detection Tool for Downstream Binaries

Hongyi Li [1,†], Daojing He [1,2,*,†], Xiaogang Zhu [3] and Sammy Chan [4]

1 Software Engineering Institute, East China Normal University, Shanghai 200062, China; 51194501009@stu.ecnu.edu.cn
2 School of Computer Science and Technology, Harbin Institute of Technology (Shenzhen), Shenzhen 518055, China
3 Department of Computer Science and Software Engineering, School of Software and Electrical Engineering, Swinburne University of Technology, Melbourne 3122, Australia; xiaogangzhu@swin.edu.au
4 Department of Electrical Engineering, City University of Hong Kong, Hong Kong SAR, China; eeschan@cityu.edu.hk
* Correspondence: djhe@sei.ecnu.edu.cn or hedaojinghit@163.com; Tel.: +86-21-6223-1233
† These authors contributed equally to this work.

Abstract: In the past decades, due to the popularity of cloning open-source software, 1-day vulnerabilities are prevalent among cyber-physical devices. Detection tools for 1-day vulnerabilities effectively protect users who fail to adopt 1-day vulnerability patches in time. However, manufacturers can non-standardly build the binaries from customized source codes to multiple architectures. The code variants in the downstream binaries decrease the accuracy of 1-day vulnerability detections, especially when signatures of out-of-bounds vulnerabilities contain incomplete information of vulnerabilities and patches. Motivated by the above observations, in this paper, we propose P1OVD, an effective patch-based 1-day out-of-bounds vulnerability detection tool for downstream binaries. P1OVD first generates signatures containing patch information and vulnerability root cause information. Then, P1OVD uses an accurate and robust matching algorithm to scan target binaries. We have evaluated P1OVD on 104 different versions of 30 out-of-bounds vulnerable functions and 620 target binaries in six different compilation environments. The results show that P1OVD achieved an accuracy of 83.06%. Compared to the widely used patch-level vulnerability detection tool ReDeBug, P1OVD ignores 4.07 unnecessary lines on average. The experiments on the *x86_64* platform and the *O0* optimization show that P1OVD increases the accuracy of the state-of-the-art tool, BinXray, by 8.74%. Besides, it can analyze a single binary in 4 s after a 20-s offline signature extraction on average.

Keywords: out-of-bounds; vulnerable detection; patch

Citation: Li, H.; He, D.; Zhu, X.; Chan, S. P1OVD: Patch-Based 1-Day Out-of-Bounds Vulnerabilities Detection Tool for Downstream Binaries. *Electronics* **2022**, *11*, 260. https://doi.org/10.3390/electronics11020260

Academic Editor: Arman Sargolzaei

Received: 13 December 2021
Accepted: 12 January 2022
Published: 14 January 2022

1. Introduction

Vulnerabilities acknowledged by vendors are called 1-day vulnerabilities and are often fixed by upstream software developers using security patches [1]. In the past decades, 1-day vulnerabilities are widely spread among cyber-physical devices due to the popularity of open-source software cloning [2]. In the Debian system alone, researchers [3] have found 145 cloned 1-day vulnerabilities. Over the last few years, various automatical 1-day vulnerability detection tools for binaries have been proposed [3–18] to protect users who fail to adopt 1-day vulnerability patches in time.

However, manufacturers usually non-standardly build binaries from customized source codes for multiple target architectures. Such code variants decrease the accuracy of 1-day vulnerability detection, especially out-of-bounds vulnerabilities. Moreover, the inaccuracy can lead to safety risks or extra manual efforts for security analysis. Due to the prevalence of 1-day out-of-bounds vulnerabilities, in this paper, we propose P1OVD, a 1-day out-of-bounds vulnerability detection tool, which has higher accuracy when code variants appear in the downstream binaries.

Code variants are common and diverse. Research shows that among 6027 counterparts of 285 Android Kernel functions, over 72% of them contain codes that are different from their mainstream versions [19]. On the one hand, the code variants that are caused by target architectures or optimization levels can be large. Although they heavily change the instructions, function basic blocks, and function CFGs (control flow graph), they hardly change function logics. On the other hand, code variants can be caused by patching vulnerabilities or unexpectedly introducing new vulnerable modules. So these code variants are critical but small. These two kinds of code variants can appear at the same time, causing two challenges.

The first challenge is that out-of-bounds vulnerability signatures can easily neglect small but important code variants. Function-level 1-day vulnerability detection tools such as Asm2Vec [16] generate vulnerability signatures from the whole vulnerable functions [4–16]. Due to their extremely large scope, they fail to capture the precise context of vulnerabilities. At the same time, patch-level 1-day vulnerability detection tools such as ReDeBug [3] and BinXray [17] generate vulnerability signatures only from patches and their signatures contain incomplete vulnerability information [3,17,18]. As a result, when the small code variants influence the vulnerability root causes, the existing tools give false predictions.

The second challenge is that large code variants can decrease the accuracy of the matching methods that depend on AST (abstract syntax tree) shaped out-of-bounds patch signature. Moreover, many existing works have difficulties in balancing the accuracy and robustness. The strict operand-based matching [20] is accurate but sensitive to unimportant code variants, while the graph-similarity-based algorithms [8,9,13,19] improve the robustness but sacrifice accuracy.

To solve the above two challenges, we propose a patch-based 1-day out-of-bounds vulnerability detection tool named P1OVD, which can automatically find 1-day out-of-bounds vulnerabilities in the downstream binaries. It first analyzes patches and outputs source signatures, which solves the first challenge. Then the signature generator maps the source signatures to binary signatures. Finally, the matching engine scans the target binary with the binary signatures. Moreover, when P1OVD matches patch signatures, it uses the novel matching algorithm to solve the second challenge. To evaluate the efficiency and effectiveness, we test P1OVD based on a dataset containing 620 binaries, which are compiled under six compilation environments from 104 different versions of 30 out-of-bounds vulnerable functions. The result shows that P1OVD has a total accuracy of 83.06% and achieves an 8.74% higher accuracythan the state-of-the-art tool BinXray. Besides, it can analyze a single binary in 4 s after a 20-s offline signature extraction on average.

We summarize our contributions as follows:

- We design an out-of-bounds vulnerability signature that mainly contains patch information and vulnerability information.
- We propose a matching algorithm that can accurately and robustly find the patch signatures in downstream binaries.
- We propose a patch-based out-of-bounds vulnerability detection method, P1OVD. P1OVD can accurately locate 1-day out-of-bounds vulnerabilities in downstream binaries even if code variants exist. We evaluate its performance on 620 binaries of 30 real-world patches in Linux Kernel [21].

The rest of this paper is organized as follows. We first summarize the challenges in Section 2. Then we describe the design of P1OVD in Section 3 and evaluate P1OVD in Section 4. Next, we review related work in Section 5. Finally, we give the conclusion in Section 6.

2. Motivation

The open-source software can be built with customized codes and non-standard building configurations to meet the needs of downstream manufacturers [19], causing code variants. We believe that there are two main challenges caused by code variants: vulnerability signature (Section 2.1) and patch signature matching (Section 2.2).

2.1. Vulnerability Signature

The first challenge is that the code variants can significantly affect the vulnerability detection results but can be hard to detect if the vulnerability signatures do not contain enough patch information and vulnerability information. Version differences are parts of the results of the third-party code customization and cause existing works to have high false rates in out-of-bounds vulnerability detection. We take the function *init_desc* of the Linux Kernel as a motivation example to figure out the severe impacts of code variants. Figure 1 shows three different versions of function *init_desc*. Red codes are the earliest version, at that time, the vulnerable statement *hash_algo_name[hash_algo]* had not existed. Then a commit removed the red codes and introduced green codes, where the out-of-bounds vulnerability is located. The parameter *hash_algo* is possibly tainted and can read the array *hash_algo_name* out of the buffer bound. The blue codes are added by the patch, they restrict the parameter *hash_algo*, and relieve the panic.

```
static struct shash_desc *init_desc(char type)
static struct shash_desc *init_desc(char type, uint8_t hash_algo)
{
        long rc;
        char *algo;
        const char *algo;
        ......
        if (type == EVM_XATTR_HMAC) {
                ......
        } else {
                if (hash_algo >= HASH_ALGO__LAST)
                        return ERR_PTR(-EINVAL);
                tfm = &hmac_tfm;
                algo = evm_hmac;
                tfm = &evm_tfm[hash_algo];
                algo = hash_algo_name[hash_algo];
```

Figure 1. Function *init_desc* in three Versions.

These three versions challenge the state-of-the-art tools because their signatures miss either patch information or vulnerability information. As Figure 2 shows, function-level vulnerability detection tools [4–15] take the whole 53-line function into concern and fail to capture the precise context of vulnerabilities. As a result, they think the functions that are similar to known vulnerable functions are vulnerable. Due to the small differences between these three versions, they think the three versions are all vulnerable, which results in high false positives. As Figure 2 shows, some patch-level vulnerability detection tools [3,17], mistakenly think the patch disappearances are the vulnerabilities and fail to include vulnerability information into their signatures. So they focus on the blue codes rather than the green codes. As a result, the red version is labeled vulnerable even if it has no vulnerable operation at all.

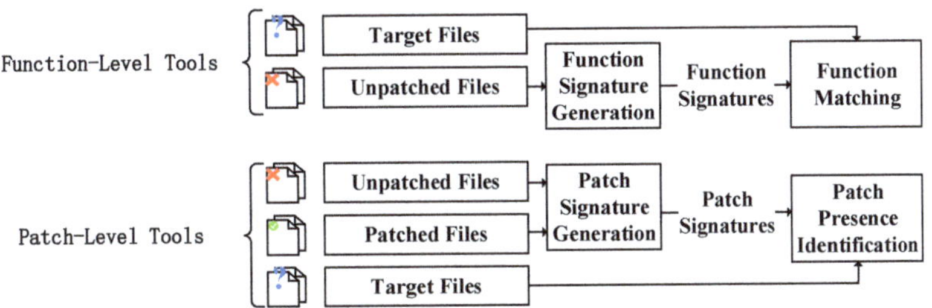

Figure 2. Function-level tools and patch-level tools.

By considering both vulnerabilities and patches, MVP [18] outperforms vulnerability detection tools. Further, researchers manually evaluate MVP's failures by comparing signatures with vulnerability root causes in case studies. Although MVP can successfully identify all three versions, its vulnerable line searching algorithm introduces a few vulnerability-irrelevant codes, e.g., the function call of *ERR_PTR*, which can be replaced by customized error handlings and harm the binary-level signatures.

2.2. Patch Signature Matching

As the solution of the first challenge, the binary-level patch information is AST-shaped. However, code variants caused by optimization levels or target architectures can influence the structures of ASTs, which is the second challenge. As Figure 3 shows, the patch checks inputs at 0x401955 (*x86_64*) and 0x17e4 (*aarch64*). The AST in *x86_64* is *[arg + 6] <= 6*, while the AST in *aarch64* can be *![arg + 6] > 6* if patched or *![arg + 6] > 8* if unpatched. The *arg* represents a function argument and now it stands for the variable *cmd*.

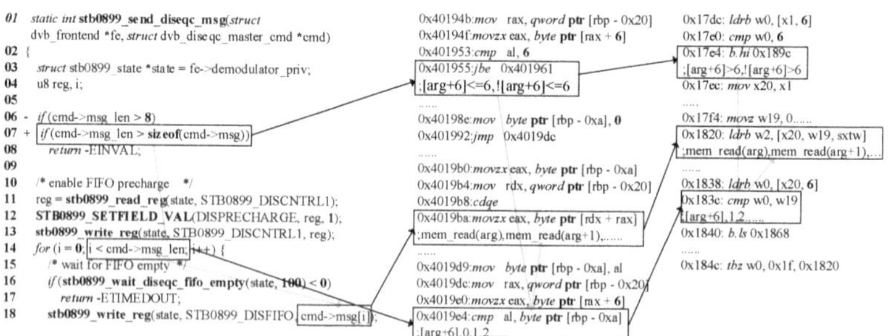

Figure 3. Patch for commit b9f62ffe, patched binary in *x86_64-O0* and target binary in *aarch64-O2*.

There are two kinds of matching algorithms. However, neither of them can balance accuracy and robustness. First, Fiber [20] performs a strict operand-based matching, while assuming that the same semantic can result in the same ASTs with few changes on the address-related immediate numbers. However, as Figure 3 shows, the *[arg + 6] <= 6* in *x86_64* can be transformed into the *![arg + 6] > 6* in *aarch64*. A strict matching can falsely think the patch signatures generated on the *x86_64* platform are different from the patch signatures generated on the *aarch64* platform. Second, Pewny et al. [8,9], Feng et al. [13] and Jiang et al. [19] match ASTs with an inaccurate graph-similarity-based structural matching to improve the robustness. However, as Figure 3 shows, the patch only changes *8* to *sizeof(cmd->msg)*, while the latter is an immediate number *6*. These tools cannot distinguish the unpatched versions from the patch versions because the patch does not cause any structural difference.

3. Design of P1OVD

In this section, we first introduce the architecture of our tool (Section 3.1), and then we will introduce the three main parts of P1OVD in detail, including patch analysis (Section 3.2), signature generator (Section 3.3), and matching engine (Section 3.4). The first challenge (Section 2.1) is solved in patch analysis, and the second challenge (Section 2.2) is solved in equation matching (Section 3.4.2).

3.1. System Architecture

Figure 4 shows that the P1OVD has four inputs, including unpatched sources, patched sources, reference binaries that are compiled from patched sources, and target binaries waiting to be checked. Since a large number of function-level binary similarity tools are currently available, e.g., Asm2Vec [16], we can obtain the address of the possibly vulnerable function in the target binary by finding out the function most similar to the vulnerable function, without requiring a symbol table.

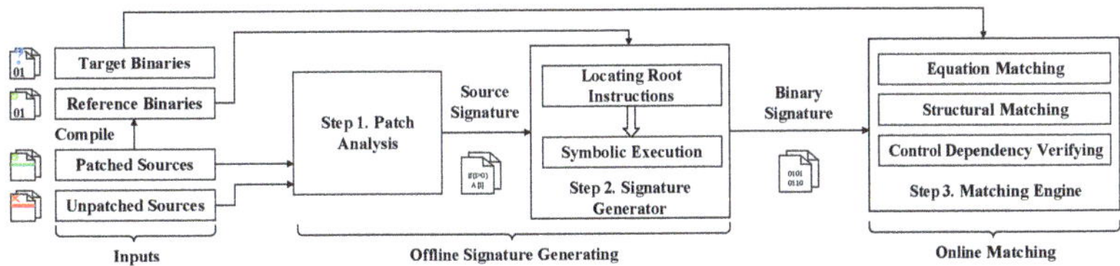

Figure 4. System Architecture.

P1OVD has three parts: patch analysis, signature generator, and matching engine. Patch analysis is designed for generating source signatures from the patched sources and unpatched sources. The generated source signatures are accurate and robust enough to overcome the first challenge. Then the signature generator maps the source signatures to binary signatures while keeping their accuracy and robustness. Patch analysis and signature generator are combined to generate binary-level signatures for out-of-bounds vulnerabilities. Finally, the matching engine searches vulnerabilities in the unknown target binaries according to the binary signatures. Especially, the novel patch signature matching algorithm matches the patch signatures accurately and robustly, while solving the second challenge.

3.2. Patch Analysis

In this step, we generate the signatures to represent vulnerabilities. As mentioned in Section 2.1, important code variants can be ignored when the vulnerability signatures incompletely contain vulnerability information or patch information. Inspired by the fact that both patch information and vulnerability information can increase the signature accuracy and root causes are widely used to evaluate vulnerability signatures [18], we define that out-of-bounds vulnerability signatures should mainly contain patch information and out-of-bounds vulnerability root cause information.

To obtain vulnerability root causes accurately, P1OVD utilizes a patch analysis tool, SID [22]. Patch analysis tools aim at removing the gap between patches and vulnerabilities. Especially, SID outperforms the statical approaches at the accurate out-of-bounds root causes locating. SID defines that the root cause of out-of-bounds vulnerabilities is memory access without proper bound checks. A branching statement, either an if statement or a loop statement that exists in the patch is regarded as a bound check. They are what out-of-bounds patches try to add or correct. Memory access always includes directly indexing arrays by subscripts or calling certain functions to access memory indirectly, which is the root cause of out-of-bounds vulnerabilities. P1OVD locates the memory access and the bound checks according to SID's security rules.

After obtaining patch information and vulnerability root cause information, P1OVD constructs a local PDG (program dependency graph), which is a subgraph of the function PDG starts at bound check and ends at memory access, linking a series of branching statements that are positioned between memory access and bound. The topology of such a local PDG reflects the relationship between patches and vulnerabilities. Compared to local CFG [20], this local PDG is more robust to code variants because compilation environments e.g., optimization levels can significantly change the CFG structures.

Example 1. *As Figure 3 shows, the if statement at line 7 is added by patch. So it is a bound check and is the start of local PDG. The variable i is used to index cmd->msg[i] at line 18. So line 18 is the memory access and is the end of local PDG. Finally, line 14 where the variable i compares with cmd->msg_len, which is important because the dissatisfaction of the bound check can make the function skip line 14 and exit directly. So line 14 is included in the local PDG and is the successor of the bound check and the predecessor of the memory access. In conclusion, we extract only three lines as a signature. With little unnecessary information and complete patch information, as well as vulnerability information, this signature can overcome code variants.*

3.3. *Binary Signature Generator*

Although the generated signatures are accurate, they are at the source level. Thus, in this step, we map the source signatures to the binary signatures by reference binaries, which are manually generated by compiling the patched sources with the *O0* optimization level to *x86_64* architecture while reserving the debugging information. P1OVD keeps the binary signatures in the form of local PDGs and only maps each node of the local PDGs from source-level to binary-level because the local PDGs remain the same even if the compilation environments change.

Theoretically, all instructions that correspond to the local PDG statement nodes can be part of the binary signature. However, Zhang et al. [20] announced that only a subset of instructions i.e., root instructions actually summarize the statement key behaviors, and the unnecessary instructions in the signatures can lead to mismatches. Hence, in this step, P1OVD accurately locates root instructions and uses the semantic information of root instruction to represent the statements.

3.3.1. Root Instructions Locating

Due to the significant difference between binary and C source codes, a statement that originally contains multiple instructions may even be divided into multiple basic blocks, during the compilation procedure. For example, an *if* statement with a logical operation, e.g., && or ||, will be separated into two multiple basic blocks. Hence, for each statement in the local PDG, P1OVD locates the root instructions accurately by taking line numbers, data dependency, variable names, and statement types into concern.

P1OVD first narrows the scope of possible root instructions by selecting the instructions corresponding to statement lines. This can be done with the help of debugging information from reference binaries.

Next, P1OVD narrows the scope of possible root instructions again by variable-based data dependency analysis because variables represent the behavior of the statement in most cases. For example, the vulnerable statement *cmd->msg[i]* contains two important variables *cmd* and *i* and they are combined to generate an out-of-bounds vulnerability. The variable names can be easily obtained by parsing source codes. However, when variables are parameters of operator *sizeof* they can be turned into a constant and disappear from binaries due to the preprocessing. For example, *sizeof(cmd->msg))* corresponds to the immediate number six in the binary. So P1OVD excludes all variables that are only used in the operator *sizeof*. After extracting variable names, P1OVD uses debugging information to map variable names to *rbp* related addresses on the stack because without optimization, the *GCC* compiler stores each local variable on the stack. Since each extracted variables are part of the original statements, the root instructions should data-depend on all extracted variables. Thus, P1OVD performs a data dependency analysis to exclude irrelevant instructions. We define an instruction data-depends on a certain variable if it directly uses the *rbp* related address or uses the result of another instruction that is data-dependent on the variable.

Finally, one statement can have multiple behaviors at the same time, while some of them are less important. For example, line 18 reads the memory and calls a function. But only reading the memory can cause the exception. Thus, P1OVD locates the root instructions that represent the key behaviors of the statements among the selected candidates. Bound checks and extra branching statements control the values of the program counters. Thus, they are compiled into PSW (program status word) writing instructions and branching instructions. Generally, they are positioned at the end of the basic blocks. We require the root instructions of bound checks are branching instructions because they reserve the important comparison operator information since out-of-bounds patches can only correct the comparison operators. But we require the root instructions of extra branching statements are PSW writing instructions. As mentioned in Section 4.2.3, the results of branching instructions can be simplified. Further, extra branching statements do not need comparison operator information. Finally, there are two kinds of memory access, including function calling and array indexing and they trigger exceptions through load or store instructions in the current functions or the callees. Hence, such behaviors are stored in the function call instructions and load or store instructions. In conclusion, Table 1 shows the type of root instruction we required.

Table 1. Mapping source statement to binary AST.

Statement	Root Instruction Type	AST
Bound Check	Branching Instruction	Branching Condition
Memory Access (Call Function)	Function Call Instruction	Access Expression (Callee and All Function Arguments)
Memory Access (Index Array)	Load or Store Instruction	Access Expression (Memory Adress)
Extra Branching Satement	Branching Instruction	PSW Write Arguments

3.3.2. Symbolic Execution

In this section, P1OVD generates sufficient information for root instructions so that they can represent the vulnerabilities and patches. Researchers [8,9,13,19,20] have demonstrated that symbolic execution results i.e., ASTs can robustly represent the operands of instructions. Thus, P1OVD symbolically executes the reference functions from their entries and extracts ASTs forass root instructions. Besides, as Table 1 shows, since the operands of different root instructions are different, P1OVD generates different ASTs for them. The extracted ASTs can represent the vulnerabilities and patches. For example, P1OVD extracts *[arg + 6] <= 6* and *![arg + 6] <= 6* for statement *cmd->msg_len>sizeof(cmd->msg)*. The *cmd* is function argument and AST uses *arg* to represent it. Then it load the member *msg _len* with offset six corresponding to *[arg + 6]*. Finally, it is compared to the constant number six and forks the basic block, as ASTs indicate. In conclusion, all statements in the source local PDGs are replaced with ASTs.

Example 2. *As Figure 5 shows. The root cause contains two variables, named cmd and i. P1OVD maps the line to instructions first. Then among these instructions, P1OVD finds that i which is located at rbp-0x20 is used at 0x409b4, while cmd is used at 0x409b8. Since the memory operation at 0x409ba uses both variables to read the memory, P1OVD thinks it is a root instruction. Moreover, by symbolic execution we generate mem_read(arg) etc. to represent the access expression it read. Similarly other nodes of local PDG can be mapped to binary-level.*

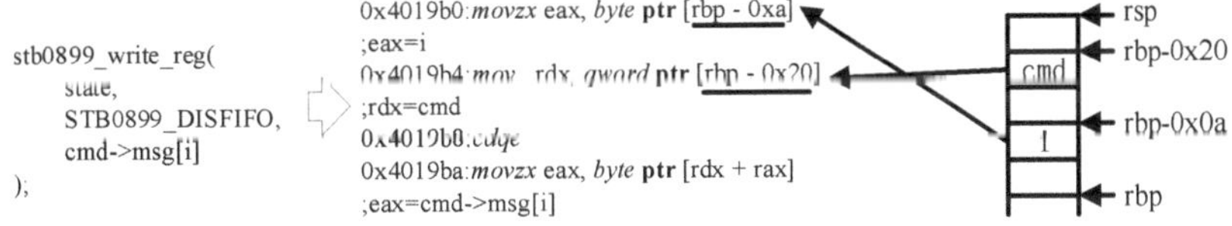

Figure 5. Locating instructions by mapping line to instructions and variable name to stack.

3.4. Matching Engine

The matching engine can judge if an unknown binary is vulnerable or not by using the binary signatures generated from reference binaries. For one binary it has four kinds of output: not vulnerable, patched, vulnerable, unable to judge. Before actually starting to match the vulnerabilities, we use the code similarity to find out the functions that may contain the vulnerabilities in the binaries and use symbolic execution to extract all ASTs of the target functions. Then we start the vulnerability matching.

We find access expressions and PSW write arguments by structural matching and find branching conditions by equation matching. Because the structural matching is faster than the equation matching but dissatisfies the high accuracy required by branching conditions. Finally, we verify the control dependencies using local PDGs.

3.4.1. Structural Matching

Structural matching finds operations with similar semantics to out-of-bounds access expressions or PSW write arguments in the target binaries by calculating the graph similarity of two ASTs. Empirically, we have found that ASTs with the same semantics may have subtle differences when extracted from binary functions of different compilation environments. For example, both *a+((b+1)«1)* and *a+2+(b«1)* can be found in binaries when statement *a[b+1]* appears in the source code. Therefore, we do not require the ASTs to be structurally identical but structurally similar. Thus, the edit-distance-based graph similarity can better reflect the similarity of ASTs and we compare the graph similarity with a predefined threshold to determine whether two ASTs are similar.

3.4.2. Equation Matching

The main task of equation matching is to find a branching condition the same as the patched bound check. However, as mentioned in Section 2.2, both structural matching and strict matching cannot overcome the second challenge.

Fiber [20] matches the same kind of nodes, e.g., immediate numbers with different algorithms according to their positions in the ASTs. Thus, we infer that different parts of branching conditions should be matched according to different precisions. We define a subtree in the AST as a data object if its root node is a memory read operation or it only contains one node that is a function return value or a parameter. Additionally, a data object should not be a subtree of another data object. After extracting data objects, remain the boolean expressions. Empirically, we learn that under different compilation environments, data objects have similar structures, while boolean expressions preserve fixed semantics. So P1OVD matches them according to their structures and semantics correspondingly.

After extracting data objects by traversing the branching conditions, P1OVD uses the structural matching presented in Section 3.4.1 to generate data object pairs, one from the target branching condition and one from the reference branching condition. The matched data object pairs in two ASTs are replaced with the same symbol. In the case that two matched data objects have different sizes, e.g., one is 64-bits long, and the other is 32-bits, P1OVD defines a symbol in the shorter size and replaces two ASTs with this symbol. To satisfy the length requirement of the longer tree, P1OVD pads zero to the left of the defined symbol. P1OVD replaces the data objects to ensure the two bool expressions have identical symbol sets and can be used for solving.

Next, we accurately solve [23] the boolean expressions. Since the boolean expressions are semantically identical, only equal or opposite boolean expressions are matched. In other words, given two boolean expressions $Expr_1$ and $Expr_2$, they are considered matched if only one of $Expr_1 = Expr_2$ or $Expr_1 =! Expr_2$ can be solved.

Example 3. *As Figure 6 shows, to match the [arg + 6] <= 6 and ![arg + 6] > 6, P1OVD first extracts data objects from them. The extracted data objects are both [arg + 6] and they are structurally similar obviously. So P1OVD replaces them with a single symbol x in their original AST. Then P1OVD checks the solving possibility of conditions* $(x \leqslant 6) =! (x > 6)$ *and* $(x \leqslant 6) =!! (x > 6)$*. Results show* $x = 0$ *satisfies the first constrain and the second constrain will never be satisfied, thus P1OVD finds that the two branching conditions are equal. So, in conclusion, the two ASTs represent the same branching condition. At the same time, both strict matching and graph-similarity-based matching cannot distinguish such changes in ASTs.*

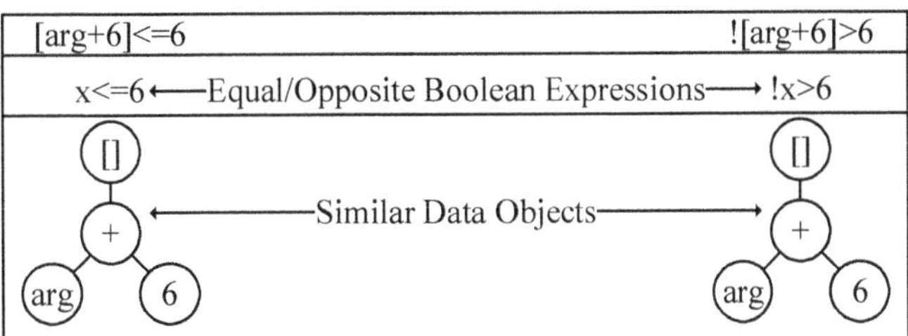

Figure 6. Two-step matching algorithm.

3.4.3. Verify Control Dependency

In this step, P1OVD matches the local PDG topologies by verifying the edge i.e. control dependency of their nodes. This step is to ensure that the bound checks actually control the memory access. Since even if both bound checks and memory access appear in the same function, vulnerabilities can still appear. For example, in Linux Kernel commit 1fa2337, the patch only moves the check of *d->msg_len* forward to secure the *d->msg[i]* used in function *printk*. P1OVD can infer statement A controls statement B from two cases:

- A is a loop statement. The loop structure is often reordered by optimization, and P1OVD requires every trace that passes the B twice or more to contain an A between every neighbor B.
- A is an *if* statement. Since the *if* statement may be in a loop, P1OVD requires each trace that reaches the B from the function entry to pass A.

4. Evaluation

We have developed P1OVD with 1200 lines of python code on top of open source libraries Angr [24], joern [25], and pyelftool [26]. P1OVD supports *aarch64*, *x86_64*, and *x86_32* target architectures as well as, *O0* and *O2* optimization levels. In this section, we first evaluate P1OVD in terms of accuracy (Section 4.2) and efficiency (Section 4.3), and then we compare P1OVD with other tools in terms of the overall performance (Section 4.4) and the effectiveness of signature generation (Section 4.5), and AST matching (Section 4.6).

4.1. Datasets

We evaluate our tool based on Linux Kernel because not only is Linux Kernel widely used [27], but also the out-of-bounds vulnerabilities in the Linux Kernel are widely analyzed [22]. Our tool has four inputs, including unpatched source code, patched source code, patched binary, and target binary. Thus, we collect three datasets.

4.1.1. Source Codes

Source codes include the patched and unpatched source codes. To ensure that the patch analysis can successfully operate, we use 30 out-of-bounds patches listed in the appendix of SID [22] while two of them are patches of CVE-2017-18379 and CVE-2019-15926. We exclude some patches that are too old that cannot be successfully compiled.

4.1.2. Reference Binaries

The reference binaries are obtained by compiling the patched source code. Still, most of the out-of-bounds vulnerabilities are located in the Linux Kernel optional modules, which are difficult to trigger by the default compilation options. We manually adjust the compilation options for each out-of-bounds patch to satisfy the constraints and apply *GCC* to compile the patched source codes while reserving the debugging information.

4.1.3. Target Binaries

The target binaries are used to prove that the P1OVD can fight against binaries built with non-standard configurations from customized codes on multiple architectures. To obtain target binaries, we compile source code into 620 different binaries. The collected binaries vary from three aspects, including versions, optimization levels, and target architectures.

We think the first way to customize source codes is to compile the source codes of different versions. When we refer to versions, we do not mean the software release version e.g., "Linux-5.0-y" because a newer software release version does not change the vulnerable function sometimes. We define a new version according to the vulnerable functions. For a vulnerable function, we regard all commits that change this function as versions and divide these versions into three categories, including not vulnerable versions, vulnerable versions, and patched versions. A function had no vulnerability at first and we think the functions in these versions are not vulnerable. And then, at a notable point, the vulnerable memory access began to appear in the function. We think these functions are vulnerable. After the vulnerability was discovered, the repository maintainer corrected the bound check using a patch. We consider these functions are patched. Totally, we obtain 104 different versions from Linux Kernel "master" branch and we classify these versions manually. Second, optimization levels are usually customized and *O0* and *O2* are the most commonly used optimization. Moreover, the reference binaries are generated based on *O0*. Thus, we evaluate P1OVD on *O0* and *O2* optimization levels. Finally, different manufacturers can build sources for different target architectures. We also evaluate P1OVD on three architectures including *x86_64*, *x86_32*, and *aarch64*. The *x86_32* uses 32-bit addresses while *aarch64* and *x86_64* use 64-bit addresses while *aarch64* has an instruction set different from *x86_64* and *x86_32*. After obtaining 620 binaries, we generate ground truths for them according to their versions because the changes of compilation environments do not affect the vulnerability detection results.

4.2. Accuracy

We evaluate the P1OVD based on the multiclass classification problem. Only the outputs of P1OVD that correctly predict both vulnerability and patch are considered correct. We use precision, recall, F-1 score, and accuracy to measure the accuracy of P1OVD. The four evaluation metrics are defined in Equation (1). P1OVD first generates signatures based on *O0* optimization and *x86_64* architecture and then scans the target binaries with the signatures.

$$\begin{aligned} Precision &= \frac{TP}{TP+FP} \\ Recall &= \frac{TP}{TP+FN} \\ F_1\ Score &= \frac{2 * Precision * Recall}{Precision + Recall} \\ Accuracy &= \frac{Correct}{Total} \end{aligned} \tag{1}$$

Table 2 shows the accuracy of P1OVD. Each row represents the compilation environments, and each column stands for the evaluation metrics of three version categories. P1OVD obtains an accuracy of 83.06%. And we manually analyzed the false predict cases and summarized the following four reasons, while the first three are common challenges for symbolic execution-based tools.

Table 2. Vulnerability Detection Accuracy Test.

Compilation Environment		Recall			Precision			F-1 Score			Accuracy
		NV	V	P	NV	V	P	NV	V	P	
O0	*aarch64*	100.00%	84.09%	86.00%	45.45%	100.00%	97.73%	0.62	0.91	0.91	86.54%
	x86_64	80.00%	95.45%	100.00%	88.89%	95.45%	98.00%	0.84	0.95	0.99	96.12%
	x86_32	80.00%	63.64%	79.59%	80.00%	100.00%	97.50%	0.80	0.78	0.88	72.82%
O2	*aarch64*	100.00%	84.09%	80.00%	47.62%	90.24%	97.56%	0.65	0.87	0.88	83.65%
	x86_64	80.00%	86.36%	95.92%	80.00%	97.44%	97.92%	0.80	0.92	0.97	90.29%
	x86_32	80.00%	61.36%	73.47%	66.67%	93.10%	97.30%	0.73	0.74	0.84	68.93%
All		86.67%	79.17%	85.81%	61.90%	95.87%	97.69%	0.72	0.87	0.91	83.06%

NV stands for not vulnerable. V stands for vulnerable. P stands for patched.

4.2.1. Function Inline

Function inline contributes most of the false rates. A function with an inline tag may not be inline during the compilation procedure. It is influenced by many factors e.g., the optimization level, and target architecture. For example, function *nvmet_fc_getqueueid* called by the function *nvmet_fc_find_target_queue* is inlined when compiled with *aarch64* and *O0*. However, function *nvmet_fc_getqueueid* is not inlined when compiled with *x86_64* and *O0*. Whether the callee is inline affects the extracted signature through symbolic execution and function inline is also the key reason why P1OVD has worse performance on *aarch64* architecture or *O2* optimization level, compared to *x86_64* and *O0*.

4.2.2. Conditional Execution Instructions

Conditional execution instructions are instructions that select whether to perform operations based on the PSW, e.g., *CSEL*. When multiple conditional branches occur continuously, *aarch64* optimizes the efficiency by replacing branch instructions with conditional execution instructions since branch instructions slow the assembly. Function *qxl_clientcap_ioctl* in commit 62c8ba7 compares *qdev->pdev->revision* with 4 and *byte* with 58 continuously, and only one branch instruction to deal with the exception for them. Since we used the AST of branching conditions for boundary check, the inability to find the correct branching conditions in the binary led to false positives.

4.2.3. Simplified Expression

P1OVD will output a false result when a patch changes the loop bound. For example commit 43622021d2e2b changes operator <= to < in the statement *for(j=0; j<HID_MAX_IDS; j++)*. Every time the executor compares the *j* and *HID_MAX_IDS*, the value of *j* is a constant. As a result, Angr will automatically optimize a boolean expression containing only constants to True or False. This prevents us from extracting branching conditions correctly and this is the key reason why the signature cannot ensure all predictions are correct on *x86_64* and *O0*.

4.2.4. Structure Dissimilar

Function *hid_register_report* in commit 43622021d, indexes an array through *report_enum->report_id_hash[id]* while *report_enum* is calculated by *device->report_enum+type*. However, all the structs have pointer members, which means that when the system address lengths change the sizes of structs change. Although the difference between the two integers is small, optimizing codes by replacing multiplication with an arithmetical left shift is often used in the addressing process, resulting in structural differences. As a result, *x86_32* has the worst performance. We believe that graph embedding is a feasible solution to this type of problem.

4.3. Performance

Experimented on intel-i7-8700 and 12GB RAM, Table 3 records the time consumption of P1OVD from three aspects of patch analysis, signature generator, and matching engine. Patch analysis and signature generator are used to extract binary signatures offline. We calculate the average time used to generate a signature for one patch. The matching engine is used to determine whether a target binary is vulnerable or not. To solve the situation that different vulnerable functions have different numbers of versions and better reflect the time consumption, we first calculate the average time used for finding a certain vulnerability in various binaries, then we calculate the overall average vulnerabilities finding times.

Table 3. Vulnerability Detection Performance Test.

	Step	Total Time	Number	Average
Offline	Patch Analyze	470.68 s	30	15.69 s
	Signature Generate	99.90 s	30	3.33 s
Online	Match	108.07 s	30	3.60 s

4.4. Accuracy Comparison with Vulnerability Detection Tools

In this section,we evaluate the accuracy of P1OVD by comparing it with the state-of-the-art vulnerability detection tools. We choose BinXray [17] and Asm2Vec [16] as references because they are both open-sourced and BinXray and Asm2Vec are the state-of-the-art patch-level and function-level vulnerability detection tools. We compare P1OVD, Asm2Vec, and BinXray from four aspects, including precision, recall, F-1 score, and accuracy. For one function, Binxray only has two kinds of outputs vulnerable or patched. So we require P1OVD to predict if the functions are vulnerable or not. Thus, we relabel the patched binaries as not vulnerable binaries. At the same time, Asm2Vec ranks the possibly vulnerable functions. So we think the function in the unknown target binary that has the highest similarity to the vulnerable function in the reference binary is vulnerable.

Table 4 shows the results of the comparison. Asm2Vec, a function-level tool, cannot distinguish three versions well and consider them are all vulnerable, which result in high false positive. Meanwhile, BinXray assumes that vulnerabilities exist when patches disappear. However, bound checks and memory access can both disappear because of code customization. When BinXray cannot detect the patches, it mistakenly believes that the vulnerabilities exist. P1OVD achieves the highest precisions due to its vulnerability signature containing both vulnerability root cause information and patch information. However, P1OVD cannot ensure that all binaries predicted safe are accurately safe. When the patch changes the loop boundary, the simplified expressions described in Section 4.2.3 can cause the results of symbolic execution to contain too little information and P1OVD cannot distinguish the patched version from the unpatched version. However, Binxray uses the patch codes in the binaries as signatures, which works well when the target binary and reference binary are under the same compilation environment.

4.5. Effectiveness of Vulnerability Signatures

In this step, we evaluate the effectiveness of our vulnerability signatures by comparing the source signatures of P1OVD with ReDeBug [3] based on the source dataset. Because ReDeBug is a widely used open-source source-level unpatched buggy code detection tool. As Table 5 shows, compared to ReDeBug, P1OVD extracts fewer lines but our signatures contain more vulnerability root cause information and patch information, which means P1OVD can generate more accurate vulnerability signatures. This is because ReDeBug only pays attention to the lines close to the patches. On the contrary, P1OVD focuses more on the statements that control the function security (vulnerability root cause information and patch information), by which P1OVD overcomes the first challenge (Section 2.1). For example, ReDebug always includes the error handling of bound checks because they are added by patches. Meanwhile, P1OVD thinks they are widely customized [19] and excludes them from signatures.

Table 4. Comparing P1OVD with BinXray and Asm2Vec.

Tool	Compilation Environment		Recall		Precision		F1-score		Accuracy
			Not Vulnerable	Vulnerable	Not Vulnerable	Vulnerable	Not Vulnerable	Vulnerable	
P1OVD (Graph Similarity)	O0	aarch64	60.00%	84.09%	83.72%	61.67%	0.70	0.71	70.19%
		x86_64	50.85%	95.45%	93.75%	59.15%	0.66	0.73	69.90%
		x86_32	52.54%	63.64%	93.94%	62.22%	0.67	0.63	57.28%
	O2	aarch64	56.67%	84.09%	85.00%	58.73%	0.68	0.69	68.27%
		x86_64	49.15%	86.36%	93.55%	57.58%	0.64	0.69	65.05%
		x86_32	50.85%	61.36%	90.91%	60.00%	0.65	0.61	55.34%
P1OVD (Two Step)	O0	aarch64	98.33%	84.09%	89.39%	100.00%	0.94	0.91	92.31%
		x86_64	96.61%	95.45%	96.61%	95.45%	0.97	0.95	96.12%
		x86_32	81.36%	63.64%	96.00%	100.00%	0.88	0.78	73.79%
	O2	aarch64	93.33%	84.09%	90.32%	90.24%	0.92	0.87	89.42%
		x86_64	94.92%	86.36%	96.55%	97.44%	0.96	0.92	91.26%
		x86_32	77.97%	61.36%	93.88%	93.10%	0.85	0.74	70.87%
BinXray	O0	x86_64	81.36%	95.45%	100.00%	84.00%	0.90	0.89	87.38%
Asm2Vec	O0	x86_64	5.08%	95.45%	60.00%	42.86%	0.09	0.59	43.69%
	O2	x86_64	16.95%	79.55%	52.63%	41.67%	0.26	0.55	43.69%

Table 5. Comparing P1OVD with ReDeBug.

	P1OVD	**ReDeBug**
Bound Check Coverage	100%	100%
Memory Access Coverage	100%	50.94%
Used Lines	2.80	6.87

4.6. Effectiveness of Two-Step AST Matching Algorithm

In this section, we evaluate the effectiveness of our two-step AST matching algorithm by comparing it with the graph-similarity-based AST matching algorithm, because the graph-similarity-based AST matching algorithm is the most widely used AST matching algorithm among vulnerability detection tools [8,9,13] and is used to enhance the robustness of Fiber [20]. To compare with it, we generate a new version of P1OVD by replacing our two-step equation matching component with the graph similarity matching. Table 4 shows that the modified P1OVD has more false positives, which demonstrates that the two-step AST matching algorithm can address the second challenge (Section 2.2). This is because the graph-similarity-based matching can only distinguish action-related nodes and condition-related nodes [13] and cannot distinguish the in-node changes. Moreover, when only operands are different, the graph-similarity-based matching algorithm remains unaware. The two-step matching algorithm splits the AST into two parts. If such changes happen in the data objects, P1OVD ignores them. On the contrary, if boolean expression semantics are changed, P1OVD is warned by the solver.

4.7. Limitation

P1OVD analyzes the out-of-bounds patches based on security rules of SID [22], which leads to one limitation. Although SID outperforms patch analysis tools, it can neither analyze the patches that involve multiple functions nor understand out-of-bounds patches that do not correct bound checks e.g., extending the array size. Thus, P1OVD cannot successfully detect all out-of-bounds vulnerabilities. To address this problem, we are considering replacing SID with other dynamic patch analysis tools e.g., PatchScope [28].

Further, memory access is a common root cause of many kinds of memory-centric vulnerabilities, e.g., use after free and correcting missing or wrong checks before the memory access can also be the key behavior of security patches. We can polish bound checks to generalize the vulnerabilities that P1OVD can detect.

Finally, the performance of P1OVD in 32-bit architecture is not as good as in 64-bit architecture. This is mainly due to the structural changes related to address length (Section 4.2.4). Thus, we try to calculate the similarity scores of two ASTs with a more robust algorithm. Many works [16,29] train neural networks to calculate the graph similarity of CFG or PDG and we think these approaches can be adapted to AST similarity calculation.

5. Related Work

This article is closely related to four branches of study, function-level 1-day vulnerability detection, patch-level 1-day vulnerability detection, patch presence test, and patch analysis. In the following four sections, we give a brief review of the works that lead to our own.

5.1. Function-Level 1-Day Vulnerability Detection

At present, many studies focus on detecting vulnerabilities in source files and binary files by judging whether the target function is similar to the vulnerable function. Usually, function-level 1-day vulnerability detection tools extract features from reference sources or binaries and match them with special algorithms.

Early algorithm using normalized source codes [7], ASTs [5], PDG [4,6], etc. to comprehensively represent the whole source vulnerable function. However, they cannot detect the 1-day vulnerabilities in binaries, due to the lack of binary semantic information.

DiscovRE [10] tries to solve this problem by extracting numerical features from the basic blocks and CFG structural features. Introduced by Genius [12], neural networks use vectors to better represent the function feature, e.g., numerical and structural information [14,15], assembly codes [16]. Some tools use tree-liked formulas to represent basic blocks [8], function IO behaviors [9], or even the high-level function semantic information [13]. These function-level 1-day vulnerability detection tools take the whole vulnerable functions as the vulnerability signatures and their extremely large scope of function-level signatures cause the first challenge (Section 2.1). When small code variants involve patches or vulnerability root causes, the function-level signatures bring much useless information and cannot give correct predictions.

5.2. Patch-Level 1-Day Vulnerability Detection

During the last decade, researchers use patches to improve the function-level 1-day vulnerability detection, which is called patch-level vulnerability detection. Early tools [3,17,30,31] believe the missing patch-added codes are the root causes of vulnerabilities. By using normalized and tokenized patches [3], patch sensitive matching algorithms [30], LSTM-embedded code vectors [31], patch modified traces [17], they enhance the patch searching rather than vulnerability searching.

Li et al. [32] think the patch-removed code is vulnerable and using concolic testing to verify the clone of vulnerable code. MVP first [18] announced that the patch and the corresponding vulnerability have different information and it thinks the deleted codes are vulnerable and the added codes are patches. So it uses CPG (code property graphs) to slice vulnerability-related codes as vulnerability signatures and patch-related codes as patch signatures. Further, researchers manually evaluate MVP's failures by comparing signatures with vulnerability root causes in case studies. Although the signatures containing both patch information and vulnerability information improve source-level 1-day vulnerability detection, they contain too much unnecessary information for binary-level vulnerability detection, which leads to the first challenge (Section 2.1).

5.3. Patch Presence Test

The concept of patch presence tests is first proposed by Zhang et al. [20]. Its main purpose is to accurately confirm whether a binary contains a particular patch or not, while we try to find vulnerabilities rather than patches. Other patch presence tests improve the Fiber [20] in terms of the diversifying source program languages [33] and robustness [19], or polish it with other dynamic tools [34].

Vulnerabilities detections need high accuracy and robustness, which is similar to patch present tests. So, we have a deep look into the Fiber and Pidff [19]. They both locate basic blocks corresponding to the patches in the patched binaries. Later, by symbolic execution, they extract the AST-shaped results of these basic blocks as signatures. However, they match their signatures with different methods. While Fiber proposes a strict operand-based patch matching algorithm with little relaxations (inter-changeable operators, address-related immediate numbers), PDiff proposed a robust but less accurate graph similarity-based matching algorithm. Unfortunately, neither of them can accurately and robustly match branching conditions in downstream binaries, which leads to the second challenge (Section 2.2). In Section 3.4.2 we benefit from both and propose a novel two-step matching algorithm.

5.4. Patch Analysis

Patch analyses are used to understand how security patches fix the vulnerabilities. At first, Corley et al. [35] links between bugs and patches while requiring an issue tracking system. Later, Spain [36] proposed binary-level patch patterns to detect unexplored vulnerabilities but limited by lacking high-level semantic information, it cannot fully understand out-of-bounds patches. SID [22] outperforms other statical tools at the accurate out-of-bounds root causes locating, by utilizing symbolic execution. PatchScop [28] dynamically analyzes the patches and gets the highest accuracy although it requires POCs (Proof of Concepts). In this work, we use the state-of-the-art statical tool, SID to locate the root causes of out-of-bounds vulnerabilities.

6. Conclusions

In this work, we have had a deep look into the 1-day vulnerability detection and identified two challenges introduced by code variants, including vulnerability signature and patch signature matching. To solve the two challenges, we have proposed P1OVD, an accurate detection method for 1-day out-of-bounds vulnerabilities in downstream binaries using patches. P1OVD analyzes the patch to get accurate vulnerability signature, generates binary signatures using debugging information and symbolic execution, and accurately matches the signatures, especially branching condition. Experiments have demonstrated that P1OVD can generate accurate and robust vulnerability signatures and match the signatures accurately. Addressing the above two challenges allows P1OVD to resist interference from code customization, non-standard building configurations, and to detect 1-day out-of-bounds vulnerabilities on multiple architectures more accurately than existing tools.

Author Contributions: Conceptualization, methodology, validation, evaluating, and writing H.L.; writing—review and editing, funding acquisition, project administration, resources D.H.; writing—review and editing S.C.; writing-review and editing X.Z. All authors have read and agreed to the published version of the manuscript.

Funding: This research was supported by the National Natural Science Foundation of China, China (Grant No. U1936120), the University Grants Committee of the Hong Kong Special Administrative Region of China (City U11201421), and the Basic Research Program of State Grid Shanghai Municipal Electric Power Company (52094019007F).

Conflicts of Interest: The authors declare no conflict of interest.

Abbreviations

The following abbreviations are used in this manuscript:

AST	Abstract Syntax Tree
PDG	Program Dependency Graph
PSW	Program Status Word
CFG	Control Flow Graph
POC	Proof of Concepts

References

1. Peng, J.; Li, F.; Liu, B.; Xu, L.; Liu, B.; Chen, K.; Huo, W. 1dVul: Discovering 1-Day Vulnerabilities through Binary Patches. In Proceedings of the 2019 49th Annual IEEE/IFIP International Conference on Dependable Systems and Networks (DSN), Portland, OR, USA, 24–27 June 2019; pp. 605–616. [CrossRef]
2. Insights into the 2.3 Billion Android Smartphones in Use Around the World. Available online: https://newzoo.com/insights/articles/insights-into-the-2-3-billion-android-smartphones-in-use-around-the-world/ (accessed on 12 December 2021).
3. Jang, J.; Agrawal, A.; Brumley, D. ReDeBug: Finding unpatched code clones in entire os distributions. In Proceedings of the 2012 IEEE Symposium on Security and Privacy, San Francisco, CA, USA, 20–23 May 2012; pp. 48–62.
4. Pham, N.H.; Nguyen, T.T.; Nguyen, H.A.; Nguyen, T.N. Detection of recurring software vulnerabilities. In Proceedings of the IEEE/ACM International Conference on Automated Software Engineering, New York, NY, USA, 20–24 September 2010; pp. 447–456.
5. Yamaguchi, F.; Lottmann, M.; Rieck, K. Generalized vulnerability extrapolation using abstract syntax trees. In Proceedings of the 28th Annual Computer Security Applications Conference, Orlando, FL, USA, 3–7 December 2012; pp. 359–368.
6. Zou, D.; Qi, H.; Li, Z.; Wu, S.; Jin, H.; Sun, G.; Wang, S.; Zhong, Y. SCVD: A New Semantics-Based Approach for Cloned Vulnerable Code Detection. In Proceedings of the International Conference on Detection of Intrusions and Malware, and Vulnerability Assessment, Bonn, Germany, 6–7 July 2017; pp. 325–344.
7. Kim, S.; Woo, S.; Lee, H.; Oh, H. Vuddy: A scalable approach for vulnerable code clone discovery. In Proceedings of the 2017 IEEE Symposium on Security and Privacy (SP), San Jose, CA, USA, 22–26 May 2017; pp. 595–614.
8. Pewny, J.; Schuster, F.; Bernhard, L.; Holz, T.; Rossow, C. Leveraging semantic signatures for bug search in binary programs. In Proceedings of the 30th Annual Computer Security Applications Conference, New Orleans, LA, USA, 8–12 December 2014; pp. 406–415.
9. Pewny, J.; Garmany, B.; Gawlik, R.; Rossow, C.; Holz, T. Cross-architecture bug search in binary executables. In Proceedings of the 2015 IEEE Symposium on Security and Privacy, San Jose, CA, USA, 17–21 May 2015; pp. 709–724.
10. Eschweiler, S.; Yakdan, K.; Gerhards-Padilla, E. discovRE: Efficient Cross-Architecture Identification of Bugs in Binary Code. In Proceedings of the NDSS, San Diego, CA, USA, 21–24 February 2016; pp. 58–79.
11. Feng, Q.; Zhou, R.; Xu, C.; Cheng, Y.; Testa, B.; Yin, H. Scalable graph-based bug search for firmware images. In Proceedings of the 2016 ACM SIGSAC Conference on Computer and Communications Security, Vienna, Austria, 24–28 October 2016; pp. 480–491.
12. Xu, X.; Liu, C.; Feng, Q.; Yin, H.; Song, L.; Song, D. Neural network-based graph embedding for cross-platform binary code similarity detection. In Proceedings of the 2017 ACM SIGSAC Conference on Computer and Communications Security, Dallas, TX, USA, 30 October–3 November 2017; pp. 363–376.
13. Feng, Q.; Wang, M.; Zhang, M.; Zhou, R.; Henderson, A.; Yin, H. Extracting conditional formulas for cross-platform bug search. In Proceedings of the 2017 ACM on Asia Conference on Computer and Communications Security, Abu Dhabi, United Arab Emirates, 2–6 April 2017; pp. 346–359.
14. Gao, J.; Yang, X.; Fu, Y.; Jiang, Y.; Sun, J. VulSeeker: A semantic learning based vulnerability seeker for cross-platform binary. In Proceedings of the 2018 33rd IEEE/ACM International Conference on Automated Software Engineering (ASE), Montpellier, France, 3–7 September 2018; pp. 896–899.
15. Liu, B.; Huo, W.; Zhang, C.; Li, W.; Li, F.; Piao, A.; Zou, W. αdiff: Cross-version binary code similarity detection with dnn. In Proceedings of the 33rd ACM/IEEE International Conference on Automated Software Engineering, Montpellier France, 3–7 September 2018; pp. 667–678.
16. Ding, S.H.; Fung, B.C.; Charland, P. Asm2vec: Boosting static representation robustness for binary clone search against code obfuscation and compiler optimization. In Proceedings of the 2019 IEEE Symposium on Security and Privacy (SP), San Francisco, CA, USA, 19–23 May 2019; pp. 472–489.
17. Xu, Y.; Xu, Z.; Chen, B.; Song, F.; Liu, Y.; Liu, T. Patch based vulnerability matching for binary programs. In Proceedings of the 29th ACM SIGSOFT International Symposium on Software Testing and Analysis, Virtual Event, USA, 18–22 July 2020; pp. 376–387.
18. Xiao, Y.; Chen, B.; Yu, C.; Xu, Z.; Yuan, Z.; Li, F.; Liu, B.; Liu, Y.; Huo, W.; Zou, W.; et al. MVP: Detecting Vulnerabilities Using Patch-Enhanced Vulnerability Signatures. Available online: https://chenbihuan.github.io/paper/sec20-xiao-mvp.pdf (accessed on 12 December 2021).

19. Jiang, Z.; Zhang, Y.; Xu, J.; Wen, Q.; Wang, Z.; Zhang, X.; Xing, X.; Yang, M.; Yang, Z. PDiff: Semantic-based Patch Presence Testing for Downstream Kernels. In Proceedings of the 2020 ACM SIGSAC Conference on Computer and Communications Security, Virtual Event, USA, 9–13 November 2020; pp. 1149–1163.
20. Zhang, H.; Qian, Z. Precise and accurate patch presence test for binaries. In Proceedings of the 27th USENIX Security Symposium (USENIX Security 18), Baltimore, MD, USA, 15–17 August 2018; pp. 887–902.
21. Linux Kernel. Available online: https://github.com/torvalds/linux (accessed on 12 December 2021).
22. Wu, Q.; He, Y.; McCamant, S.; Lu, K. Precisely characterizing security impact in a flood of patches via symbolic rule comparison. In Proceedings of the Network and Distributed System Security Symposium (NDSS), San Diego, CA, USA, 23–26 February 2020.
23. Z3Prover/z3: The Z3 Theorem Prover. Available online: https://github.com/Z3Prover/z3 (accessed on 12 December 2021).
24. Shoshitaishvili, Y.; Wang, R.; Salls, C.; Stephens, N.; Polino, M.; Dutcher, A.; Grosen, J.; Feng, S.; Hauser, C.; Kruegel, C.; et al. SoK: (State of) The Art of War: Offensive Techniques in Binary Analysis. In Proceedings of the 2016 IEEE Symposium on Security and Privacy (SP), San Jose, CA, USA, 22–26 May 2016.
25. Yamaguchi, F.; Golde, N.; Arp, D.; Rieck, K. Modeling and discovering vulnerabilities with code property graphs. In Proceedings of the 2014 IEEE Symposium on Security and Privacy, San Jose, CA, USA, 18–21 May 2014; pp. 590–604.
26. Parsing ELF and DWARF in Python. Available online: https://github.com/eliben/pyelftools (accessed on 12 December 2021).
27. Hall, C. Survey Shows Linux the Top Operating System for Internet of Things Devices. Available online: https://www.itprotoday.com/iot/survey-shows-linux-top-operating-system-internet-things-devices (accessed on 12 December 2021).
28. Zhao, L.; Zhu, Y.; Ming, J.; Zhang, Y.; Zhang, H.; Yin, H. Patchscope: Memory object centric patch diffing. In Proceedings of the 2020 ACM SIGSAC Conference on Computer and Communications Security, Virtual Event, USA, 9–13 November 2020; pp. 149–165.
29. Chandramohan, M.; Xue, Y.; Xu, Z.; Liu, Y.; Cho, C.Y.; Tan, H.B.K. Bingo: Cross-architecture cross-os binary search. In Proceedings of the 2016 24th ACM SIGSOFT International Symposium on Foundations of Software Engineering, Seattle, WA, USA, 13–18 November 2016; pp. 678–689.
30. Li, Z.; Zou, D.; Xu, S.; Jin, H.; Qi, H.; Hu, J. VulPecker: An automated vulnerability detection system based on code similarity analysis. In Proceedings of the 32nd Annual Conference on Computer Security Applications, Los Angeles, CA, USA, 5–9 December 2016; pp. 201–213.
31. Li, Z.; Zou, D.; Xu, S.; Ou, X.; Jin, H.; Wang, S.; Deng, Z.; Zhong, Y. Vuldeepecker: A deep learning-based system for vulnerability detection. *arXiv* **2018**, arXiv:1801.01681.
32. Li, H.; Kwon, H.; Kwon, J.; Lee, H. A scalable approach for vulnerability discovery based on security patches. In Proceedings of the International Conference on Applications and Techniques in Information Security, Melbourne, Australia, 26–28 November 2014; pp. 109–122.
33. Dai, J.; Zhang, Y.; Jiang, Z.; Zhou, Y.; Chen, J.; Xing, X.; Zhang, X.; Tan, X.; Yang, M.; Yang, Z. BScout: Direct Whole Patch Presence Test for Java Executables. In Proceedings of the 29th USENIX Security Symposium (USENIX Security 20), Boston, MA, USA, 12–14 August 2020; pp. 1147–1164.
34. Sun, P.; Garcia, L.; Salles-Loustau, G.; Zonouz, S. Hybrid firmware analysis for known mobile and iot security vulnerabilities. In Proceedings of the 2020 50th Annual IEEE/IFIP International Conference on Dependable Systems and Networks (DSN), Valencia, Spain, 29 June–2 July 2020; pp. 373–384.
35. Corley, C.S.; Kraft, N.A.; Etzkorn, L.H.; Lukins, S.K. Recovering traceability links between source code and fixed bugs via patch analysis. In Proceedings of the 6th International Workshop on Traceability in Emerging Forms of Software Engineering, Waikiki, HI, USA, 23 May 2011; pp. 31–37.
36. Xu, Z.; Chen, B.; Chandramohan, M.; Liu, Y.; Song, F. Spain: Security patch analysis for binaries towards understanding the pain and pills. In Proceedings of the 2017 IEEE/ACM 39th International Conference on Software Engineering (ICSE), Buenos Aires, Argentina, 20–28 May 2017; pp. 462–472.

 electronics

Article

Detection and Isolation of DoS and Integrity Cyber Attacks in Cyber-Physical Systems with a Neural Network-Based Architecture

Carlos M. Paredes [1], Diego Martínez-Castro [1], Vrani Ibarra-Junquera [2] and Apolinar González-Potes [2,3,*]

1 Departamento de Automática y Electrónica, Universidad Autónoma de Occidente, Cali 760030, Colombia; cmparedes@uao.edu.co (C.M.P.); dmartinez@uao.edu.co (D.M.-C.)
2 Laboratorio de Agrobiotecnología, Universidad de Colima, Colima 28400, Mexico; vij@ucol.mx
3 Facultad de Ingeniería Mecánica y Eléctrica, Universidad de Colima, Colima 28400, Mexico
* Correspondence: apogon@ucol.mx

Abstract: New applications of industrial automation request great flexibility in the systems, supported by the increase in the interconnection between its components, allowing access to all the information of the system and its reconfiguration based on the changes that occur during its operations, with the purpose of reaching optimum points of operation. These aspects promote the Smart Factory paradigm, integrating physical and digital systems to create smarts products and processes capable of transforming conventional value chains, forming the Cyber-Physical Systems (CPSs). This flexibility opens a large gap that affects the security of control systems since the new communication links can be used by people to generate attacks that produce risk in these applications. This is a recent problem in the control systems, which originally were centralized and later were implemented as interconnected systems through isolated networks. To protect these systems, strategies that have presented acceptable results in other environments, such as office environments, have been chosen. However, the characteristics of these applications are not the same, and the results achieved are not as expected. This problem has motivated several efforts in order to contribute from different approaches to increase the security of control systems. Based on the above, this work proposes an architecture based on artificial neural networks for detection and isolation of cyber attacks Denial of Service (DoS) and integrity in CPS. Simulation results of two test benches, the Secure Water Treatment (SWaT) dataset, and a tanks system, show the effectiveness of the proposal. Regarding the SWaT dataset, the scores obtained from the recall and F1 score metrics was 0.95 and was higher than other reported works, while, in terms of precision and accuracy, it obtained a score of 0.95 which is close to other proposed methods. With respect to the interconnected tank system, scores of 0.96, 0.83, 0.81, and 0.83 were obtained for the accuracy, precision, F1 score, and recall metrics, respectively. The high true negatives rate in both cases is noteworthy. In general terms, the proposal has a high effectiveness in detecting and locating the proposed attacks.

Keywords: anomaly detection; anomaly isolation; artificial neural networks; Cyber Physical System

Citation: Paredes, C.M.; Martínez-Castro, D.; Ibarra-Junquera, V.; González-Potes, A. Detection and Isolation of DoS and Integrity Cyber Attacks in Cyber-Physical Systems with a Neural Network-Based Architecture. *Electronics* **2021**, *10*, 2238. https://doi.org10.3390/electronics10182238

Academic Editor: Arman Sargolzaei

Received: 30 July 2021
Accepted: 31 August 2021
Published: 12 September 2021

1. Introduction

Cyber Physical Systems (CPSs) emerge from the attempts to unify the emerging applications of embedded computers and communication technologies used to monitor, control, as well as generate actions on physical elements to fulfill with a specific task [1], and they have an important impact on different sectors [2].

The different parts of the system are usually interconnected using communication networks to share information and data that interact with each other and, sometimes, cloud computing services [3–5]. CPSs can be represented in layers, as shown in Figure 1. The first is the physical layer, where the physical infrastructure of the system, sensors, and actuators are located, with the objective of monitoring and controlling physical processes.

The second is the network layer, which implements the transmission data and allows the interaction between the physical layer and the cybernetic layer. Finally, a cybernetic layer allows the abstractions of the received data, as well as the interaction between networks, devices, and the physical infrastructure [6].

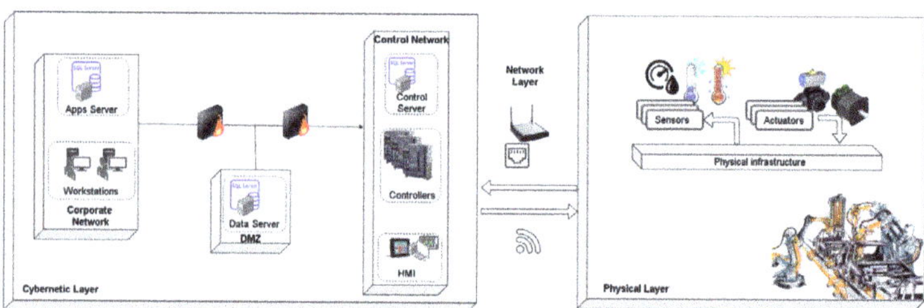

Figure 1. Architecture of a CPS.

Society currently relies on multiple automatic systems supported by CPSs. These applications are focused in contexts, such as industrial, health, and environmental, among others [7,8]. Security and reliability are fundamental requirements in these systems. Cyber attacks can generate inappropriate behaviors and catastrophic impacts on the physical world, causing damage to both the system infrastructure and industrial products and even threaten human lives [9]. Examples, such as attacks on smart grids, aviation systems, water plants, chemical plants, and oil and natural gas distribution systems, are becoming increasingly high [10–14]. The above has caused this research area to be active in recent years.

Therefore, there must be mechanisms to detect the occurrence anomalies to avoid exploiting vulnerabilities in the devices connected to the system network. Real-time detection is very important in order to ensure reliability and security in these systems, where sensors are prone to malicious attacks. For this reason, detection systems are often used, such as Intrusion Detection Systems (IDS), which monitor data traffic to identify and protect systems from these eventualities. Based on detailed analysis of network traffic and device usage, IDSs seek to evaluate this information to identify unwanted events. IDSs do this by carrying out three stages: monitoring, analysis, and detection. Monitoring relies on a sensor network or host-based sensors, the analysis stage is based on any method to identify and extract features, and the detection stage relies on anomaly detection [15,16].

Within these can be highlighted: [17] the methods based on traditional information technologies, where network traffic analysis is used to detect anomalies [18–26]; and model-based methods, where detection is performed by comparing the system actual output with an expected value [4,27–31].

According to reference [16,32], host-based IDS methods operate on the data collected from the individual parts of the computer systems and can detect internal changes and determine which processes and/or users are involved in malicious activities, which can be not significant with some devices; thus, this method sometimes fails. Whereas a network-based IDS will detect malicious packets as they enter your network or unusual behavior on your network, such as flooding attacks, more traditional IDS can do it on one channel or across the network. These monitor the entire network traffic to detect known or unknown attacks using techniques based on anomalies, signatures, and specifications [16,33,34]. Hence, IDSs help to avoid critical consequences and assist in making appropriate decisions when system events occur by performing two main tasks: attack detection, which decides whether or not an anomaly has occurred; and attack isolation, which decides which elements of the system are being affected by the unwanted.

In such a way, the purpose of this research is to present the design of an architecture that allows detecting and isolating attacks that may occur between the elements of the physical layer and the controller, generating alerts that allow detection and localization of the origin of the cyber attacks. For this, a new architecture was proposed for the detection and isolation of attacks using techniques based on artificial intelligence. The proposal integrates two approaches: regression and classification. The first approach allows generating models that describe the behavior of the real process to estimate the outputs by using process input data, obtaining in this way the model to be compared with the real process values in order to detect and isolate the attack. The second approach allows generating detection systems to carry out the detection and isolation of attacks. The proposal was subjected to two test benches, obtaining better results than those reported in previous works. The contributions of this paper are as follows:

- The design of an architecture using one-dimensional convolutional neural networks to detect and isolate cyber-attacks that involve the elements of the physical layer and the controller of a CPS, generating alerts to detect and locate the origin of the cyber attack.
- The architecture proposed is an architecture based on the process information, where the dynamic properties of the process are covered, in order to evaluate the possibility of a cyber attack occurring in different parts of the system, without the need to define a threshold that allows separating normal situations with events where a cyber attack is possibly occurring.
- The design of the architecture allows detecting and locating the occurrence of cyber attacks occurring simultaneously in different parts of the system, even when the attacks are of different types.

The remaining sections are structure as follows. Section 2 presents related works. Section 3 describes the problem statement. In Section 4, the attack detection and isolation method is proposed. Section 5 exposes the results obtained using the method proposed in two test benches. Finally, in the last section, we present the conclusions.

2. Related Works

Protection systems in industrial processes have used strategies that have presented good performance in other environments, such as office environments. However, the characteristics of these applications are not the same, so the results obtained are not as expected. This is because the availability of equipment in industrial systems is very high; so, in many cases, a simple solution in corporate environments, such as patching, simply does not work because the machine is not available to shut down until a planned outage. It is also difficult to predict how a newly introduced patch will affect the operation of a control system, especially if the patch is not rigorously tested, increasing the organization's reluctance to act on potential threats. The implementation of security patches can affect application performance and, therefore, the stability, availability, and real-time behavior of machines. Something equivalent occurs with the impact on data traffic through the communications network associated with solutions that evaluate network traffic, which can affect delays in control strategies and, in turn, the performance of control loops [35]. This problem has motivated different projects with the purpose of contributing from different approaches to increase the security of control systems. In this section, the related works are described.

Some ongoing projects to improve security in these systems have included methods to provide aspects, such as data confidentiality and authentication, access control, within the network, and privacy and reliability of applications, as well as the inclusion of security and privacy policies [36]. Even so, CPSs are vulnerable to multiple attacks aimed at disrupting the network and modifying process variables, altering its operation. For this reason, new defense mechanisms designed to detect cyber attacks have been generated. One of the best known mechanisms is IDS. IDS approaches may be classified as signature-based, anomaly-based, or specification-based [33].

The signature-based method only detects records that are inside of a database, and it is highly accurate and effective against known threats, consumes more power, and does not detect new events [33]. The anomaly-based method is efficient in detecting new attacks [16] since it compares the system activities in a moment against an usual behavior profile and generates alerts whenever a threshold defined by the system's normal behavior is cross [34].

However, anything that does not match normal behavior is considered an intrusion, and learning all normal behavior is not an easy task. Therefore, this method generally has high false-positive rates. On the other hand, the methods based on specifications use a set of rules and thresholds that define the expected behavior of the different components of the network. It has the same purpose as anomaly-based methods, with the difference that this method is specified manually by an expert who determines the specifications. Manually defined specifications typically provide low false-positive rates versus anomaly-based detection and do not require training steps because it can be used immediately. However, these methods cannot be adapted to different environments and can be time-consuming to adjust and error-prone [33].

Other authors have developed state observers for detection, such as the Luenberger Observer (LO), while the isolation process is realized by structured residues generated using Unknown Input Observers (UIOs) [37–40]. These methods present drawbacks because the detection of anomalies is realized by a comparison of a fixed threshold defined by a historical data of normal behavior, with the difference between the variables of the actual process and the values generated by an estimated model. Then, it can lead to a considerable rate of false positives and false negatives. The above is because, for the design of the observer banks, the knowledge of the parameters and the dynamics of the system is used, which sometimes can be significantly different of the real system performance. So, both proposals are limited by the knowledge of the process, such as the definition of the threshold, which, in real situations, it may not be easy to model accurately.

In the last few years, data-driven methods have been employed to detect cyber attacks [18–23,25,41]. These methods have presented good performance to find models of processes that even present quite pronounced non-linear dynamics. Machine learning technology is one of the data-driven methods emerging as a method to detect attacks in these systems [23,26,42–50].

Random Forest-based algorithms have been employed recently to detect malicious behavior by using databases; in this case, binary classification is applied to classify whether the content of a packet is malicious or not. This method reduces computational cost but does not guarantee high accuracy [51]. In this way, it is not possible to identify which task transmitted the packet, and it does not allow specifying the type of attack [15,16]. From another point of view, in Reference [52], a scheme was proposed to protect remote patient monitoring systems against DoS attacks. An attack detection model was established by developing mechanical learning using decision trees. The model could help to locate various types of attacks, focusing mainly on flooding attacks, and could be appropriate to devices with limited memory and processing resources, such as sensors and healthcare devices. As future work, they propose the possibility of identifying other types of attacks and even developing a mechanism to block a wide range of attacks.

Other approaches have used different artificial intelligence techniques, such as Support Vector Machines (SVMs), genetic algorithms [32], self-organized networks of ant colonies, and extreme learning machines, which provide models with very high accuracies applied in the context of security in computer networks, and especially in the detection of intrusions. The purpose of these techniques is to achieve better intrusion recognition rates, but it is still noticeable that the false positive rate remains the problem to be approached in all these studies. Although some technique can reduce the false positive rate, it increases the training time and classification, which is a relevant index for real-time detection [53].

In Reference [18], an SVM-based algorithm was used to classify normal and abnormal behavior of data traffic that may be subjected to DoS attacks. This algorithm reaches good attacks predictions rate with less training time. In Reference [19], a method based on Principal Component Analysis (PCA) and SVM to detect DoS attacks was presented. The paper analyzes the effects of DoS attacks in a network using TCP protocol. The PCA algorithm is used in order to filter the interference of the environment to extract the main features effectively and reduce the dimensioning of information without losing information from the original data. The results show that the algorithm has high accuracy and a low false positive and false negative rate (FPR and FNR). In the same context, an SVM using a radial basis kernel function is proposed in Reference [20] to detect attacks in networked automotive systems. This proposal aims to avoid drawbacks associated with cases in which there is not an events dataset, or it is probably not sufficiently representative because many of the possible situations of a system are unknown. However, these techniques are not suitable for detecting mutations from various attacks.

Advanced techniques, such as Deep Belief Networks (DBN) and Deep Convolutional Neural Networks (Deep CNN) [54,55], are trained to extract low-dimensional features and are used to discriminate usual and hacking packets. In Reference [56], an anomaly detector based on a neural network recurrent Long short-term memory (LSTM) was proposed to detect attacks with low false alarm rates. These methods have had the best response in these environments, although the computational costs sometimes are high [20,55]. Thus, applying machine learning and other artificial intelligent techniques is a challenge because it requires more memory and computational cost that can affect the performance of the system.

In addition, to validate the proposal, two test benches were used. For the selection of these datasets, a search was performed that included keywords, such as security in industrial control systems, detection of faults, anomalies and cyber attacks in control systems, and design of secure CPSs. From this search, we considered the publications that had a publication time of less than 5 years, as well as the number of times that the datasets had been used to evaluate the security on CPSs. We also considered the type of attacks that were implemented, since our approach was to address different types of attacks, including those with the highest frequency and impact on the control systems found in the CPSs (integrity and DoS attacks).

The first one corresponds to the SWaT dataset, which provides real data from a simplified version of a real world water treatment plant. This dataset allows researchers to design and evaluate defense mechanisms for CPSs and contains both network traffic and data concerning the physical properties of the system [57]. On the other hand, there is another test bed which consists of three interconnected tanks [58] that has allowed the validation of different types of detection methods for cyber attacks on CPS. These two test benches have made it possible to validate different proposals focused on techniques that allow us, in one way or another, to analyze the detection of cyber attacks [37,42,59–69] and have made it possible to direct this research to improve the proposed proposals.

Based on this review, Table 1 summarizes each of the related reports to a set of characteristics in order to highlight the issues that need to be addressed to improve the strategies and proposals in the future.

Table 1. Summary of related works.

Reference	Main Domain	Technique	Type of Anomaly	Advantages	Limitations	Evaluation
[18]	Mobile networks	SVM, signature and anomaly based methods	DoS attacks	High accuracy to detect normal and anomalous behavior	Only detects DoS attacks	Dataset KDD
[19]	Mobile networks	PCA-SVM	Low rate DoS attacks	High detection rate and low FPR and FNR	Only detects DoS attacks	Simulation
[20]	In-vehicle networks	One-class SVM	Possible errors in the recordings	The proposed methodology could be applied to several fields	TNR below 77% and precision below 76%.	Dataset from a real vehicle
[23]	Mobile networks	MLP for intrusion detection	DoS attacks	High accuracy to detect normal and anomalous behavior	Only detects DoS attacks	Dataset KDD
[25]	Heavy duty vehicle system	Gaussian radial basis function neural network	Deception attacks	Can be applied to a variety of nonlinear CPSs	Attacks occur in only one part of the system	Simulation
[26]	Solar Farms	Multilayer LSTM network	Integrity attacks	Accuracy, recall, precision and F1 score are above 90%	Attacks occur in only one part of the system	Simulation
[37]	Three-tank system	Luenberger Observers (LOs) and Unknown Input Observers (UIOs)	Integrity attacks	Possibility to mitigate the effect of the attack	Attacks occur in only one part of the system. Dependence on threshold selection.	Simulation
[38]	Smart grids	Unknown Input Observers (UIOs)	Integrity attacks	Possibility to mitigate the effect of the attack.	Attacks occur in only one part of the system Dependence on threshold selection.	Simulation
[39]	Power systems	Unknown Input Observers (UIOs)	Integrity attacks	Possibility to mitigate the effect of the attack	Attacks occur in only one part of the system. Dependence on threshold selection	Simulation
[40]	Power systems	Luenberger Observers (LOs) and Unknown Input Observers (UIOs)	Integrity and DoS attacks	Platform for simulating different types of cyber attacks	Detection depends on the selection of the threshold.	Emulation and simulation
[41]	Automotive Brake Systems	Recurrente neural networks	Integrity attacks	High accuracy	The attacks are applied on the same part.	Experimental
[42]	Industrial Control Systems	1D CNN and GRU	Integrity attacks	High precision and F1 score.	False alarm rate needs to decrease.	SWaT Dataset
[50]	Automated Vehicles	LSTM and CNN	Various	Detecting different single anomaly types.	In some cases the TPR is low.	Experimental.
[55]	Heavy-duty gas turbines of combined cycle power plants	Stacked denoising autoencoder	Various	Real time detection, high TPR and low FPR.	Only detects, does not locate.	Simulation and data from real plants.
[56]	Automobile Control Network Data	LSTM	Integrity attacks	High TPR and low FPR.	It is required to achieve a practical level to reliably detect anomalies.	Simulation.
[59]	Industrial Control System	Genetic algorithms and neural network	Various	High accuracy to locate the sensor under attack.	Metrics, such as F1 score and recall, must be improved.	SWaT Dataset.
[60]	Industrial Control System	Deep Neural Networks	Various	Successfully detects the vast majority of attacks with a low level of false positives.	Metrics, such as F1 score and recall, must be improved.	SWaT Dataset.
[61]	Industrial Control System	Graphical model-based	Various	High precision.	Metrics, such as F1 score and recall, must be improved.	SWaT Dataset.
[62]	Industrial Control System	SVM and Deep Neural Networks	Various	High precision.	Metrics, such as F1 score and recall, must be improved.	SWaT Dataset.
[63]	Industrial Control System	LSTM and CNN	Various	High precision.	Metrics, such as F1 score and recall, must be improved.	SWaT Dataset.
[64]	Industrial Control System	Lightweight Neural Networks and PCA	Various	Good precision.	Metrics, such as F1 score and recall, must be improved.	SWaT, BATADAL, and WADI Dataset.
[65]	Networked Control (Three-tank system)	Resilient Tracking Control	Deception and DoS attacks	A combination of attacks can be taken into account to form a sophisticated and stealthy attack model.	High dependence on knowledge of system parameters.	Simulation and experimental.
[66]	Three-tank system	Model-based fault/attack tolerant	Integrity attacks	Determines when the control input is to be updated again, depending on the occurrence of the anomaly.	High dependence on knowledge of system parameters.	Simulation and experimental.

Based on the review of the related works, it became evident that there are still challenges concerning the detection of cyber attacks within the control systems found in the CPSs. On the one hand, methods must be sought to decrease both the false positive and false negative rates, and to increase the true positive and true negative rates. This will improve the overall performance of these detection systems. It is also evident that the phenomenon of simultaneous attacks has not been addressed in the design of cyber attack detection systems, which is worrying because these situations can occur very often in the real world. Is important to clarify that, within a CPS, there are many points where a cyber-attack can occur and that can cause different consequences in the system. The emphasis of this work seeks to design an architecture that allows detecting and locating attacks that occur between the elements of the physical layer and the controller of a CPS, precisely in attacks that modify or interrupt the sending of data from one element to another. In this way, this paper presents the design of an architecture that explores the potential of convolutional neural networks to extract features and, thus, to determine whether there is an event related to the possibility of a cyber attack occurring. This approach may have a closer approach to the implementation in real cases in which there is a high degree of uncertainty in the process models, since, on many occasions, the way to detect an anomaly or not is done under a process of comparison between estimated values and the real values of the process, which is subsequently evaluated by a threshold. In our proposal, this evaluation is carried out in an intrinsic way by the architecture based on convolutional neural networks, generating a better performance than current works, as well as shows promising results in the detection and isolation of simultaneous attacks.

3. Problem Statement

Several control applications supported in these systems can be labeled as safety critical in relation to the fulfillment of strict real time deadlines, associated with the generation of actions from the interaction between the computational systems and the physical systems related to the application, because the non-fulfillment of these requirements can cause irreparable damage to the physical system being controlled, as well as to the people depending on it [70]. Additionally, measurements and control actions can be altered while being transmitted through communication networks, thus requiring new control algorithms or design architectures, which, in the presence of adverse situations, can bring the system to safe and stable states [71,72]. The proposal presented in this work focuses in the detection and isolation of DoS and integrity cyber attacks on CPSs, specifically on the exchange of information between sensors, actuators, and controllers. The approach realized is based in the fault detection and isolation systems for what anomalies are represented as a variation of the system parameters [58]. Then, any control system where its control signals and/or measured variables are susceptible to be attacked can be modeled as a combination of the two models defined in (1) and (2).

$$x(k+1) = Ax(k) + Bu(k) + F_a f_a(k), \tag{1}$$

$$y(k) = Cx(k) + F_s f_s(k), \tag{2}$$

where $x(k)$ represents the state vector, $x(k) \in \mathbb{R}^{n\times 1}$, $y(k)$ is the output vector, $y(k) \in \mathbb{R}^{p\times 1}$, $u(k)$ is the control action, $u(k) \in \mathbb{R}^{m\times 1}$, matrix A is the state matrix, $A \in \mathbb{R}^{n\times n}$, B is the input matrix, $B \in \mathbb{R}^{n\times m}$, C is the output matrix, $C \in \mathbb{R}^{p\times n}$, D is the feedthrough matrix, $D \in \mathbb{R}^{p\times m}$, $F_a = B$, and $f_a = (\Gamma - I)U + U_{f0}$. ΓU and U_{f0}, represent the effect of a multiplicative anomaly and an additive effect in the control action, respectively. DoS and integrity attacks are visible as anomalies on the control action. If the i-th control action is attacked, then the matrix F_a corresponds to the i-th column of the matrix B, and f_a corresponds to the magnitude of the attack that directly affects the controller.

Similarly, if the i-th sensor is attacked, the matrix F_s is the i-th row of the matrix C, and the vector of attacks is f_s, which represents the magnitude of the effect produced in the i-th sensor.

The problem with traditional methods based on mathematical models that describe the behavior of the system is that these models are dispensable of the complete knowledge of the system parameters, and the adaptation in real conditions can cause the overall performance to decrease. Because of this, we intend to address this problem from models based on artificial neural networks, precisely in one-dimensional convolutional neural networks, which have shown very promising results in fields where patterns are sought to identify a class.

Modeling of the Cyber Attack

Measurements of process signals and control action values are critical to the proper functioning of a control system, and its modification by cyber attacks can produce instability in the control system [73,74]. A cyber attack by data manipulation is called an integrity attack, modeled by (3), and an attack that results in a prolonged loss of these signals is called a type DoS attack, which is modeled by (4).

$$\overline{y}_i(k) = y_i(k) + y_i(k)^a, \tag{3}$$

$$\overline{y}_i(k) = y_i(k)_{t_{s-1}}, \tag{4}$$

where $\overline{y}_i(k)$ corresponds to the sensor measurement that reaches the controller in the k-time, $y_i(k)$ corresponds to the sensor measurement before being transmitted to the controller in the k-time, and $y_i(k)^a$ is a vector injected by the attackers which changes the $y_i(k)$ measure in the k-time. $y_i(k)_{t_{s-1}}$ corresponds to the measurement before the start of the DoS attack. The time interval for the occurrence of the attack is defined by $\tau_a = [t_s \; t_e]$.

For the development of the proposal, it is assumed that any sensor can be affected by any type of attack, integrity, or DoS. Additionally, the attacks may occur at any time in various parts of the system. The last premise is significant to note because simultaneous attacks are less discussed in previous works; thus, depending on the type of attack carried out on the system, output (2) may take the form of (3) and/or (4).

4. Attack Detection and Isolation Method

In the context of this work, most cyber attack detection methods use the available data to develop a model that determines the usual behavior of the system. Then, by a comparison between the estimated outputs of the model and the actual process outputs, determination of if the behavior of the system is normal or if a cyber attack is taking place. To isolate the attack, which is nothing more than locating the part of the system that is being affected directly by the cyber attack, decoupled models of the system are developed that are susceptible only to cyber attacks that occur in specific parts of the system.

The procedure to perform this task can be grouped into three steps. Firstly, the generation of a residual signal is realized, and this process consists of comparing the measured output with an estimated output. This signal is denoted as residual signal, $res(k)$, this is described in (5).

$$res(k) = y(k) - \hat{y}(k), \tag{5}$$

where $y(k)$ are the set of output measures of the actual process, and $\hat{y}(k)$ are the set of outputs estimated. The second step corresponds to the evaluation of the residual; in this case, a comparison of the residuals is made with a predefined threshold, as is shown in (6).

$$|res(k)| > \tau_{thresholds}. \tag{6}$$

The thresholds are obtained from data in which the attacks have been presented, thus allowing their detection and isolation. Finally, a decision-making process is carried out through indicators.

These steps involve the use of residuals that should take values close to 0 in situations where the system is not being attacked. On the other hand, when an attack is present, the residual signals must have values other than 0.

Although a single residual signal can alert or detect a cyber attack, a set of residuals is required to isolate it. Then, to locate the origin of the cyber attack, it is necessary that some residues be sensitive only for a particular part of the system. The above implies that the set of residuals must be independent of other cyber attacks defined. In this way, to isolate a cyber attack, a structured set of residuals is considered, where each residual vector can be used to detect a cyber attack in a specific place of the system.

In the architecture model proposed in this work, it is emphasized that second step will be an implicit step because the architecture based on artificial neural networks will interpret the input data generating intrinsic characteristics that will allow the evaluation to detect and isolate the attacks.

Architecture Proposed for the Detection and Isolation of Cyber Attacks in CPS

The architecture proposed is presented in Figure 2. This architecture includes a prediction model which uses an input dataset $x_0, x_1, \ldots, x_{k-1}$ to estimate outputs $\hat{y}_1, \hat{y}_2, \ldots, \hat{y}_k$ (these datasets will depend specifically on the type of data available from the process), and these values are used to obtain the residual signal $res(k)$, as is shown in (5). These signals are used by a classifier to detect anomalies presents in the process.

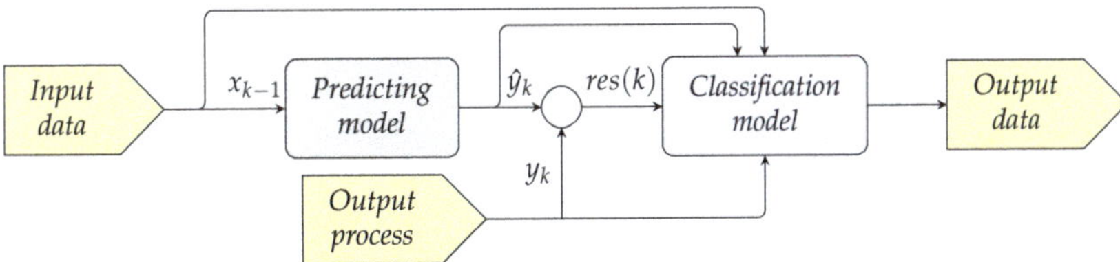

Figure 2. General architecture model to detect and isolate cyber attack.

As the characteristics of the signals in a specific process are different then values with different magnitude could affect the classifier training procedure, therefore, all input data to the classifier are normalized using its mean and standard deviation to obtain a z-score for each one as is shown in (7).

$$z = |x - \mu| / \sigma, \tag{7}$$

where x are the input data, μ is the mean, and σ is the standard deviation.

Although the architecture presented is general, it is a base for selecting different types of machine learning for the prediction and classification stages. The idea is to use deep neural networks to extract patterns that allow the detection of cyber attacks (such as LSTM or CNN 1-dimensional). As was not included a method to find spatial-temporal correlations to detect cyber attacks, it is expected that neural networks will be able to carry out this task implicitly.

The architecture can be detailed as follows for a specific CPS, shown in Figure 3. A model of the dynamics of the process generates the outputs signals $x(k)_s$ which correspond to the reconstruction of all the states (it is assumed that the outputs are the process states or some linear combination of them, although it can be extended to non-linear cases). In order to isolate the attack, there is a set of neural network models that relate the process states with their respective control actions for generate states that are decoupled from each other $(x(k)_{d_{1,2,\ldots x}})$; in this way, it is possible to isolate the attack in a way equivalent to the use of UIOs, but with the advantage that neural networks allow addressing the uncertainty in the representations. With this set of neural networks, $res(k)$ is generated.

Detection and isolation functions are implemented using artificial neural networks, which use the process states $x(k)$, the control actions $u(k)$, the reference signals $r(k)$, the residual signals $res(k)$, and the signals generated by the predicting model.

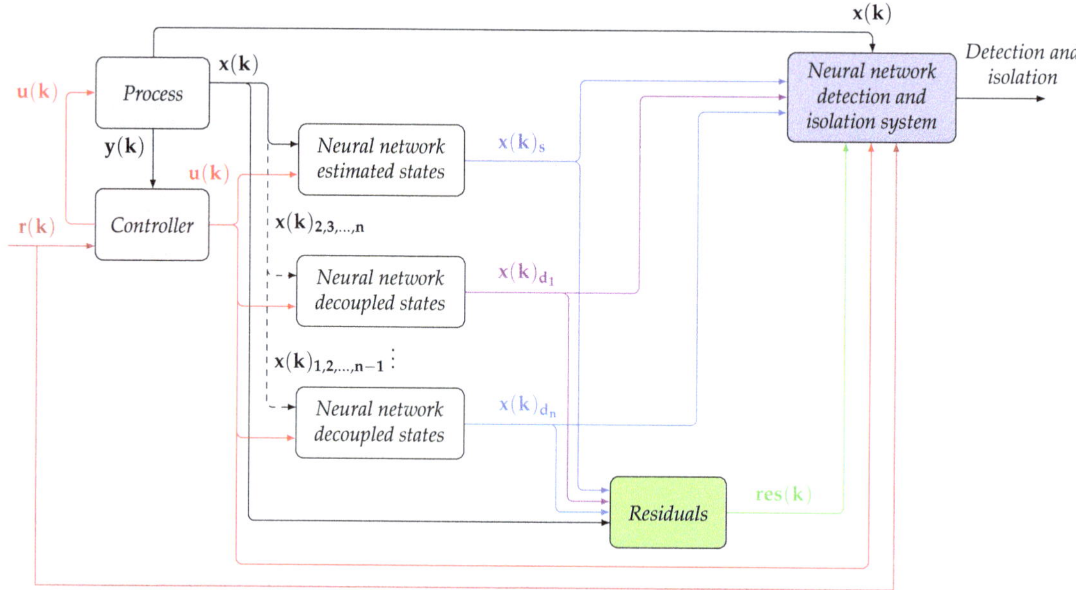

Figure 3. Architecture based on neural networks for the detection and isolation of the cyber attack.

Mean squared error (MSE) [75] is adopted as the model's loss function to train the predicting model.

$$MSE = \frac{1}{n}\sum_{i=1}^{n}(x_i - \hat{x}_i)^2, \tag{8}$$

where n is the amount of data, x_i is the real state, and $\hat{x}_i$ is the estimated state. For the classifier, the cost function categorical crossentropy (CCE) is used [76] because it is a single-label multi-class classification problem.

$$J_{CCE} = -\sum_{q=1}^{l}\sum_{k=1}^{p} d_{qk}\log(y_{qk}). \tag{9}$$

With p classes, training data size of l, the input of x_q, where $q = 1, 2, \ldots, l$ and y_{qk} $(0 \le y_{qk} \le 1)$,$k = 1, 2, \ldots, p$ is the estimated probability that belongs to class k, and d_{qk} (0 or 1) becomes the given label (9).

5. Case Studies and Results Analysis

Two test benches were used to evaluate the performance of the proposed architecture, the SWaT dataset [77,78] and an interconnected tank [58].

5.1. Secure Water Treatment Dataset-SWaT

This dataset was completed by the Singapore University of Technology and Design to provide researchers with data collected from a complex and realistic ICS environment. The testbed is a fully operational scale water treatment plant that produces purified water. SWaT is composed of six main processes corresponding to the physical and control components of the water treatment plant; each stage (from P1 to P6) is equipped with a certain number of sensors and actuators. The sensors include flow meters, water level meters, conductivity, and pH analyzers, among others, while the actuators consist of pumps that transfer water from one stage to another, pumps that dose chemicals, and valves that control inlet flow. The process is not circular, and P6 water is removed. Sensors and actuators in each stage are connected to the corresponding PLC (programmable logic controller). This process is shown in Figure 4.

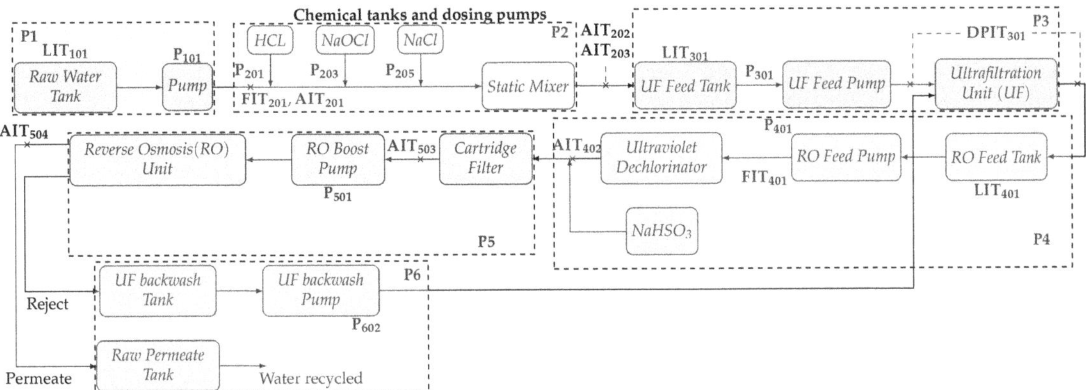

Figure 4. SWaT testbed processes overview [57].

Stage P1 controls the flow of raw water by opening or closing a motorized valve that is connected to the inlet of tank T101. By means of the P101 pump, water starts flowing from T101 through the chemical dosing station in stage P2 and is followed by the ultrafiltration (UF) process located in stage P3, which eliminates unwanted materials. Similarly, the feed pump in stage P3 is responsible for supplying the water to the ultrafiltration unit. In the P5 stage, inorganic impurities are separated by a reverse osmosis process. The output of the reverse osmosis process is stocked in the permeate tank of stage P6 for its distribution. The P6 stage is also controlling the cleaning of the ultrafiltration membranes in P3 by the backwashing process. Every certain period of time, the backwash process is triggered by turning on the backwash pump and is accomplished in under one minute. The backwash process can alternatively be started by a PLC when the differential pressure sensor value increases above 0.4, which means that the UF membranes are choked [57,78].

5.1.1. Dataset Description

Training Dataset 1 and Training Dataset 2 were used. The first one corresponds to data collected under normal operating conditions. This dataset was released on November 20, 2016 and was generated from a one-year long simulation. The second dataset corresponds to situations when attack scenarios are generated. This dataset with partially labeled data was released on 28 November 2016. The dataset is around six months long and contains several attacks, as shown in Table 2.

Table 2. Attacks featured in Training dataset 2 [78].

ID	Duration (Hours)	Attack Description	SCADA Concealment
1	50	Attackers change L_T7 thresholds (which controls PU10/PU11) by altering SCADA transmision to PLC9. Low levels in T7.	Replay attack on L_T7.
2	24	Like Attack # 1.	Like Attack # 1 but replay attack extended to PU10/PU11 flow and status.
3	60	Attackers alter L_T1 readings sent by PLC2 to PLC1, which reads a constant low level and keeps pumps PU1/PU2 ON. Overflow in T1	Polyline to offset L_T1 increase.

Table 2. *Cont.*

ID	Duration (Hours)	Attack Description	SCADA Concealment
4	94	Like Attack # 3.	Replay attack on L_T1, PU1/PU2 flow and status, as well as pressure at pumps outlet.
5	60	Working speed of PU7 reduced to 0.9 of nominal speed causes lower later levels in T4.	
6	94	Like Attack # 5, but speed reduced to 0.7.	L_T4 drop concealed with replay attack.
7	110	Like Attack # 6.	Replay attack on L_T1, as well as PU1/PU2 flow and status.

5.1.2. Data Preparation and Model Training

The data from the first dataset is used to generate a model corresponding to the "Predicting model" block shown in Figure 2. The architecture proposed in this case is based on a 1D CNN model, as shown in Figure 5.

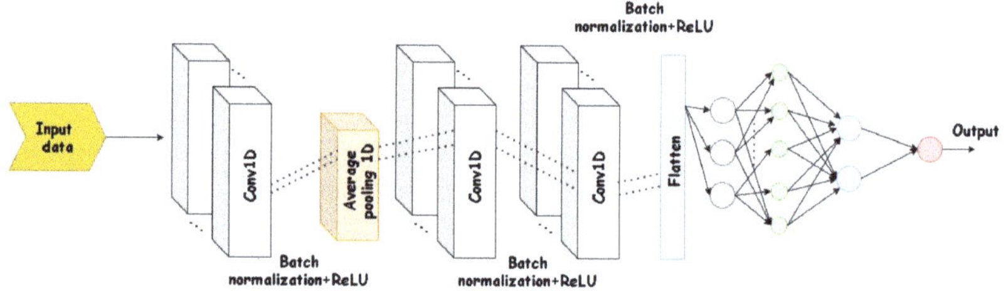

Figure 5. Prediction model for SWaT dataset.

The input data is composed of 43 characteristics compounded mainly by sensors measurements, states of the pumps, and valves positions. The first convolution layer consists of 2 filters, and the kernel size is 3. The 1D average pooling layer has a stride of 2 and the same padding; the second convolution layer has 20 filters and a kernel size of 20; the last convolution layer is composed of 10 filters and a kernel size of 5. Finally, a fully connected layer is used with a 43 neurons layer and a neuron in the output layer, all with linear activation functions. Additionally, the batch normalization layer is added with ReLU activation in various parts of the network. The loss function used was MSE, and the optimizer was the stochastic gradient descent with momentum. For training, a maximum of 40 epochs was available with an initial learning rate of 0.001. In this case, 30% of the data was used to validate, and 70% of the data to train.

The parameters of the layers for this network were found in such a way that the lowest possible MSE will be achieved. Increasing the number of layers, neurons, filter size, or number of filters did not correspond to a significant improvement performance.

The second dataset was used for the classification process; it is composed of 4177 data, of which 3685 data correspond to normal operating conditions, 50 belong to the first attack scenario, 24 correspond to the second attack scenario, 60 to third and fifth attack, 94 to fourth and sixth attack, and 110 to the seventh scenario. As can be seen in Figure 6a, this dataset is unbalanced and would then generate problems to the classifier. The bar centered at 0 corresponds to normal operating conditions, while the other corresponds to the different attack scenarios which are shown in the ID column of Table 2. It could affect the algorithms in relation to the minority classes. To address this situation, initially, methods, such as Random Oversampling and Undersampling, were used for imbalanced classification without obtaining satisfactory results. For this reason, the approach shown in Reference [79] was followed. This proposal is a modification of temporal data determined by optimal sequences that are aligned with the original data, thus generating new time-

synthesized data to the training dataset. The distribution of the different classes for the new dataset to be used is shown in Figure 6b. Although it is observed that it is an unbalanced dataset, the amount of data generated from the attack scenarios was increased, and the performance was improved.

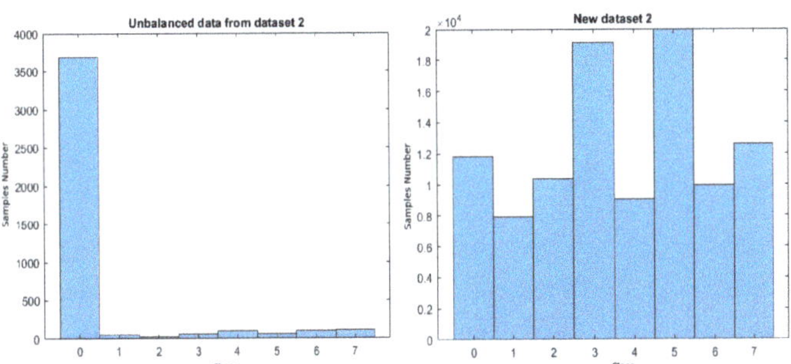

(**a**) Original SWAT dataset distribution. (**b**) New SWAT dataset distribution.
Figure 6. SWAT dataset distribution.

This new dataset was used to estimate the outputs using the architecture shown in Figure 5, which were compared with the usual process variables to obtain the residual signal.

The input data for the classifier whose architecture is shown in Figure 7 are: the estimated outputs, the process variables, and the corresponding residuals. This corresponds to the "Classification model" block shown in Figure 2 and was implemented by a group of cascaded convolutional layers with a batch normalization layer with ReLU activation function between them. The number of convolutional layers selected was five, obtaining a higher accuracy than 90%. The number of filters implemented from the input to the fully connected layer were 128, 64, 32, 16, and 8, respectively. The kernel size in each one was 10. The fully connected layer is composed of eight neurons in its input layer with linear activation function, while the last layer has eight neurons with softmax activation functions corresponding to the 7 attacks and the usual operation scenarios.

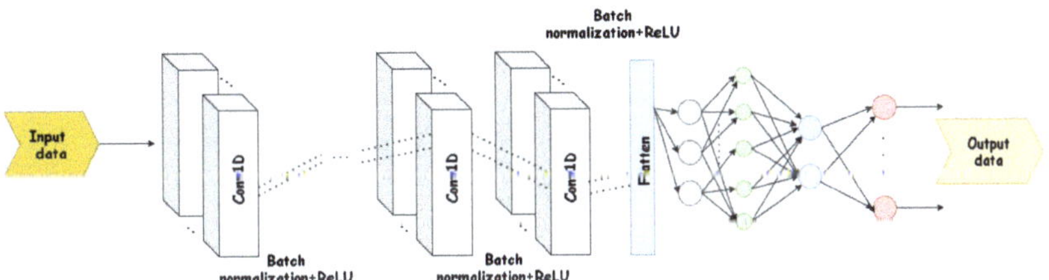

Figure 7. Classification model for SWaT dataset.

The loss function used was CCE, and the optimizer used was stochastic gradient descent with momentum. For the training, a maximum of 4 epochs was available, with an initial learning rate of 0.0001. For training, a maximum of 4 epochs was available, with an initial learning rate of 0.0001, and 30% of the dataset was used to validate, while 70% was used to train.

5.1.3. Evaluation Metrics

The metrics considered in this work were true positives (*TP*), false positives (*FP*), true negatives (*TN*), and false negatives (*FN*). In order to evaluate the performance of the architecture proposed, the following metrics were used: precision, accuracy, recall (sensitivity or TPR), F1 score, and true negative rate or specificity (*TNR*). These metrics were calculated as follows:

$$Precision = \frac{TP}{TP + FP}, \tag{10}$$

$$Accuracy = \frac{TP + TN}{TP + TN + FP + FN}, \tag{11}$$

$$Recall = \frac{TP}{TP + FN}, \tag{12}$$

$$F1\ Score = \frac{2TP}{2TP + FP + FN} = 2\frac{Precision \times Recall}{Precision + Recall}, \tag{13}$$

$$TNR = \frac{TN}{FP + TN}. \tag{14}$$

Additionally, the ROC (Receiver Operating Characteristics) and Precision-Recall Curves were considered.

5.1.4. Analysis of Results of SWaT Case

The results obtained for this dataset are shown in this section. The training and recovering results are carried out in MATLAB software. Figure 8 shows the confusion matrix for each of the available classes. From these results, the metrics defined in the previous section are obtained and are presented in Table 3.

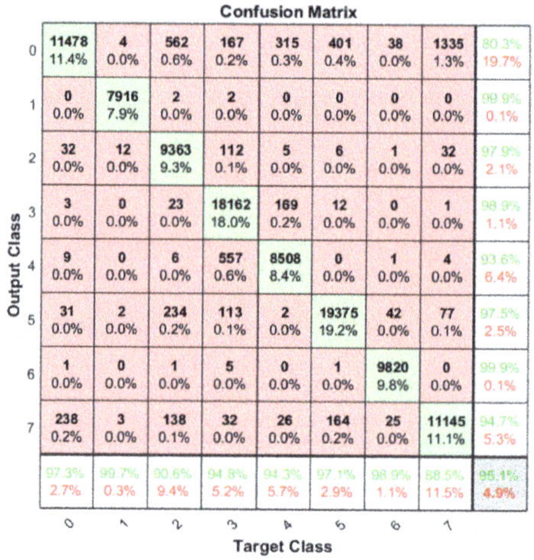

Figure 8. Confusion matrix for SWaT dataset.

Table 3. Summary of metrics.

	Accuracy	Precision	Recall	F1 Score	TNR
Class 0	0.97	0.81	0.97	0.88	0.98
Class 1	0.99	0.99	0.99	0.99	0.99
Class 2	0.99	0.98	0.91	0.94	0.99
Class 3	0.99	0.99	0.95	0.97	0.98
Class 4	0.99	0.94	0.94	0.94	0.99
Class 5	0.99	0.98	0.97	0.97	0.98
Class 6	0.99	0.99	0.99	0.99	0.99
Class 7	0.98	0.95	0.89	0.92	0.98

Class 0 corresponds to the usual operation, while class 1 to 7 are the different attacks scenarios shown in Table 2. It is observed that accuracy is high in all cases. The above shows a high percentage ratio of samples correctly classified by our model. On the other hand, for precision, it is observed that all attack scenarios present a score above 0.94, which means that a lot of data was correctly classified in the different attack scenarios. Similarly, the recall scores are above 0.91 in the majority of classes, which allows minimizing the false alarm rate. Finally, the F1 score shows scores above 0.92. The high rate of TNR in each of the classes is highlighted, which means that FPR is low.

The ROC and Precision-Recall Curves shown in Figure 9a,b present an appropriate performance, indicating that the model has a good capability to distinguish different classes.

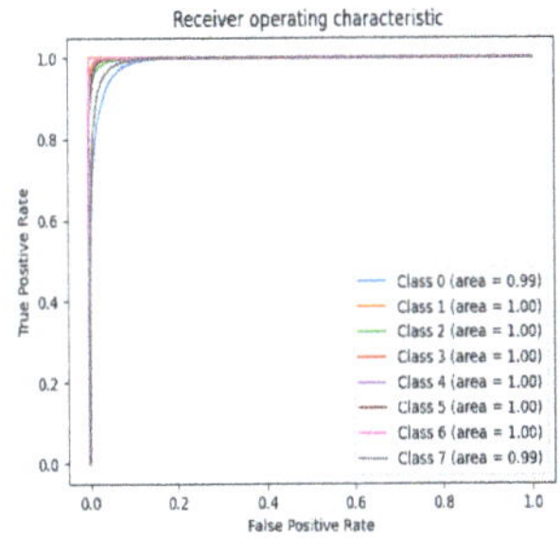

(**a**) ROC curve.

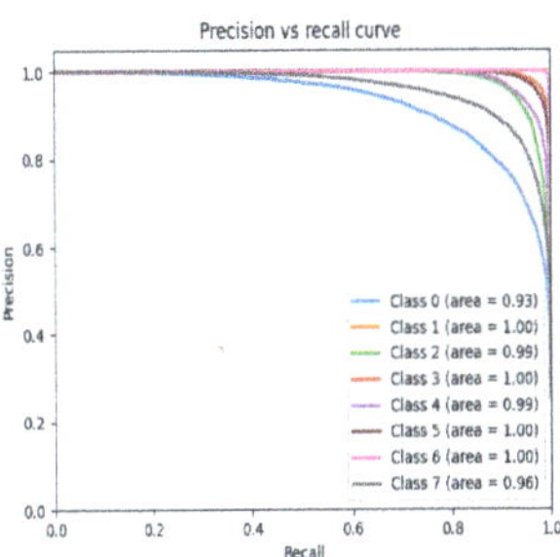

(**b**) Precision-Recall Curve.

Figure 9. ROC and Precision-Recall Curves for SWaT dataset.

Table 4 presents a comparison of the proposal presented in this is paper with other methods. In the recall and F1 score metrics, the proposed method presents a better performance related to the other methods. For values of precision and accuracy, the proposed method is above in almost all cases, except for the last two methods, which exceed it by a score margin of 0.04. However, the performance of the F1 score metric is high, indicating that a satisfactory and reliable class detection was obtained.

Table 4. Summary of the results and performance comparison on the SWaT dataset.

Method	Accuracy	Precision	Recall	F1 Score
Proposed	0.95	0.95	0.95	0.95
SVM [59]	-	0.93	0.70	0.79
RNN [59]		0.91	0.70	0.80
1D CNN [60]	-	0.96	0.80	0.87
TABOR [61]	0.95	0.86	0.79	0.82
STAE-AD [63]	-	0.96	0.82	0.88
AE [64]	-	0.89	0.80	0.84
AE Frequency [64]	-	0.92	0.83	0.87
LSTM [62]	-	0.98	0.68	0.88
One Class SVM [62]	-	0.93	0.70	0.80
SDA+1D CNN+ LSTM [42]	0.99	0.99	0.85	0.91
SDA+1D CNN+ GRU [42]	0.99	0.99	0.85	0.92

5.2. *Interconnected Tank Testbed*

This testbed has been used extensively to test proposals to detect anomalies [37,65–69]. The hydraulic system consists of three identical cylindrical tanks with equal cross-sectional area S, as shown in Figure 10. These tanks are connected by two cylindrical pipes of the same cross-sectional area, denoted S_n, and have the same outflow coefficient, denoted μ_{13} and μ_{32}. The nominal outflow located at tank 2 has the same cross-sectional area as the coupling pipe between the cylinders, but a different outflow coefficient, denoted μ_{20}. The inlet flow of the tanks comes from two pumps, with a flow rate, q_1 and q_2. A digital/analog converter is used to control each pump. A piezo-resistive differential pressure sensor carries out the necessary level measurement. The idea of the system is to maintain the height levels of the fluid stored in tanks 1 and 2 at a particular operating point.

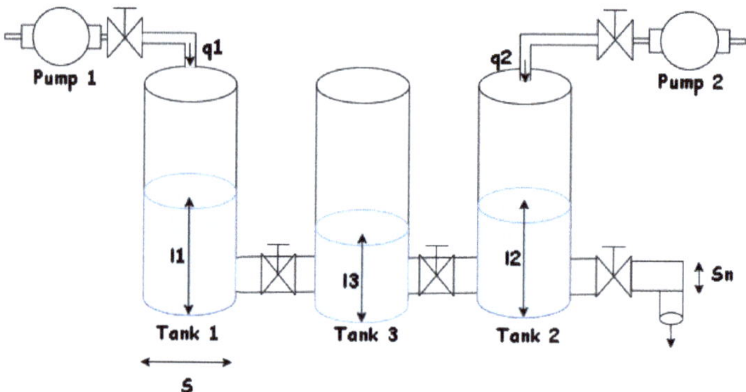

Figure 10. Schematic diagram of the three-tank system.

The parameters are shown in Table 5, and the mathematical model is presented in (15)–(17) [58].

$$\begin{aligned} \frac{dl_1(t)}{dt} &= (q_1(t) - q_{13}(t))/S \\ \frac{dl_2(t)}{dt} &= (q_2(t) + q_{32}(t) - q_{20}(t))/S, \\ \frac{dl_3(t)}{dt} &= (q_{13}(t) - q_{32}(t))/S \end{aligned} \quad (15)$$

$$q_{mn}(t) = \mu_{mn} S_p sign(l_m(t) - l_n(t)) \sqrt{2g|l_m(t) - l_n(t)|} \; (m, n = 1, 2, 3 \; \forall \; m \neq n), \quad (16)$$

$$q_{20}(t) = \mu_{20} S_p \sqrt{2g l_2(t)}. \quad (17)$$

Table 5. Parameters value of the three-tank system.

Variable	Symbol	Value
Tank cross sectional area	S	0.0154 m^2
Inter tank cross sectional area	S_n	5×10^{-5} m^2
Outflow coefficient	$\mu_{13} = \mu_{32}$	0.05
Outflow coefficient	μ_{20}	0.675
Maximum flow rate	$q_{imax}(i \in [1\ 2])$	1.2×10^{-4} m^3/s
Maximum level	$l_{jmax}(j \in [1\ 2\ 3])$	0.62

5.2.1. Dataset Generation

Assuming that $l_1 > l_2 > l_3$, a linear approximation was established around an equilibrium point (U_0, Y_0) using Taylor series expansion. The linearized system is described by a discrete state space representation with a sampling period of $T_s = 1s$. This is shown in (18).

$$\begin{aligned} x(k+1) &= Ax(k) + Bu(k) \\ y(k) &= Cx(k) \end{aligned} \quad . \quad (18)$$

The states $x(k)$ correspond to the fluid level of the tanks.

The purpose of this study is to control system around the operating point (U_0, Y_0), as is shown in (19).

$$\begin{aligned} Y_0 &= [0.4\ 0.2\ 0.3]^T (m) \\ U_0 &= [0.35 \times 10^{-4}\ 0.375 \times 10^{-4}]^T (m^3/s) \end{aligned} \tag{19}$$

A tracking control problem was considered in this study case, where the desired outputs $y = [l_1\ l_2]^T$ are required to track references. The state feedback pole assignment technique was used. Thus, a feedback gain matrix K was designed, so that the closed-loop eigenvalues of the augmented system are equal to $[0.92\ 0.97\ 0.9\ 0.95\ 0.94]$. MATLAB software was used to find the matrices A and B, as well as the controller gains. The values can be observed in (20)–(22).

$$A = \begin{bmatrix} 0.9888 & 0.0001 & 0.0112 \\ 0.0001 & 0.9781 & 0.0111 \\ 0.0112 & 0.0111 & 0.9776 \end{bmatrix}, \tag{20}$$

$$B = \begin{bmatrix} 64.5687 & 0.0014 \\ 0.0014 & 64.2202 \\ 0.3650 & 0.3637 \end{bmatrix}, \tag{21}$$

$$K = [K_1\ |K_2] = 10^{-4}\left[\begin{pmatrix} 21.6 & 3 & -5 \\ 2.9 & 19 & -4 \end{pmatrix} | \begin{pmatrix} -0.95 & -0.32 \\ -0.3 & -0.91 \end{pmatrix}\right]. \tag{22}$$

In order to construct the dataset for detecting attacks, the scheme shown in Figure 11 was implemented, which has modules to obtain measurements of the process variables, as well as the control actions applied by the actuators. An Ethernet was used as a control network. This representation is equivalent to boxes "Process" and "Controller" in the architectures presented in Figure 3.

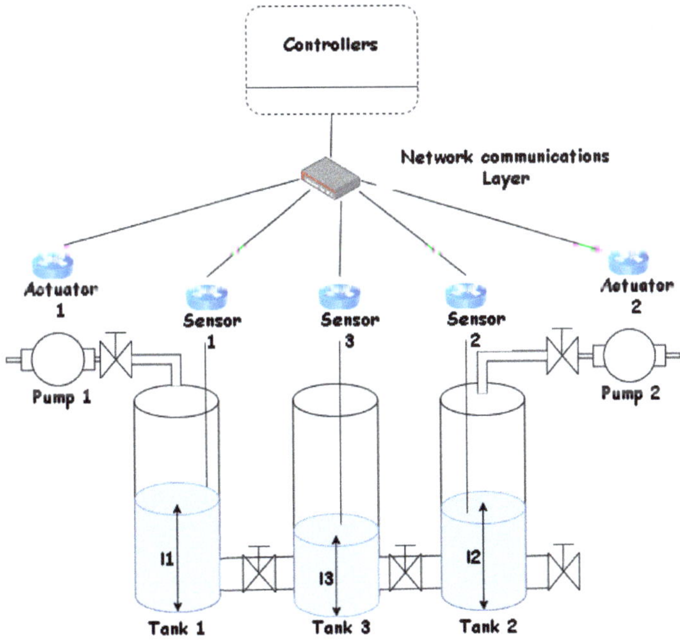

Figure 11. Interconnected tank testbed.

Two datasets were generated. The first one is a dataset in normal operations to determine a model that estimates the system outputs. The second one includes cyber attacks on sensors 1 and 2. These cyber attacks can be integrity or DoS attacks.

In both cases, 499,000 samples were generated. The system references range between 0.35 m and 0.45 m for l_1, and between 0.185 m and 0.25 m for l_2. The time intervals were defined randomly with a uniform distribution and reference changes every 500 s to 850 s.

The cases are shown in Table 6. Case 0 corresponds to operation without attacks. The following cases correspond to situations in which integrity or DoS cyber attacks can be generated on any sensor, following the models described by the Equations (3) and (4). In cases 1 to 4, only one cyber attack is generated every time, while cases 5 to 8 correspond to simultaneous attacks.

Table 6. Cases raised.

Case	Description
Case 0	Normal operation
Case 1	Integrity attack on sensor 1
Case 2	Integrity attack on sensor 2
Case 3	DoS attack on sensor 1
Case 4	DoS attack on sensor 2
Case 5	Integrity attack on sensor 1 and DoS on sensor 2
Case 6	Integrity attack on sensor 2 and DoS on sensor 1
Case 7	Integrity attack on sensor 1 and 2
Case 8	DoS attack on sensor 1 and 2

The time intervals in which cyber attacks occur were defined such that the dataset was balanced, so it were defined randomly and uniformly distributed. The integrity attacks were implemented by changing the modified variable in a range of 5% to 8% of its measured value. This range of values depends on the sensitivity of the system since there will be particular processes where the effect of the variation of the measurements in a given range does not has as much impact as in others. All cases presented correspond to the classes that the classifier will identify. The distribution of these data is shown in Figure 12.

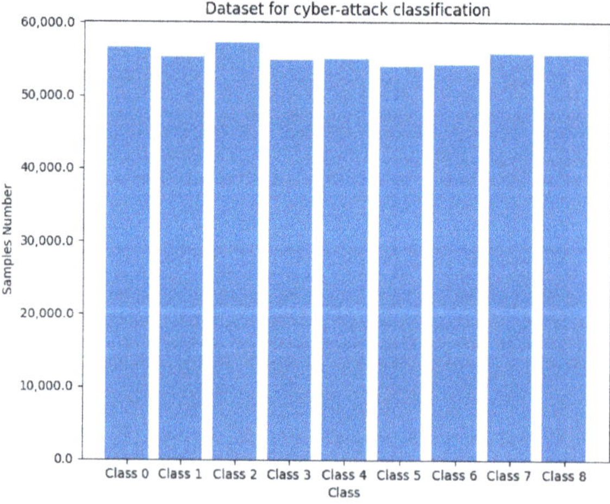

Figure 12. Dataset for cyber attack classification.

5.2.2. Model Training

Figure 3 presents the architecture implemented. The first model generates the process states estimate, while two more models were obtained to reconstruct independent states x_1 and x_2, according to those states susceptible to cyber attack.

The first network has the architecture shown in Figure 13. Its input data is composed of five characteristics, which are composed of the measurements of the sensors and the control actions corresponding to vector (23):

$$\begin{aligned} input\ data = &[x_1(1), \ldots, x_1(k-1), x_2(1), \ldots, x_2(k-1), x_3(1), \ldots, x_3(k-1), \\ &u_1(1), \ldots, u_1(k-1), u_2(1), \ldots, u_2(k-1)]^T \end{aligned} \quad (23)$$

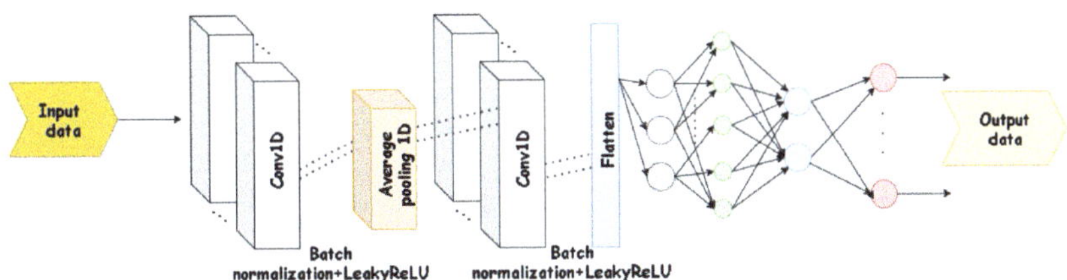

Figure 13. Model to estimate all states.

The model has three outputs corresponding to the states of the process. The vector to be reconstructed is (24):

$$\begin{aligned} output\ data\ 1 &= \hat{x}_1 = [x_1(2), \ldots, x_1(k)]^T \\ output\ data\ 2 &= \hat{x}_2 = [x_2(2), \ldots, x_2(k)]^T, \\ output\ data\ 3 &= \hat{x}_3 = [x_3(2), \ldots, x_3(k)]^T \end{aligned} \quad (24)$$

where k is the number of samples. This model has two convolutional layers, one average pooling 1D layer between the convolutional layers, and one fully connected layer. The first convolutional layer has a kernel size of 5 and has eight filters, while the second layer has a kernel size of 3 with 16 filters. Each of these layers has hyperbolic tangent activation function. Between previous layers, there is an average pooling 1D layer with a pool size of 2 and strides of 2 with same padding. Between the convolutional layers and the fully connected layer, there is a batch normalization layer with Leaky ReLU type activation function. In the fully connected layer, there is an input layer of 48 neurons and an output layer composed of 3 neurons with a linear activation function to estimate the corresponding states. The loss function used was MSE, and the optimizer used was Adam. For training, a maximum of 4 epochs and a batch size of 10 were available with initial learning rate of 0.01. To train the model, 30% of the data was used to validate, and 70% to train. The various parameters of the layers of this network were found in such a way that the lowest possible MSE will be achieved, it was 0.000067. Increasing the number of layers, neurons, filter size, or number of filters did not correspond to a significant improvement to the proposed architecture.

The second and third networks have the architecture shown in the Figure 14.

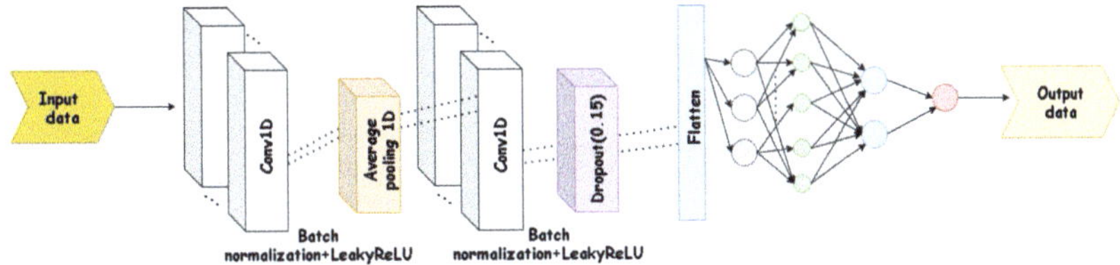

Figure 14. Model to estimate the decoupled states.

The input data for the second architecture is composed of four characteristics corresponding to the measurements of the sensors and the control actions, as is presented in vector (25):

$$\begin{aligned} \textit{input data} =& [x_2(1), \ldots, x_2(k-1), x_3(1), \ldots, x_3(k-1), \\ & u_1(1), \ldots, u_1(k-1), u_2(1), \ldots, u_2(k-1)]^{T^{\cdot}} \end{aligned} \tag{25}$$

The model generates an estimated uncoupled output for the first state as (26):

$$\textit{output data} = \hat{x}_{1d} = [x_1(2), \ldots, x_1(k)]^T, \tag{26}$$

where k is the number of samples. This model has two convolutional layers and one fully connected layer. The first convolutional layer has a kernel size of 4 and has 8 filters, while the second layer has a kernel size of 2 with 16 filters. Each of these layers has the hyperbolic tangent activation function. Between these layers, there is an average pooling 1D layer with a pool size of 2 and a stride of 2 with the same padding. Between the convolutional layers and the fully connected layer, there is a batch normalization layer and a Leaky ReLU type activation function. Before the fully connected layer, a dropout layer (0.15) was added. In the fully connected layer, there is an input layer of 32 neurons and an output layer composed of 1 neuron with linear activation function to estimate the corresponding state. The loss functions used was MSE, and the optimizer used was Adam. For training, a maximum of 4 epochs and a batch size of 10 was available with initial learn rate of 0.01. Seventy percent of the data was used to train the model, and 30% to validate it. The various parameters of the layers of this network were found in such a way that the lowest possible MSE will be achieved, and it was 0.00047. Increasing the number of layers, neurons, filter size, or number of filters did not correspond to a significant improvement to the proposed architecture.

Finally, the structure used to estimate the second uncoupled state of the process is shown in (27) and (28). The respective MSE for this case was 0.000031.

$$\begin{aligned} \textit{input data} =& [x_1(1), \ldots, x_1(k-1), x_3(1), \ldots, x_3(k-1), \\ & u_1(1), \ldots, u_1(k-1), u_2(1), \ldots, u_2(k-1)]^{T'} \end{aligned} \tag{27}$$

$$\textit{output data} = \hat{x}_{2d} = [x_2(2), \ldots, x_2(k)]^T \tag{28}$$

The architecture proposed for the classifier of the cyber attack is similar to that shown in Figure 7. It is composed of three convolutional layers whose activation function is hyperbolic tangent. The first convolutional layer has a kernel size of 15 with several 80 filters. The second and third convolutional layers have the same kernel size, but the number of filters is 60 and 30, respectively. There is also a batch normalization layer with Leaky ReLU activation function. Finally, a fully connected layer is used with an input layer of 25 neurons and an output layer with nine neurons corresponding to the established classes above. The last layer uses the softmax function. The loss function used was CCE, and the optimizer used was stochastic gradient descent with momentum. For training, a maximum of 1000 epochs was established, with a batch size of 10 and initial learning rate of 0.0001. For model training, 30% of the data was used to validate, and 70% to train. The input data is (29):

$$\textit{input data} = [x_1, x_2, x_3, \hat{x}_1, \hat{x}_2, \hat{x}_3, \hat{x}_{1d}, \hat{x}_{2d}, q_1, q_2, res, res_1, res_2]^T, \tag{29}$$

where x_1, x_2, x_3 correspond to the real variables of the process; $\hat{x}_1, \hat{x}_2, \hat{x}_3$ are the outputs estimated by the architecture shown in Figure 13; $\hat{x}_{1d}$ and $\hat{x}_{2d}$ correspond to the decoupled states estimated by the architecture of the Figure 14, q_1 and q_2 are the process references, and res, res_1, and res_2 are the residual signals obtained by comparing the real process states with the estimated states, and the individual comparison between the first two real process states and the estimated decoupled states, respectively.

Figure 15a,b present the evolution of the cost function and the accuracy metric obtained during the classifier training procedure.

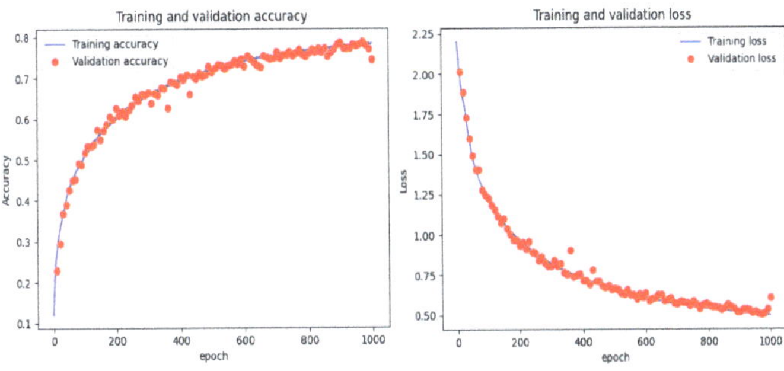

(**a**) Accuracy. (**b**) Loss function.

Figure 15. Training and Validation: accuracy and loss function through epochs.

Python programming language and the Keras library were used for training and obtaining the results. With the purpose of evaluating the performance of this architecture, the same metrics of the previous case were used.

5.2.3. Performance Analysis of the Three-Tank System Testbed

Indices values obtained for this case are presented in Figures 16 and 17a,b and Table 7. For accuracy, the best scores were obtained when simultaneous attacks happened with values above 0.97. The above is an important result because this situation has been little explored. In terms of recall, class 7 has a slightly fair score, while the other situations have scores above 0.83. Additionally, the F1 score also has high values. The scores show that the proposed architecture allows for high specificity and high sensitivity.

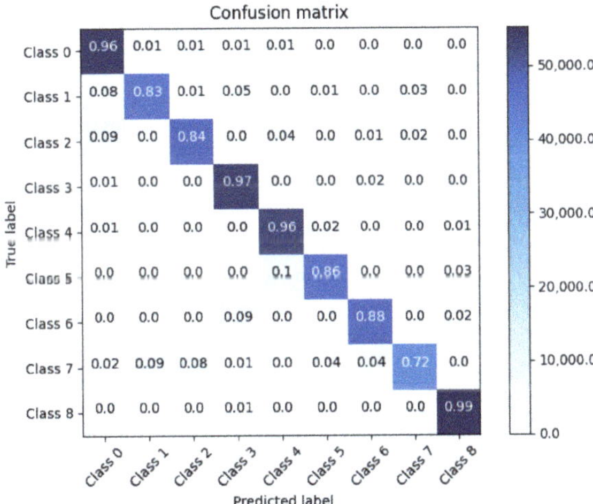

Figure 16. Confusion matrix for the three-tank system.

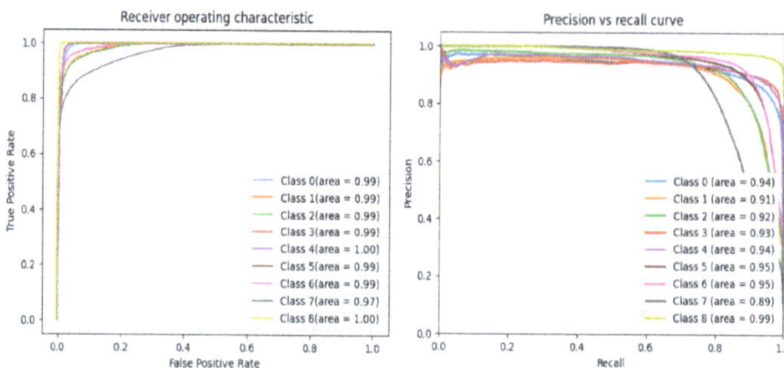

(**a**) ROC Curve. (**b**) Precision-Recall Curve.

Figure 17. ROC and Precision-Recall Curves for Interconnected tank.

Table 7. Summary of metrics.

	Accuracy	Precision	Recall	F1 Score	TNR
Class 0	0.97	0.83	0.96	0.89	0.98
Class 1	0.96	0.89	0.83	0.86	0.99
Class 2	0.97	0.89	0.84	0.87	0.99
Class 3	0.98	0.86	0.97	0.91	0.98
Class 4	0.98	0.87	0.96	0.91	0.98
Class 5	0.98	0.91	0.86	0.89	0.99
Class 6	0.98	0.92	0.88	0.90	0.99
Class 7	0.96	0.94	0.72	0.81	0.99
Class 8	0.99	0.94	0.99	0.97	0.99

The alarm indicator was implemented from the classifier in order to know the process state. Since the classifier provides the probability to classify an input data in particular class, the alarm signal is generated taking in to account the maximum value obtained from the classifier. In Figure 18, the alarm indicator is 1 when sensor 1 or 2 is under attack, and 0 when it is not. Additionally, it is discriminated if the attack is DoS or integrity type. The response of the process when it is attacked is shown in Figure 19. Boxes indicate the time instance when the attack occurs in both sensors, according to the alarm signals generated. Red boxes correspond to DoS attacks, and black boxes correspond to integrity attacks.

Additionally, the effect is different, depending on whether it is DoS or integrity attack. The system proposed in this work performed appropriately to detect the occurrence of the cyber attack, as well as the location and type of the attack. As results obtained using convolutional networks were better than those employing RNN or LSTM networks, convolutional networks were then chosen for this proposal.

In summary, the key steps for using the proposed architecture are as follows:

1. Generate an estimated output of the process under a regression model.
2. Generate a residual signal under the comparison of the measured process outputs with estimated outputs.
3. Use a classification model that from some system characteristics, such as control actions, estimated outputs, measured process outputs, and residual signals, allows evaluating if there is an attack in any part of the system.
4. From the detected class, generate alarm signals to report the occurrence of a cyber attack to define the type of attack and the part of the system that is being affected by it.

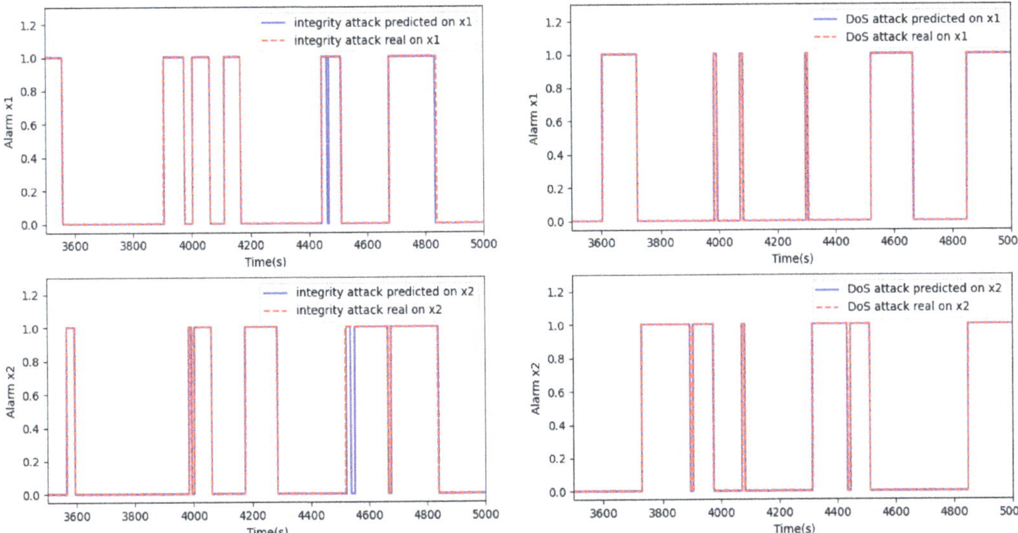

Figure 18. Alarm generation.

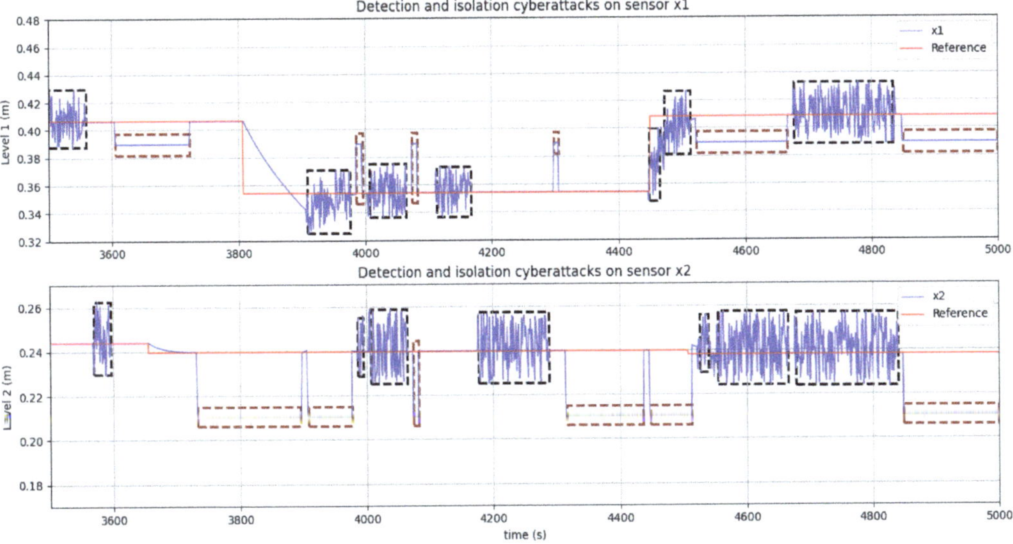

Figure 19. Temporal response.

6. Conclusions

New applications of industrial automation request great flexibility in the systems, which is supported by the increase in the interconnection between its components. At the same time, it generates a large gap that affects the security of control systems. Current solutions are oriented mainly to avoid the occurrence of attacks, but, regardless, the problems appear; so, recently, the interest in developing new proposals that contribute to detect attacks has grown.

In this work, a new architecture for DoS and integrity cyber attacks detection and isolation in Cyber Physical Systems using one-dimensional Convolutional Neural Networks was presented, thereby overcoming other models that are based on machine learning and model-based methods, such as the use of Unknown Input Observers. This architecture involves a series of steps to achieve its purpose. The first step was to generate an estimated output of the process under a regression model. The next step was to generate a residual signal under the comparison of the measured process outputs with estimated outputs. Then, a classification model was added whose input data are different characteristics, such as control actions, estimated outputs, measured process outputs, and residual signals. This model allowed for detection and isolation of different eventualities that were defined in classes. Finally, from the detected class, alarm signals were generated that are used to report the occurrence of a cyber attack, allowing to define the type of attack and the part of the system that is being affected by the attack.

The architecture proposed does not use threshold information to detect and isolate attacks, as is the case with model-based methods, such as Unknown Input Observers, which often use this information. These models require an exhaustive selection of these thresholds, which can cause both false detections and anomalous situations that go undetected, and the proposed architecture provides shows advantages over this.

The performance of the proposed architecture was validated by two test benches obtaining satisfactory results compared to other methods. The results on the SWaT dataset allowed observing that, in terms of precision and accuracy, the indexes are very close to the highest scores of other works, and these obtained a score of 0.95. In terms of recall and F1Score metrics, it presented a score of 0.95, which outperforms the previously proposed methods by a good margin. Overall, the proposed system has a high true positive rate and a low false positive rate. On the other hand, the ability of the system to be able to detect and isolate cyber attacks that may occur simultaneously is highlighted, which was presented in the three-tank system testbed. In the defined classes, the accuracy presents scores above 0.96, and the precision is above 0.83, in cases where attacks occur in a single part of the system, while the score is higher than 0.91 in cases where simultaneous attacks occur. In terms of the F1 score metric, the scores are above 0.81, which is a very promising result. Finally, with respect to the recall metric, the scores are above 0.83, in most cases. With the cases presented in this testbed, it was possible to demonstrate the ability of the proposed architecture to detect and locate attacks that may occur simultaneously. This is interesting because these types of experiments are rarely performed, let alone provide evidence of systems that can detect these types of situations, which are not alien to eventualities that may occur in reality. In both cases highlighted, there was a high rate of TNR in each of the classes, ranging between 0.98 and 0.99.

Author Contributions: Conceptualization, C.M.P.; methodology, C.M.P. and D.M.-C.; software, C.M.P.; validation, C.M.P. and D.M.-C.; formal analysis, C.M.P.; investigation C.M.P. ; resources, D.M.-C.; data curation, C.M.P.; writing—original draft preparation, C.M.P.; writing—review and editing, C.M.P, D.M.-C., V.I.-J. and A.G.-P.; visualization, D.M.-C. and A.G.-P; supervision, D.M.-C.; project administration, D.M.-C. All authors have read and agreed to the published version of the manuscript.

Funding: This research received no external funding.

Data Availability Statement: Interested partis can contact the first author about the availability of datasets.

Conflicts of Interest: The authors declare no conflict of interest.

Abbreviations

The following abbreviations have been used in this manuscript:

AE	Autoencoder
CCE	Categorical crossentropy
CPS	Cyber Physical System
CNN	Convolutional Neural Network
DBN	Deep Belief Networks
DoS	Denial of Service

FNR	False negative rate
FPR	False positive rate
GRU	Gated recurrent unit
ICS	Industrial Control System
IDS	Intrusion Detection System
LO	Luenberger Observer
LSTM	Long short-term memory
MSE	Mean squared error
PCA	Principal Component Analysis
PLC	Programmable Logic Controller
ROC	Receiver Operating Characteristics
RNN	Recurrent Neural Network
SDA	Stacked denoising autoencoder
STAE-AD	Spatio-Temporal Autoencoder for Anomaly Detection
SVM	Support Vector machine
SWaT	Secure Water Treatment
TNR	True negative rate
TPR	True positive rate
UIO	Unknown Input Observer

References

1. Vale, Z.A.; Morais, H.; Silva, M.; Ramos, C. Towards a future SCADA. In Proceedings of the 2009 IEEE Power Energy Society General Meeting, Calgary, AB, Canada, 26–30 July 2009; pp. 1–7. [CrossRef]
2. Humayed, A.; Lin, J.; Li, F.; Luo, B. Cyber-Physical Systems Security—A Survey. *IEEE Internet Things J.* **2017**, *4*, 1802–1831. [CrossRef]
3. Cardenas, A.; Amin, S.; Lin, Z.S.; Huang, Y.; Huang, C.Y.; Sastry, S. Attacks against process control systems: Risk assessment, detection, and response. In Proceedings of the 6th ACM Symposium on Information, Computer and Communications Security, Hong Kong, China, 22–24 March 2011; pp. 355–366. [CrossRef]
4. Shoukry, Y.; Chong, M.; Wakaiki, M.; Nuzzo, P.; Sangiovanni-Vincentelli, A.L.; Seshia, S.A.; Hespanha, J.P.; Tabuada, P. SMT-Based Observer Design for Cyber-Physical Systems under Sensor Attacks. In Proceedings of the 2016 ACM/IEEE 7th International Conference on Cyber-Physical Systems (ICCPS), Vienna, Austria, 11–14 April 2016; pp. 1–10. [CrossRef]
5. Fawzi, H.; Tabuada, P.; Diggavi, S. Security for control systems under sensor and actuator attacks. In Proceedings of the 2012 IEEE 51st IEEE Conference on Decision and Control (CDC), Maui, HI, USA, 10–13 December 2012; pp. 3412–3417. [CrossRef]
6. Januário, F.; Cardoso, A.; Gil, P. A Distributed Multi-Agent Framework for Resilience Enhancement in Cyber-Physical Systems. *IEEE Access* **2019**, *7*, 31342–31357. [CrossRef]
7. Challa, S.; Das, A.K.; Gope, P.; Kumar, N.; Wu, F.; Vasilakos, A.V. Design and analysis of authenticated key agreement scheme in cloud-assisted cyber–physical systems. *Future Gener. Comput. Syst.* **2020**, *108*, 1267–1286. [CrossRef]
8. Kim, K.D.; Kumar, P. An Overview and Some Challenges in Cyber-Physical Systems. *J. Indian Inst. Sci.* **2013**, *93*, 341–352.
9. Hink, R.C.B.; Goseva-Popstojanova, K. Characterization of Cyberattacks Aimed at Integrated Industrial Control and Enterprise Systems: A Case Study. In Proceedings of the 2016 IEEE 17th International Symposium on High Assurance Systems Engineering (HASE), Orlando, FL, USA, 7–9 January 2016; pp. 149–156. [CrossRef]
10. Ahmadian, M.M.; Shajari, M.; Shafiee, M.A. Industrial control system security taxonomic framework with application to a comprehensive incidents survey. *Int. J. Crit. Infrastruct. Prot.* **2020**, *29*, 100356. [CrossRef]
11. Slay, J.; Miller, M. *Lessons Learned from the Maroochy Water Breach*; Goetz, E., Shenoi, S., Eds.; Critical Infrastructure Protection; Springer: Boston, MA, USA, 2008; pp. 73–82.
12. Moore, D.; Paxson, V.; Savage, S.; Shannon, C.; Staniford, S.; Weaver, N. Inside the Slammer worm. *IEEE Secur. Priv.* **2003**, *1*, 33–39. [CrossRef]
13. Nicholson, A.; Webber, S.; Dyer, S.; Patel, T.; Janicke, H. SCADA security in the light of Cyber-Warfare. *Comput. Secur.* **2012**, *31*, 418–436. [CrossRef]
14. Daniela, T. Communication security in SCADA pipeline monitoring systems. In Proceedings of the 2011 RoEduNet International Conference 10th Edition: Networking in Education and Research, Iasi, Romania, 23–25 June 2011; pp. 1–5. [CrossRef]
15. Mitchell, R.; Chen, I.R. A Survey of Intrusion Detection Techniques for Cyber-Physical Systems. *ACM Comput. Surv. (CSUR)* **2014**, *46*. [CrossRef]
16. Deorankar, A.V.; Thakare, S.S. Survey on Anomaly Detection of (IoT)- Internet of Things Cyberattacks Using Machine Learning. In Proceedings of the 2020 Fourth International Conference on Computing Methodologies and Communication (ICCMC), Erode, India, 11–13 March 2020; pp. 115–117. [CrossRef]
17. Miao, K.; Shi, X.; Zhang, W.A. Attack signal estimation for intrusion detection in industrial control system. *Comput. Secur.* **2020**, *96*, 101926. [CrossRef]

18. Justin, V.; Marathe, N.; Dongre, N. Hybrid IDS using SVM classifier for detecting DoS attack in MANET application. In Proceedings of the 2017 International Conference on I-SMAC (IoT in Social, Mobile, Analytics and Cloud) (I-SMAC), Palladam, India, 10–11 February 2017; pp. 775–778.
19. Zhang, D.; Tang, D.; Tang, L.; Dai, R.; Chen, J.; Zhu, N. PCA-SVM-Based Approach of Detecting Low-Rate DoS Attack. In Proceedings of the 2019 IEEE 21st International Conference on High Performance Computing and Communications, Zhangjiajie, China, 10–12 August 2019; pp. 1163–1170.
20. Theissler, A. Anomaly detection in recordings from in-vehicle networks. *Big Data Appl.* **2014**, *23*, 26.
21. Karimipour, H.; Dinavahi, V. Robust Massively Parallel Dynamic State Estimation of Power Systems Against Cyber-Attack. *IEEE Access* **2018**, *6*, 2984–2995. [CrossRef]
22. Dudek, D. Collaborative detection of traffic anomalies using first order Markov chains. In Proceedings of the 2012 Ninth International Conference on Networked Sensing (INSS), Antwerp, Belgium, 11–14 June 2012; pp. 1–4.
23. Alpaño, P.V.S.; Pedrasa, J.R.I.; Atienza, R. Multilayer perceptron with binary weights and activations for intrusion detection of Cyber-Physical systems. In Proceedings of the Tencon 2017—2017 IEEE Region 10 Conference, Penang, Malaysia, 5–8 November 2017; pp. 2825–2829. [CrossRef]
24. Khan, I.A.; Pi, D.; Khan, Z.U.; Hussain, Y.; Nawaz, A. HML-IDS: A Hybrid-Multilevel Anomaly Prediction Approach for Intrusion Detection in SCADA Systems. *IEEE Access* **2019**, *7*, 89507–89521. [CrossRef]
25. Farivar, F.; Haghighi, M.S.; Jolfaei, A.; Alazab, M. Artificial Intelligence for Detection, Estimation, and Compensation of Malicious Attacks in Nonlinear Cyber-Physical Systems and Industrial IoT. *IEEE Trans. Ind. Inform.* **2020**, *16*, 2716–2725. [CrossRef]
26. Li, F.; Li, Q.; Zhang, J.; Kou, J.; Ye, J.; Song, W.; Mantooth, H.A. Detection and Diagnosis of Data Integrity Attacks in Solar Farms Based on Multilayer Long Short-Term Memory Network. *IEEE Trans. Power Electron.* **2021**, *36*, 2495–2498. [CrossRef]
27. Wan, Y.; Cao, J.; Chen, G.; Huang, W. Distributed Observer-Based Cyber-Security Control of Complex Dynamical Networks. *IEEE Trans. Circuits Syst. I Regul. Pap.* **2017**, *64*, 2966–2975. [CrossRef]
28. Pasqualetti, F.; Dörfler, F.; Bullo, F. Attack Detection and Identification in Cyber-Physical Systems. *IEEE Trans. Autom. Control* **2013**, *58*, 2715–2729. [CrossRef]
29. Do, V.L. Statistical detection and isolation of cyber-physical attacks on SCADA systems. In Proceedings of the IECON 2017—43rd Annual Conference of the IEEE Industrial Electronics Society, Beijing, China, 29 October–1 November 2017; pp. 3524–3529.
30. Lopez Rodriguez, V.M.; Cheng, A.M.K.; Doan, B. Work-in-Progress: Combining Two Security Methods to Detect Versatile Integrity Attacks in Cyber-Physical Systems. In Proceedings of the 2019 IEEE Real-Time Systems Symposium (RTSS), Hong Kong, China, 3–6 December 2019; pp. 596–599. [CrossRef]
31. Boudehenn, C.; Cexus, J.; Boudraa, A.A. A Data Extraction Method for Anomaly Detection in Naval Systems. In Proceedings of the 2020 International Conference on Cyber Situational Awareness, Data Analytics and Assessment (CyberSA), Dublin, Ireland, 15–19 June 2020; pp. 1–4. [CrossRef]
32. Cao, L.; Jiang, X.; Zhao, Y.; Wang, S.; You, D.; Xu, X. A Survey of Network Attacks on Cyber-Physical Systems. *IEEE Access* **2020**, *8*, 44219–44227. [CrossRef]
33. Zarpelão, B.B.; Miani, R.S.; Kawakani, C.T.; de Alvarenga, S.C. A survey of intrusion detection in Internet of Things. *J. Netw. Comput. Appl.* **2017**, *84*, 25–37. [CrossRef]
34. Tan, S.; Guerrero, J.M.; Xie, P.; Han, R.; Vasquez, J.C. Brief Survey on Attack Detection Methods for Cyber-Physical Systems. *IEEE Syst. J.* **2020**,*14*, 5329–5339. [CrossRef]
35. Siemens. *Primer for Cybersecurity in Industrial Automation*; Siemens: Munich, Germany, 2019.
36. Sicari, S.; Rizzardi, A.; Grieco, L.; Coen-Porisini, A. Security, privacy and trust in Internet of Things: The road ahead. *Comput. Netw.* **2015**, *76*, 146–164. [CrossRef]
37. Cómbita, L.F.; Cárdenas, A.A.; Quijano, N. Mitigation of sensor attacks on legacy industrial control systems. In Proceedings of the 2017 IEEE 3rd Colombian Conference on Automatic Control (CCAC), Cartagena, Colombia, 18–20 October 2017; pp. 1–6. [CrossRef]
38. Li, Y.; Li, J.; Luo, X.; Wang, X.; Guan, X. Cyber Attack Detection and Isolation for Smart Grids via Unknown Input Observer. In Proceedings of the 2018 37th Chinese Control Conference (CCC), Wuhan, China, 25–27 July 2018; pp. 6207–6212. [CrossRef]
39. Wang, Z.; Zhao, Y.; Yang, K.; Yao, J.; Ding, Z.; Zhang, K. UIO-based Cyber Attack Detection and Mitagation Scheme for Load Frequency Control System. In Proceedings of the 2019 3rd International Conference on Electronic Information Technology and Computer Engineering (EITCE), Xiamen, China, 18–20 October 2019; pp. 1257–1262. [CrossRef]
40. Valencia, C.M.P.; Alzate, R.E.; Castro, D.M.; Bayona, A.F.; García, D.R. Detection and isolation of DoS and integrity attacks in Cyber-Physical Microgrid System. In Proceedings of the 2019 IEEE 4th Colombian Conference on Automatic Control (CCAC), Medellin, Colombia, 15–18 October 2019; pp. 1–6. [CrossRef]
41. Shin, J.; Baek, Y.; Lee, J.; Lee, S. Cyber-Physical Attack Detection and Recovery Based on RNN in Automotive Brake Systems. *Appl. Sci.* **2019**, *9*, 82. [CrossRef]
42. Xie, X.; Wang, B.; Wan, T.; Tang, W. Multivariate Abnormal Detection for Industrial Control Systems Using 1D CNN and GRU. *IEEE Access* **2020**, *8*, 88348–88359. [CrossRef]
43. Siegel, B. Industrial Anomaly Detection: A Comparison of Unsupervised Neural Network Architectures. *IEEE Sens. Lett.* **2020**, *4*, 1–4. [CrossRef]

44. Bernieri, G.; Conti, M.; Turrin, F. Evaluation of Machine Learning Algorithms for Anomaly Detection in Industrial Networks. In Proceedings of the 2019 IEEE International Symposium on Measurements Networking (M N), Catania, Italy, 8–10 July 2019; pp. 1–6. [CrossRef]
45. Sokolov, A.N.; Pyatnitsky, I.A.; Alabugin, S.K. Research of Classical Machine Learning Methods and Deep Learning Models Effectiveness in Detecting Anomalies of Industrial Control System. In Proceedings of the 2018 Global Smart Industry Conference (GloSIC), Chelyabinsk, Russia, 13–15 November 2018; pp. 1–6. [CrossRef]
46. Elmrabit, N.; Zhou, F.; Li, F.; Zhou, H. Evaluation of Machine Learning Algorithms for Anomaly Detection. In Proceedings of the 2020 International Conference on Cyber Security and Protection of Digital Services (Cyber Security), Dublin, Ireland, 15–19 June 2020; pp. 1–8. [CrossRef]
47. Vinayakumar, R.; Alazab, M.; Soman, K.P.; Poornachandran, P.; Al-Nemrat, A.; Venkatraman, S. Deep Learning Approach for Intelligent Intrusion Detection System. *IEEE Access* **2019**, *7*, 41525–41550. [CrossRef]
48. Kim, D.E.; Gofman, M. Comparison of shallow and deep neural networks for network intrusion detection. In Proceedings of the 2018 IEEE 8th Annual Computing and Communication Workshop and Conference (CCWC), Las Vegas, Nevada, USA, 8–10 January 2018; pp. 204–208. [CrossRef]
49. Nie, J.; Ma, P.; Wang, B.; Su, Y. A Covert Network Attack Detection Method Based on LSTM. In Proceedings of the 2020 IEEE 5th Information Technology and Mechatronics Engineering Conference (ITOEC), Chongqing, China, 12–14 June 2020; pp. 1690–1693. [CrossRef]
50. Javed, A.R.; Usman, M.; Rehman, S.U.; Khan, M.U.; Haghighi, M.S. Anomaly Detection in Automated Vehicles Using Multistage Attention-Based Convolutional Neural Network. *IEEE Trans. Intell. Transp. Syst.* **2020**. [CrossRef]
51. Checkoway, S.; McCoy, D.; Kantor, B.; Anderson, D.; Shacham, H.; Savage, S.; Koscher, K.; Czeskis, A.; Roesner, F.; Kohno, T. Comprehensive Experimental Analyses of Automotive Attack Surfaces. In Proceedings of the 20th USENIX Conference on Security, San Francisco, CA, USA, 8–12 August 2011; p. 6.
52. Liagkou, V.; Kavvadas, V.; Chronopoulos, S.K.; Tafiadis, D.; Christofilakis, V.; Peppas, K.P. Attack Detection for Healthcare Monitoring Systems Using Mechanical Learning in Virtual Private Networks over Optical Transport Layer Architecture. *Computation* **2019**, *7*, 24. [CrossRef]
53. da Costa, K.A.; Papa, J.P.; Lisboa, C.O.; Munoz, R.; de Albuquerque, V.H.C. Internet of Things: A survey on machine learning-based intrusion detection approaches. *Comput. Netw.* **2019**, *151*, 147–157. [CrossRef]
54. Tabassum, A.; Erbad, A.; Guizani, M. A Survey on Recent Approaches in Intrusion Detection System in IoTs. In Proceedings of the 2019 15th International Wireless Communications Mobile Computing Conference (IWCMC), Tangier, Morocco, 24–28 June 2019; pp. 1190–1197. [CrossRef]
55. Yan, W.; Mestha, L.K.; Abbaszadeh, M. Attack Detection for Securing Cyber Physical Systems. *IEEE Internet Things J.* **2019**, *6*, 8471–8481. [CrossRef]
56. Taylor, A.; Leblanc, S.; Japkowicz, N. Anomaly Detection in Automobile Control Network Data with Long Short-Term Memory Networks. In Proceedings of the 2016 IEEE International Conference on Data Science and Advanced Analytics (DSAA), Montreal, QC, Canada, 17–19 October 2016; pp. 130–139. [CrossRef]
57. Goh, J.; Adepu, S.; Junejo, K.; Mathur, A. A Dataset to Support Research in the Design of Secure Water Treatment Systems. In *International Conference On Critical Information Infrastructures Security*; Springer: Paris, France, 2016.
58. Noura, H.; Theilliol, D.; Ponsart, J.C.; Chamseddine, A. *Fault-Tolerant Control Systems: Design and Practical Applications*; Springer Science & Business Media: Berlin/Heidelberg, Germany, 2009; doi:10.1007/978-1-84882-653-3. [CrossRef]
59. Shalyga, D.; Filonov, P.; Lavrentyev, A. Anomaly Detection for Water Treatment System based on Neural Network with Automatic Architecture Optimization. *arXiv* **2018**, arXiv:1807.07282.
60. Kravchik, M.; Shabtai, A. *Detecting Cyber Attacks in Industrial Control Systems Using Convolutional Neural Networks*; CPS-SPC '18; Association for Computing Machinery: New York, NY, USA, 2018; pp. 72–83. [CrossRef]
61. Lin, Q.; Adepu, S.; Verwer, S.; Mathur, A. TABOR: A Graphical Model-Based Approach for Anomaly Detection in Industrial Control Systems. In Proceedings of the 2018 on Asia Conference on Computer and Communications Security, Incheon Republic, Korea, 4 June 2018; pp. 525–536. [CrossRef]
62. Inoue, J.; Yamagata, Y.; Chen, Y.; Poskitt, C.M.; Sun, J. Anomaly Detection for a Water Treatment System Using Unsupervised Machine Learning. In Proceedings of the 2017 IEEE International Conference on Data Mining Workshops (ICDMW), New Orleans, LA, USA, 18–21 November 2017; pp. 1058–1065. [CrossRef]
63. Macas, M.; Wu, C. An Unsupervised Framework for Anomaly Detection in a Water Treatment System. In Proceedings of the 2019 18th IEEE International Conference On Machine Learning And Applications (ICMLA), Boca Raton, FL, USA, 16–19 December 2019; pp. 1298–1305. [CrossRef]
64. Kravchik, M.; Shabtai, A. Efficient Cyber Attacks Detection in Industrial Control Systems Using Lightweight Neural Networks. *arXiv* **2019**, arXiv:1907.01216.
65. Mousavinejad, E.; Ge, X.; Han, Q.L.; Yang, F.; Vlacic, L. Resilient Tracking Control of Networked Control Systems Under Cyber Attacks. *IEEE Trans. Cybern.* **2021**, *51*, 2107–2119. [CrossRef]
66. Bezzaoucha Rebaï, S.; Voos, H.; Darouach, M. Attack-tolerant control and observer-based trajectory tracking for Cyber-Physical Systems. *Eur. J. Control* **2019**, *47*, 30–36. [CrossRef]

67. Rebaï, S.B.; Voos, H. Chapter 13—Observer-Based Event-Triggered Attack-Tolerant Control Design for Cyber-Physical Systems. In *New Trends in Observer-Based Control*; Boubaker, O., Zhu, Q., Mahmoud, M.S., Ragot, J., Karimi, H.R., Dávila, J., Eds.; Emerging Methodologies and Applications in Modelling; Academic Press: Cambridge, MA, USA, 2019; pp. 439–462. [CrossRef]
68. Prajapati, A.K.; Roy, B. Multi-fault Diagnosis in Three Coupled Tank System using Unknown Input Observer. *IFAC-PapersOnLine* **2016**, *49*, 47–52. [CrossRef]
69. Zhang, Y.; Wang, Z.; Ma, L.; Alsaadi, F.E. Annulus-event-based fault detection, isolation and estimation for multirate time-varying systems: Applications to a three-tank system. *J. Process Control* **2019**, *75*, 48–58. [CrossRef]
70. Ge, H.; Yue, D.; Xie, X.; Deng, S.; Zhang, Y. Analysis of cyber physical systems security via networked attacks. In Proceedings of the 2017 36th Chinese Control Conference (CCC), Dalian, China, 26–28 July 2017; pp. 4266–4272.
71. Jin, X.; Haddad, W. An Adaptive Control Architecture for Leader-Follower Multiagent Systems with Stochastic Disturbances and Sensor and Actuator Attacks. In Proceedings of the 2018 Annual American Control Conference (ACC), Milwaukee, WI, USA, 27–29 June 2018; pp. 980–985.
72. Rebaï, S.; Voos, H.; Darouach, M. A contribution to cyber-security of networked control systems: An event-based control approach. In Proceedings of the 2017 3rd International Conference on Event-Based Control, Communication and Signal Processing (EBCCSP), Funchal, Portugal, 24–26 May 2017; pp. 1–7.
73. Wang, D.; Wang, Z.; Shen, B.; Alsaadi, F.E.; Hayat, T. Recent advances on filtering and control for cyber-physical systems under security and resource constraints. *J. Frankl. Inst.* **2016**, *353*, 2451–2466. [CrossRef]
74. Sridhar, S.; Manimaran, G. Data integrity attacks and their impacts on SCADA control system. In Proceedings of the IEEE PES General Meeting, Minneapolis, MN, USA, 30 September 2010; pp. 1–6.
75. Stone, M. Cross-Validatory Choice and Assessment of Statistical Predictions. *J. R. Stat. Soc. Ser. B (Methodol.)* **1974**, *36*, 111–147. [CrossRef]
76. Rousseeuw, P.; Leroy, A. Robust Regression & Outlier Detection *J. Educ. Stat.* **1988**, *13*, 358–364. [CrossRef]
77. Taormina, R.; Galelli, S.; Tippenhauer, N.O.; Salomons, E.; Ostfeld, A.; Eliades, D.G.; Aghashahi, M.; Sundararajan, R.; Pourahmadi, M.; Banks, M.K.; et al. Battle of the Attack Detection Algorithms: Disclosing Cyber Attacks on Water Distribution Networks. *J. Water Resour. Plan. Manag.* **2018**, *144*, 04018048. [CrossRef]
78. BATADAL-Datasets. Available online: http://www.batadal.net/data.html (accessed on 15 March 2021).
79. Kamycki, K.; Kapuscinski, T.; Oszust, M. Data Augmentation with Suboptimal Warping for Time-Series Classification. *Sensors* **2020**, *20*, 98. [CrossRef]

 *electronics* MDPI

Article

REFUZZ: A Remedy for Saturation in Coverage-Guided Fuzzing

Qian Lyu [1], Dalin Zhang [1,*], Rihan Da [1] and Hailong Zhang [2,*]

[1] School of Software Engineering, Beijing Jiaotong University, Beijing 100044, China ; 20126318@bjtu.edu.cn (Q.L.); 20126291@bjtu.edu.cn (R.D.)
[2] Department of Computer and Information Sciences, Fordham University, Bronx, NY 10458, USA
* Correspondence: dalin@bjtu.edu.cn (D.Z.); hzhang285@fordham.edu (H.Z.)

Abstract: Coverage-guided greybox fuzzing aims at generating random test inputs to trigger vulnerabilities in target programs while achieving high code coverage. In the process, the scale of testing gradually becomes larger and more complex, and eventually, the fuzzer runs into a saturation state where new vulnerabilities are hard to find. In this paper, we propose a fuzzer, REFUZZ, that acts as a complement to existing coverage-guided fuzzers and a remedy for saturation. This approach facilitates the generation of inputs that lead only to covered paths by omitting all other inputs, which is exactly the opposite of what existing fuzzers do. REFUZZ takes the test inputs generated from the regular saturated fuzzing process and continue to explore the target program with the goal of *preserving* the code coverage. The insight is that coverage-guided fuzzers tend to underplay already covered execution paths during fuzzing when seeking to reach new paths, causing covered paths to be examined insufficiently. In our experiments, REFUZZ discovered tens of new unique crashes that AFL failed to find, of which nine vulnerabilities were submitted and accepted to the CVE database.

Keywords: remedial testing; greybox fuzzing; vulnerability detection; enhanced security

Citation: Lyu, Q.; Zhang, D.; Da, R.; Zhang, H. REFUZZ: A Remedy for Saturation in Coverage-Guided Fuzzing. *Electronics* **2021**, *10*, 1921. https://doi.org/10.3390/electronics10161921

Academic Editor: Arman Sargolzaei

Received: 21 June 2021
Accepted: 8 August 2021
Published: 10 August 2021

1. Introduction

Software vulnerabilities are regarded as a significant threat in information security. Programming languages without a memory reclamation mechanism (such as C/C++) have the risk of memory leaks, which may expose irreparable risks [1]. With the increase in software complexity, it is impractical to reveal all abnormal software behaviors manually. Fuzz testing, or *fuzzing*, is a (semi-) automated technology to facilitate software testing. A fuzzing tool, or *fuzzer*, feeds random inputs to a target program and, meanwhile, monitors unexpected behaviors during software execution to detect vulnerabilities [2]. Among all fuzzers, coverage-guided greybox fuzzers (CGF) have become one of the most popular ones due to their high deployability and scalability, e.g., AFL [3] and LibFuzzer [4]. They have been successfully applied in practice to detect thousands of security vulnerabilities in open-source projects [5].

Coverage-guided greybox fuzzing relies on the assumption that *more run-time bugs could be revealed if more program code is executed*. To find bugs as quickly as possible, AFL and other CGFs try to maximize the code coverage. This is because a bug at a specific program location can only be triggered unless that location is covered by some test inputs. A CGF utilizes light-weight program transformation and dynamic program profiling to collect run-time coverage information. For example, AFL instruments the target program to record transitions at the basic block level. The actual fuzzing process starts with an initial corpus of seed inputs provided by users. AFL generates a new set of test inputs by randomly mutating the seeds (such as bit flipping). It then executes the program using the mutated inputs and records those that cover new execution paths. AFL continually repeats this process, but starts with the mutated inputs instead of user-provided seed inputs. If there are any program crashes and hangs, for example, caused by memory errors, AFL would also report the corresponding inputs for further analysis.

When a CGF is applied, the fuzzing process does not terminate automatically. Practically, users need to decide when to end this process. In a typical scenario, a user sets a timer for each CGF run and the CGF stops right away when the timer expires. However, researchers have discovered empirically that, within a fixed time budget, exponentially more machines are needed to discover each new vulnerability [6]. With a limited number of machines, the CGF could rapidly reach a *saturation* state in which, by continuing the fuzzing, it is difficult to find new unique crashes (where exponentially more time is needed). Then, *what can we do to improve the capability of CGF to find bugs with constraints on time and CPU power?* In this work, we try to provide one solution to this question.

Existing CGFs are biased toward test inputs that can explore new program execution paths. These inputs are prioritized in subsequent mutations. Inputs that do not discover new coverage are considered unimportant and are not selected for mutation. However, in practice, this extensive coverage-guided path exploration may *hinder the discovery of or even overlook potential vulnerabilities on specific paths*. The rationale is that an execution path in one successful run may not be bug-free in all runs. Simply dumping "bad" inputs may cause insufficient testing of their corresponding execution paths. Rather, special attention should be paid to such inputs and paths. Intuitively, an input covering a path is more likely to cover the same path after mutation than any other arbitrary inputs. Although an input cannot trigger, in one execution, the bug in its path, it is possible that the input can do so after a few fine-grained mutations. In short, by focusing on the new execution paths, the CGFs can discover an amount of vulnerabilities in a fixed time, but they also omit some vulnerabilities, which need to be repeatedly tested on the specific execution path multiple times to be found.

Based on this, we propose a lightweight extension of CGF, REFUZZ, that can effectively find tens of new crashes within a fixed amount of time on the same machines. The goal of REFUZZ is not to achieve as high code coverage as possible. Instead, it aims to *detect new unique crashes on already-covered execution paths in a limited time*. In REFUZZ, test inputs that do not explore new paths are regarded as favored. They are prioritized and mutated often to examine the same set of paths repeatedly. All other mutated inputs are omitted from execution. As a prototype, we implement REFUZZ on top of AFL. In our experiments, it successfully triggered 37, 59, and 54 new crashes in our benchmarks that were not found by AFL, using three different experimental settings, respectively. Finally, we discovered nine vulnerabilities accepted to the CVE database.

In particular, REFUZZ incorporates two stages. Firstly, in the *initial* stage, AFL is applied as usual to test the target program. The output of this stage is a set of crash reports and a corpus of mutated inputs used during fuzzing. In addition, we record the code coverage of this corpus. Secondly, in the *exploration* stage, we use the corpus and coverage from the previous stage as seed inputs and initial coverage, respectively. During the testing process, instead of rewarding inputs that cover new paths, REFUZZ only records and mutates those that converge to the initial coverage, i.e., they contribute no new coverage. To further improve the performance, we also review the validity of each mutated input before execution and promote non-deterministic mutations, if necessary. In practice, the second stage may last until the fuzzing process becomes saturated.

Note that REFUZZ is not designed to replace CGF but as a *complement* and a *remedy* for saturation during fuzzing. In fact, the original unmodified AFL is used in the initial stage. The objective of the exploration stage is to verify whether new crashes can be found on execution paths that have already been covered by AFL and whether AFL and CGFs, in general, miss potential vulnerabilities on these paths while seeking to maximize code coverage.

We make the following contributions.

- We propose an innovative idea in which, though the input cannot trigger a bug over one execution time, it is possible that the input can do so after a few fine-grained mutations.

- We propose a lightweight extension of CGF, REFUZZ, that can effectively find tens of new crashes within a fixed amount of time on the same machines.
- We develop various supporting techniques, such as reviewing the validity of each mutated input before execution, and promote non-deterministic mutations if necessary to further improve the performance.
- We propose a new mutation strategy on top of AFL. If the input does not cover a new execution path, it is regarded as valuable, which will help to cover a specific execution path over multiple times.
- We evaluate REFUZZon four real-world programs collected from prior related work [7]. It successfully triggered 37, 59 and 54 new unique crashes in the three different experimental configurations and discovered nine vulnerabilities accepted to the CVE database.

The rest of the paper is organized as follows. Section 2 introduces fuzzing and AFL, as well as a motivating example to illustrate CGFs mutating strategy limitations. Section 3 describes the design details of REFUZZ. We report the experimental results and discussion in Sections 4 and 5. Section 6 discusses the related work, and finally, Section 7 concludes our work.

2. Background

2.1. Fuzzing and AFL

Fuzzing is a process of automatic test generation and execution with the goal of finding bugs. Over the past two decades, security researchers and engineers have proposed a variety of fuzzing techniques and developed a rich set of tools that helped to find thousands of vulnerabilities (or more) [8]. Blackbox fuzzing randomly mutates test inputs and examines target programs with these inputs. Whitebox fuzzing, on the other hand, utilizes advanced, sophisticated program analyses, e.g., symbolic execution [9], to systematically exercise all possible program execution paths. Greybox fuzzing sits in between the former two techniques. The testing is guided by run-time information gathered from program execution. Due to its high scalability and ease of deployment, coverage-guided greybox fuzzing gains popularity in both the research community and industry. Specifically, AFL [3] and its derivations [10–14] have received plenty of attention.

Algorithm 1 shows the skeleton of the original AFL algorithm. (The algorithm does not distinguish between deterministic and non-deterministic—totally random mutations for simplicity.) Given a program under test and a set of initial test inputs (i.e., the seeds), AFL instruments each basic block of the program to collect *block transitions* during the program execution and runs the program with mutated inputs derived from the seeds. The generation of new test inputs is guided by the collected run-time information. More specifically, if an input contributes no crash or new coverage, it is regarded as useless and is discarded. On the other hand, if it covers new state transitions, it is added as a new entry in the queue to produce new inputs since the likelihood of these resulting inputs achieving new coverage is heuristically higher, compared to other arbitrary inputs. However, this coverage-based exploration strategy leads to strong bias toward such inputs, making already explored paths probabilistically less inspected. In our experiments, we found that these paths actually contained a substantial number of vulnerabilities, causing programs to crash.

AFL mutates an input at both a coarse-grained level, which incorporates the changing bulks of bytes, and a fine-grained level, which involves byte-level modifications, insertions and deletions [15]. In addition, AFL adopts two strategies to apply the mutation, i.e., deterministic mutation and random mutation. In fuzzing, AFL maintains a seed queue that stores the *initial* test seeds provided by users and new test cases screened by the fuzzer. For one input in the seed queue, which has applied deterministic mutations, it will no longer be mutated through deterministic mutation in subsequent fuzzing. The deterministic mutation, including bitflip, arithmetic, interest, and dictionary methods, is one in which a new input is obtained by modifying the content of the input at a specific byte position and every input is mutated in the same way. In particular, during the interest and dictionary mutation stages, some special contents and tokens automatically generated or provided by users are replaced or inserted into the original input. On the contrary, the havoc and splice called random mutations would always be applied until the fuzzing stops. In the havoc stage, a random number is generated as the mutation combination number. According to the number, one random mutation method is selected each time, and then applied to the file in turn. In the next stage, called splice, a new input is produced by splicing two seed inputs, and the havoc mutation is continued on the file.

Algorithm 1: ORIGINALAFL

```
Input: The target program P; the initial set of seed inputs initSeeds.
queue ← initSeeds
crashes ← ∅
while in fuzzing loop do
    foreach input ∈ queue do
        foreach mutation ∈ allMutations do
            newInput ← MUTATE(input, mutation)
            result ← RUN(P, newInput)
            if CRASH(result) then
                crashes ← crashes ∪ {result}
            else if NEWCOVERAGE(result) then
                queue ← queue ∪ {newInput}
            end
        end
    end
end
return queue, crashes
```

Note that AFL is unaware of the structure of inputs. For example, it is possible that a MP3 file is generated from a PDF file because the magic number is changed by AFL. It is inefficient to test a PDF reader with a MP3 file since the execution will presumably terminate early, as the PDF parser does not accept non-PDF files, causing the major components not to be tested. Our implementation of REFUZZ tackles this problem by adding an extra check of validness of newly generated test inputs, as discussed in Section 3.

2.2. Motivating Example

Figure 1a shows a code snippet derived from the pdffonts program, which analyzes and lists the fonts used in a Portable Document Format (PDF) file. Class `Dict` defined at line 1–10 stores an array of entries. Developers can call the `find` function defined at line 12 to retrieve the corresponding entry by a key. In the experiments, we test this program by running both AFL and REFUZZ with the AddressSanitizer [16] to detect memory errors. Figure 1b shows the crashing trace caused by a heap buffer overflow error found only by REFUZZ. The crash is caused by accessing the `entries` array during the iteration at line 14–17 in Figure 1a. The root cause of this error is inappropriate destruction of the dictionary in the `XRef` and `Object` classes when pdffonts attempts to reconstruct the *cross-reference table* (xref for short, which internally uses a dictionary) for locating objects in the PDF file, e.g., bookmarks and annotations. The crash is triggered when the xref table of the test input is mostly valid (including the most important entries, such as "Root", "size", "Info", and "ID") but cannot pass the extra check to investigate whether the PDF file is encrypted. When the program issues a search of key "Encrypt", the dictionary has already been destructed by a previous query that checks for the validness of the xref table. A correct implementation should make a copy of the dictionary after the initial check.

```
class Dict {
public:
  ...
private:
  XRef *xref; // the xref table for this PDF file
  DictEntry *entries; //array of entries
  int length; //number of entries in dictionary
  ...
  DictEntry *find(char *key);
};
...
inline DictEntry *Dict::find(char *key){
  int i;
  for (i = 0; i < length; ++i) {
    if (!strcmp(key, entries[i].key))
      return &entries[i];
  }
  return NULL;
}
...
```

(**a**) Code derived from pdffonts

Function	File and Line
`main`	/Xpdf/pdffonts.cc:117
`PDFDoc::PDFDoc`	/Xpdf/PDFDoc.cc:96
`PDFDoc::setup`	/Xpdf/PDFDoc.cc:120
`XRef::XRef`	/Xpdf/XRef.cc:107
`XRef::checkEncrypted`	/Xpdf/XRef.cc:459
`Object::dictLookup`	/Xpdf/Object.h:252
`Dict::lookup`	/Xpdf/Dict.cc:72
`Dict::find`	/Xpdf/Dict.cc:56

(**b**) The crashing trace caused by a heap buffer overflow

Figure 1. The motivating example.

It is relatively expensive to find this vulnerability using AFL, compared to REFUZZ. In our experiments, by running AFL for 80 h, AFL failed to trigger this bug, even with the help of the AddressSanitizer tool. The major reason is that the check for validness of xref and the check for encryption of the PDF file are the first step when pdffonts parses an arbitrary file—that is, they are presumably regarded as "old" paths for most cases. When using AFL, if a test input does not cover a new execution path, the chance of mutating this input is low. In other words, the execution path covered by the input is less likely to be covered again (or is covered but by less "interesting" inputs) and the examination of the the two checks might not be enough to reveal subtle bugs, such as the one in Figure 1b.

To tackle this problem, REFUZZ does not aim at high code coverage. On the contrary, we want to detect new vulnerabilities residing in covered paths and to verify that AFL ignores possible crashes in such paths while paying attention to coverage. REFUZZ utilizes the corpus obtained in the initial stage (which runs the original AFL) as the seeds for the exploration stage. It only generates test inputs that linger on the execution paths that are covered in the first stage but not investigated sufficiently. In the next section, we provide more details about the design of REFUZZ.

3. Design of REFUZZ

3.1. Overview

We propose REFUZZ to further test the program under test with inputs generated by AFL to trigger unique crashes that were missed by AFL. REFUZZ consists of two stages, i.e., the *initial* stage and the *exploration* stage. In the initial stage, the original AFL is applied. The initial seed inputs are provided by the user. The output is an updated seed queue, including both the seed inputs and the test inputs covered new execution paths during fuzzing. In the exploration stage, REFUZZ uses this queue as the initial seed input, applying a novel mutation strategy designed for investigating previously executed paths to generate new test inputs. Moreover, only inputs that passed the extra format check are added to the seed queue and participate in subsequent mutations and testing. Figure 2 shows the workflow of REFUZZ.

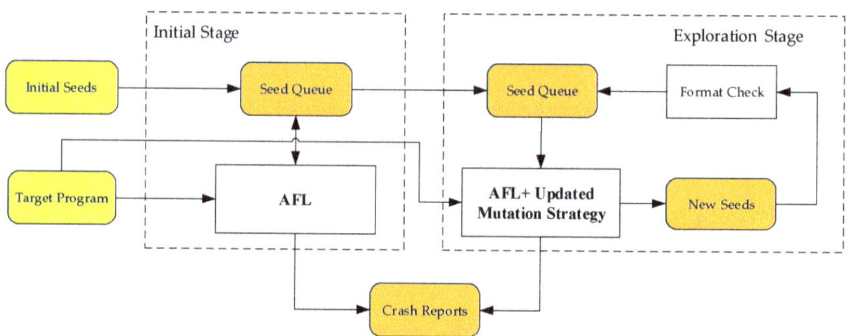

Figure 2. REFUZZ overview.

Algorithm 2 depicts the algorithmic sketch of REFUZZ. (Our implementation skips duplicate deterministic mutations of inputs in the MUTATE function.) The highlighted lines are new, compared to the original AFL algorithm. The REFUZZ algorithm takes two additional parameters besides *P* and *initSeeds*: *et*, the time allowed for the initial stage, and *ct*, the time limit for performing deterministic mutations. We discuss *ct* in the next subsection. At line 6 in Algorithm 2, when the elapsed time is less than *et*, REFUZZ is in the initial stage, and the original AFL algorithm is applied. When the elapsed time is greater than or equal to *et* (line 8–24), the testing enters the exploration stage. REFUZZ uses in this stage the input corpus queue obtained in the initial stage and applies a novel mutation strategy to generate new test inputs. If a new input passes the format check, it would be fed to the target program. The input that preserved the code coverage (i.e., did not trigger new paths) would be added to the queue. In the experiments, we set *et* to various values to evaluate the effectiveness of REFUZZ under different settings.

3.2. Mutation Strategy in Exploration Stage

REFUZZ adopts the same set of mutation operators as in AFL, including bitflip, arithmetic, value overwrite, injection of dictionary terms, havoc, and splice. The first four methods are *deterministic* because of their slight destructiveness to the seed inputs. The latter two methods will significantly damage the structure of an input, which are *totally random*. To facilitate the discovery of crashes, as shown in Algorithm 2, we introduce a parameter *ct* to limit the time since the last crash during the fuzzing process for deterministic mutations. If an input is undergoing deterministic mutation operations and no new crashes are found for a long time (>*ct*), REFUZZ will skip the current mutation operation and perform the next random mutation (line 11 of Algorithm 2). In the experiments, we initialize *ct* to 60 min and set it incrementally for each deterministic mutation. Specifically, the *n*-th deterministic mutation is skipped if there no crash is triggered in the past *n* hours by mutating an input. REFUZZ will try other more destructive mutations to facilitate the efficiency of fuzzing.

As introduced in Section 1, REFUZZ does not aim at high code coverage. Instead, it generates inputs that converge to the existing execution paths. During the initial stage, AFL saves the test inputs that have explored new execution paths in the input queue. An execution path consists of a series of *tuples*, where each tuple records the run-time transition between two basic blocks in the program code. A path is new when the input results in (1) the generation of new tuples or (2) changing of the *hit count* (i.e., the frequency) of an existing tuple. Instead, the PRESERVECOVERAGE function in Algorithm 2 checks whether new tuples are covered and returns false if this is the case. It returns true if any hit count along a path is updated. We add test inputs that preserves the coverage into the queue to participate in the next round of mutation as seeds. Using this mutation strategy, REFUZZ can effectively attack specific code areas that have been covered but are not well-tested and find vulnerabilities.

Algorithm 2: REFUZZ

Input: The target program *P*; the initial set of seed inputs *initSeeds*; the time to enter the exploration stage *et*; the time limit for performing deterministic mutations *ct*.

```
queue ← initSeeds
crashes ← ∅
lastCrash ← 0
while in fuzzing loop do
    if elapsedTime < et then                                        // Initial stage
        queue, crashes ← ORIGINALAFL(P, queue)
    else                                                            // Exploration stage
        foreach input ∈ queue do
            foreach mut ∈ allMutations do
                if ct < elapsedTime − lastCrash ∧ ISDETERMINISTIC(mut) then
                    continue
                end
                newInput ← MUTATE(input, mut)
                if FORMATCHECK(newInput) then
                    result ← RUN(P, newInput)
                    if CRASH(result) then
                        crashes ← crashes ∪ {result}
                        lastCrash ← elapsedTime
                    else if PRESERVECOVERAGE(result) then
                        queue ← queue ∪ {newInput}
                    end
                end
            end
        end
    end
end
return queue, crashes
```

3.3. Input Format Checking

Blindly feeding random test inputs to the target program leads to low performance of the fuzzer since they are likely to fail the initial input validation [8]. For instance, it is better to run a audio processing program with a MP3 file instead of an arbitrary file. Since AFL is unaware of the expected input format for each program under test, it is usual that the structure of an input is changed by random mutation operations. We propose to add an extra, light-weight format check before each program run to reduce the unnecessary overhead caused by invalid test inputs. As an exemplar, in the experiments, we check whether each input is a PDF file when testing a PDF reader and discard those that do not conform to the PDF format during testing. Specifically, in our implementation, REFUZZ takes an extra command-line argument, indicating the expected format of inputs. For each mutated input, REFUZZ checks the magic number of each input file and only adds it to the queue for further mutation if it passes the check.

4. Evaluation

4.1. Experimental Setup

To empirically evaluate the REFUZZ and its performance in finding vulnerabilities, we implement REFUZZ on top of AFL and conduct experiments on a Ubuntu V16.04.6 LTS machine with 16-core Xeon E7 2.10 GHz CPU and 32 GB RAM, using 4 programs that were also used by prior, related work [7] . Table 1 shows the details of the subjects used in our experiments. Columns "Program" and "Version" show the program names and versions. Columns "#Files" and "#LOC" list the number of files and lines of code in each program, respectively.

Table 1. Experimental subjects.

Program	Version	#Files	#LOC
pdftotext	xpdf-2.00	133	51,399
pdftopbm	xpdf-2.00	133	51,399
pdffonts	xpdf-2.00	133	51,399
MP3Gain	1.5.2	39	8701

4.2. Vulnerability Discovery

A crucial factor in evaluating a fuzzer's performance is its ability to detect vulnerabilities. We configure REFUZZ to run three different experiments for 80 h with identical initial corpus by modifying et al. Table 2 describes the time for the initial stage and the exploration stage. In the first stage, the original AFL is applied without the additional test input format checking. Then, REFUZZ takes the input queue as the initial corpus for the second stage and uses an extra parameter to pass the expected input type to the target program, e.g., PDF.

Table 2. Experimental setup.

#	Init Time	Expl Time	Total
1	60 h	20 h	80 h
2	50 h	30 h	80 h
3	40 h	40 h	80 h
4	80 h	0 h	80 h

During the fuzzing process, the fuzzer records the information of each program crash along with the input that caused the crash. To avoid duplicates in the results, we use the *afl-cmin* [17] tool in the AFL toolset to minimize the final reports by eliminating redundant crashes and inputs. Tables 3–5 show the statistics of unique crashes triggered by REFUZZ. Note that the numbers in column "Init+Expl" are not exactly the sum of the numbers in columns "Init" and "Expl". This is because REFUZZ discovers duplicate crashes in the initial stage and the exploration stage. Additionally, the numbers in column "New" are discovered by REFUZZ but not AFL. After applying afl-cmin, only the unique crashes are reported.

We also run AFL for 80 h and report the number of crashes in Table 6. The total number in column "Init" is less than the number in column "Init+Expl" in Tables 3 and 5. This indicates that REFUZZ can find more unique crashes within 80 h. In Table 4, the data in column "Init" are much fewer than the other two experimental configurations, so they are fewer than the total number of crashes in Table 6. As described in Table 7, we compare the average and variance data of the unique crashes obtained though the four programs under three different experimental configurations. The data in column "Variance" have a large deviation, which reflects the randomness of the fuzzing.

From the Tables 3–5, we can see that new unique crashes are detected during the exploration stage in all three experimental settings, except for pdftopbm, which has 0 new crashes, shown in Table 4. By applying the novel mutation strategy in the exploration stage and input format checking, REFUZZ discovers 37, 59, and 54 new unique crashes that are not discovered by AFL. These crashes are hard to find if we simply focus on achieving high code coverage since they reside in already covered paths and are not examined sufficiently with various inputs in that some vulnerabilities are detected by relying on plenty of special-type inputs.

Table 3. Number of unique crashes (60 + 20).

Program	REFUZZ			New
	Init	Expl	Init+Expl	
pdftotext	29	4	30	1
pdftopbm	33	14	40	7
pdffonts	154	74	164	10
MP3Gain	92	55	111	19
Total	308	147	345	37

Table 4. Number of unique crashes (50 + 30).

Program	REFUZZ			New
	Init	Expl	Init+Expl	
pdftotext	11	9	18	7
pdftopbm	8	1	8	0
pdffonts	153	81	164	11
MP3Gain	74	76	115	41
Total	246	167	305	59

Table 5. Number of unique crashes (40 + 40).

Program	REFUZZ			New
	Init	Expl	Init+Expl	
pdftotext	22	6	25	3
pdftopbm	32	25	41	9
pdffonts	148	88	164	16
MP3Gain	101	61	127	26
Total	303	180	357	54

Table 6. Number of unique crashes (80 + 0).

Program	Init
pdftotext	35
pdftopbm	39
pdffonts	171
MP3Gain	96
Total	341

Table 7. Average and variance of unique crashes.

Program	60 + 20	50 + 30	40 + 40	Average	Variance
pdftotext	30	18	25	24.3	24.2
pdftopbm	40	8	41	29.7	234.9
pdffonts	164	164	164	164	0
MP3Gain	111	115	127	117.7	46.2

Figure 3 shows the proportion of newly discovered unique crashes among all crashes that are triggered by REFUZZ in the exploration stage. For example, for pdftotext, the number of new unique crashes is greater than half of the total number of unique crashes (in the "40 + 40" setting). We can see that by preserving the code coverage and examining covered execution paths more, we can discover a relatively large number of new vulnerabilities that might be neglected by regular CGF, such as AFL. Note that this does not mean that AFL and others cannot find such vulnerabilities. It just implies that they have a lower chance of finding the vulnerabilities within a fixed amount of time, while REFUZZ is more likely to trigger these vulnerabilities, given the same amount of time.

In addition, we set up 12 extra experiments. The corpus obtained by running AFL for 80 h is used as the initial input of the exploration stage; then, the target programs are tested by REFUZZ for 16 h. The purpose is to verify whether REFUZZ can always find new unique crashes when AFL is saturated. The experimental data are recorded in Table 8. The column "Number of experiments" records the number of new unique crashes found by REFUZZ in 12 experiments. It can be proved that when the same initial inputs are provided, REFUZZ can always find new crashes that are not repeated with AFL, even though the fuzzing is random.

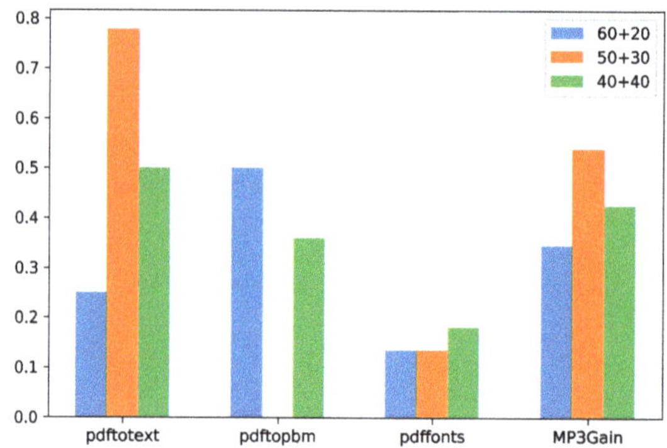

Figure 3. Proportion of newly discovered unique crashes in the exploration stage of REFUZZ.

Table 8. Number of new unique crashes (80 + 16).

Program	Number of Experiments												Average
	1	2	3	4	5	6	7	8	9	10	11	12	
pdftotext	1	7	3	2	4	4	0	0	3	5	7	4	3.3
pdftopbm	3	6	9	2	2	5	5	2	3	4	8	3	4.3
pdffonts	18	15	19	11	8	13	11	18	19	19	7	22	15
MP3Gain	1	6	15	8	7	13	7	7	8	22	14	8	9.7

We have submitted our findings in the target programs to the CVE database. Table 9 shows a summary of nine new vulnerabilities that were found by REFUZZ in our experiments. We are working on analyzing the rest crashes and will release more details in the future.

4.3. Code Coverage

As described earlier, the goal of REFUZZ is to test whether new and unique crashes can be discovered on covered paths after regular fuzzing in a limited time, instead of aiming at high code coverage. We collected the code coverage information during the execution of REFUZZ and found that the coverage for each target program remained the same during the exploration stage, which is to be expected. The results also show that

AFL only achieved slightly higher coverage compared to REFUZZ in the exploration stage, which implies that AFL ran into a saturation state, which signifies a demand for new strategies to circumvent such scenarios. REFUZZ is one such remedy, and our experimental results show its effectiveness in finding new crashes.

Table 9. Submitted vulnerabilities.

ID	Description
CVE-2020-25007	double-free vulnerability that may allow an attacker to execute arbitrary code
CVE-2020-27803	double-free on a certain position in thread
CVE-2020-27805	heap buffer access overflow in XRef::constructXRef() in XRef.cc
CVE-2020-27806	SIGSEGV in function scanFont in pdffonts.cc
CVE-2020-27807	heap-buffer-overflow in function Dict::find(char *) in Dict.cc
CVE-2020-27808	heap-buffer-overflow in function Object::fetch(XRef *, Object *) in Object.cc
CVE-2020-27809	SIGSEGV in function XRef::getStreamEnd(unsigned int, unsigned int *) in XRef.cc
CVE-2020-27810	heap-buffer-overflow in function Dict::find(char *) in Dict.cc
CVE-2020-27811	SIGSEGV in function XRef::readXRef(unsigned int *) in XRef.cc

5. Discussion

REFUZZ is effective at finding new unique crashes that are hard to discover using AFL. This is because some execution paths need to be examined multiple times with different inputs to find hidden vulnerabilities. The coverage-first strategy in AFL and other CGFs tends to overlook executed paths, which may hinder further investigation of such paths. However, such questions as "when should we stop the initial stage in REFUZZ and enter the exploration stage to start the examination of these paths", and "how long should we spend in the exploration stage of REFUZZ" remain unanswered.

How long should the initial stage take? As described in Section 4, we performed three different experiments with *et* set to 60, 50, and 40 h to gather empirical results. The intuition is that the effect of using the original AFL to find bugs would be the best when *et* is 60 h since it is to be expected that more paths could be covered and more unique crashes could be triggered if we apply the fuzzer for a longer time in the initial stage. However, our experimental results in Tables 3–5 show that the fuzzing process is unpredictable. The total number of unique crashes triggered in the initial stage of 60 h is close to 40 h (308 vs. 303), while the number obtained in 50 h is less than that of 40 h (246 vs. 303). In Algorithm 2, as well as our implementation of the algorithm, we allow the user to decide when to stop the initial stage and set *et* based on their experience and experiments. Generally, regarding the appropriate length of the initial stage, we suggest that users should pay attention to the dynamic data in the fuzzer dashboard. The code coverage remains stable, the color of the cycle numbers (*cycles done*) transforms from purple to green, or the last discovered unique crashes (*last uniq crash time*) have passed a long time, which indicates that continuing to test will not bring new discoveries. The best rigorous method is to combine these pieces of reference information to determine whether the initial stage should be paused.

How long should the exploration stage take? We conducted an extra experiment using REFUZZ with the corpus obtained from the 80-h run of AFL. We ran REFUZZ for 16 h and recorded the number of unique crashes per hour. In the experiment, each program was executed with REFUZZ for 12 trials. The raw results are shown in Figure 4 and the mean of the 12 trials are shown in Figure 5. In both figures, the x-axes show the number of bugs (i.e., unique crashes) and the y-axes show the execution time in hours. We can see that given a fixed corpus of seed inputs, the performance of REFUZZ in the exploration stage varies a lot in the 12 trials. This is due to the nature of random mutations. Overall, we can see from the figures that in the exploration stage, REFUZZ follows the empirical rule that finding a new vulnerability requires exponentially more time [6]. However, this does not negate the effectiveness of REFUZZ in finding new crashes. We suggest that the best test time to terminate the remedial testing is still when the exploration reaches saturation, and the relevant guidelines at the initial stage can be considered here.

Is REFUZZ effective as remedy for CGF? Many researchers have proposed remedial measures to CGFs. Driller [18] combines fuzzing and symbolic execution. When a fuzzer becomes stuck, symbolic execution can calculate the valid input to explore deeper bugs. T-Fuzz [19] detects whenever a baseline mutational fuzzer becomes stuck and no longer produces inputs that extend the coverage. Then, it produces inputs that trigger deep program paths and, therefore, find vulnerabilities (hidden bugs) in the program. The main cause of the saturation is due to the fact that AFL and other CGFs strongly rely on random mutation to generate new inputs to reach more execution paths. Our experimental results suggest that new unique crashes can actually be discovered if we leave code coverage aside and continue to examine the already covered execution paths by applying mutations (as shown in Tables 3–5). They also show that it is feasible and effective to use our approach as a remedy and an extension to AFL, which can easily be applied to other existing CGFs. While this conclusion may not hold for programs that we did not use in the experiments, our evaluation shows the potential of remedial testing based on re-evaluation of covered paths.

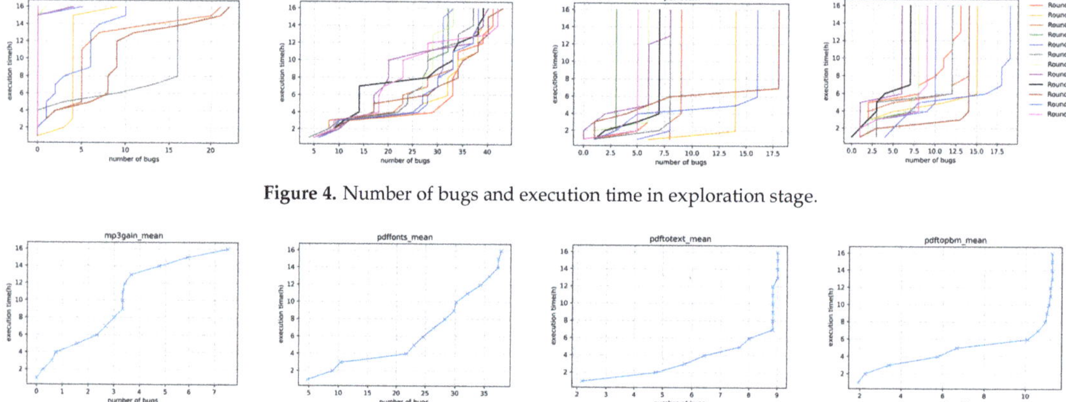

Figure 4. Number of bugs and execution time in exploration stage.

Figure 5. Average number of bugs and execution time in exploration stage.

6. Related Work

The mutation-based fuzzer uses actual inputs to continuously mutate the test cases in the corpus during the fuzzing process, and continuously feeds the target program. The code coverage is used as the key to measure the performance of the fuzzer. AFL [3] uses compile-time instrumentation and genetic algorithms to find interesting test cases, and can find new edge coverage based on these inputs. VUzzer [20] uses the "intelligent" mutation strategy based on data flow and control flow to generate high-quality inputs through the result feedback and by optimizing the input generation process. The experiments show that it can effectively speed up the mining efficiency and increase the depth of mining. FairFuzz [21] increases the coverage of AFL by identifying branches (rare branches) performed by a small amount of input generated by AFL and by using mutation mask creation algorithms to make mutations that tend to generate inputs that hit specific rare branches. AFLFast [12] proposes a strategy to make AFL geared toward the low-frequency path, providing more opportunities to the low-frequency path, which can effectively increase the coverage of AFL. LibFuzzer [4] uses SanitizerCoverage [22] to track basic block coverage information in order to generate more test cases that can cover new basic blocks. Sun et al. [23] proposed to use the ant colony algorithm to control seed inputs screening in greybox fuzzing. By estimating the transition rate between basic blocks, we can determine which the seed input is more likely to be mutated. PerfFuzz [24] generates inputs through feedback-oriented mutation fuzzing generation, can find various inputs with different hot spots in the program, and escapes local maximums to have higher execution path length inputs. SPFuzzs [25] implement three mutation strategies, namely, head, content and sequence mutation strategies. They cover more paths by driving the fuzzing process, and provide a method of randomly assigning weights through messages and strategies. By continuously updating and improving the mutation strategy, the above research effectively improves the efficiency of fuzzing. As far as we know, in our experiment, if there are no new crashes for a long time (>*ct*), and it is undergoing the deterministic mutation operations at present, then it performs the next deterministic mutation or to enter the random mutation stage directly, which reduces unnecessary time consumption to a certain extent.

The generation-based fuzzer is significant for having a good understanding of the file format and interface specification of the target program. By establishing the model of the file format and interface specification, the fuzzer generates test cases according to the model. Dam et al. [26] established the Long Short-Term memory model based on deep learning, which automatically learns the semantic and grammatical features in the code, and proves that its predictive ability is better than the state-of-the-art vulnerability prediction models. Reddy et al. [27] proposed a reinforcement learning method to solve the diversification guidance problem, and used the most advanced testing tools to evaluate the ability of RLCheck. Godefroid et al. [28] proposed a machine learning technology based on neural networks to automatically generate grammatically test cases. AFL++ [29] provides a variety of novel functions that can extend the blurring process over multiple stages. With it, variants of specific targets can also be written by experienced security testers. Fioraldi et al. [30] proposed a new technique that can generate and mutate inputs automatically for the binary format of unknown basic blocks. This technique enables the input to meet the characteristics of certain formats during the initial analysis phase and enables deeper path access. You et al. [31] proposed a new fuzzy technology, which can generate effective seed inputs based on AFL to detect the validity of the input and record the input corresponding to this type of inspection. PMFuzz [32] automatically generates high-value test cases to detect crash consistency bugs in persistent memory (PM) programs. These efforts use syntax or semantic learning techniques to generate legitimate inputs. However, our work is not limited to using input format checking to screen legitimate inputs during the testing process, and we can obtain high coverage in a short time by using the corpus obtained by AFL test as the initial corpus in the exploration phase. Symbolic execution is an extremely effective software testing method that can generate inputs [33–35]. Symbolic execution can analyze the program to obtain input for the execution of a specific code area. In other words, when using symbolic execution to analyze a program, the program uses symbolic values as input instead of the specific values used in the general program execution. Symbolic execution is a heavyweight software testing method because the possible input of the analysis program needs to be able to obtain the support of the target source code. SAFL [36] is augmented with qualified seed generation and efficient coverage-directed mutation. Symbolic execution is used in a lightweight approach to generate qualified initial seeds. Valuable exploration directions are learned from the seeds to reach deep paths in program state space earlier and easier. However, for large software projects, it takes a lot of time to analyze the target source code. As ReFuzz is a lightweight extension of AFL, in order to be able to repeatedly reach the existing execution path, we choose to add the test that fails to generate a new path to the execution corpus to participate in subsequent mutations.

7. Conclusions

This paper designs and implements a remedy for saturation during greybox fuzzing, called ReFuzz. Using the corpus of the initial stage as the seed test inputs of the exploration stage, ReFuzz can explore the same set of execution paths extensively to find new and unique crashes along those paths within a limited time. The AFL directly feeds the input obtained by the mutation into the target program for running, which causes many noncompliant seeds to be unable to explore deeper paths. In this paper, we proposed an input format checking algorithm that can filter the file conformed to the input format, which is beneficial to enhance the coverage depth of the execution path. At the same time, the mutation strategy we proposed can transition to the random mutation stage to continue testing when the deterministic mutation stage is stuck, which significantly accelerates the testing efficiency of fuzzing. We evaluated REFUZZ , using programs from prior related work. The experimental results show that REFUZZ can find new unique crashes that account for a large portion among the total unique crashes. Specifically, we discovered and submitted nine new vulnerabilities in the experimental subjects to the CVE database. We are in the process of analyzing and reporting more bugs to the developers.

In the future, in order to make our prototype tool better serviced in the real world, we will study how to combine machine learning to improve the efficiency of input format checking and design more complex automatic saturation strategies to strengthen the linkability of the tool. We will continue to improve REFUZZ to help increase the efficiency of fuzzers in the saturation state using parallel mode and deep reinforcement learning. We are planning to develop more corresponding interfaces and drivers to explore more vulnerabilities of IoT terminals for enhanced security of critical infrastructures.

Author Contributions: Conceptualization, D.Z. and H.Z.; methodology, D.Z.; software, Q.L.; validation, R.D.; writing—original draft preparation, Q.L. and H.Z.; writing—review and editing, Q.L.; supervision, D.Z. All authors have read and agreed to the published version of the manuscript.

Funding: This research was funded by the Fundamental Research Funds for the Central Universities through the project(No.2021QY010).

Conflicts of Interest: The authors declare no conflict of interest.

References

1. Frighetto, A. Coverage-Guided Binary Fuzzing with REVNG and LLVM Libfuzzer. Available online: https://www.politesi.polimi.it/bitstream/10589/173614/3/2021_04_Frighetto.pdf (accessed on 5 May 2021).
2. Sutton, M.; Greene, A.; Amini, P. *Fuzzing: Brute Force Vulnerability Discovery*; Addison-Wesley Professional: Boston, MA, USA, 2007.
3. American Fuzzy Lop. Available online: https://lcamtuf.coredump.cx/afl (accessed on 1 September 2020).
4. libFuzzer: A Library for Coverage-Guided Fuzz Testing. Available online: http://llvm.org/docs/LibFuzzer.html (accessed on 1 September 2020).
5. OSS-Fuzz: Continuous Fuzzing for Open Source Software. Available online: https://github.com/google/oss-fuzz (accessed on 1 October 2020).
6. Böhme, M.; Falk, B. Fuzzing: On the exponential cost of vulnerability discovery. In Proceedings of the 28th ACM Joint Meeting on European Software Engineering Conference and Symposium on the Foundations of Software Engineering, New York, NY, USA, 8–13 November 2020 ; pp. 713–724.
7. Li, Y.; Ji, S.; Lv, C.; Chen, Y.; Chen, J.; Gu, Q.; Wu, C. V-Fuzz: Vulnerability-Oriented Evolutionary Fuzzing. CoRR. 2019. Available online: http://xxx.lanl.gov/abs/1901.01142 (accessed on 15 October 2020).
8. Godefroid, P. Fuzzing: Hack, art, and science. *Commun. ACM* **2020**, *63*, 70–76. [CrossRef]
9. Godefroid, P.; Klarlund, N.; Sen, K. DART: Directed automated random testing. In Proceedings of the 2005 ACM SIGPLAN Conference on Programming Language Design and Implementation, Chicago, IL, USA, 12–15 June 2005; pp. 213–223.
10. Böhme, M.; Pham, V.T.; Nguyen, M.D.; Roychoudhury, A. Directed greybox fuzzing. In Proceedings of the ACM SIGSAC Conference on Computer and Communications Security, New York, NY, USA, 30 October–3 November 2017; pp. 2329–2344.
11. Chen, P.; Chen, H. Angora: Efficient fuzzing by principled search. In Proceedings of the 2018 IEEE Symposium on Security and Privacy (SP), San Francisco, CA, USA, 20–24 May 2018; pp. 711–725.
12. Böhme, M.; Pham, V.T.; Roychoudhury, A. Coverage-based greybox fuzzing as markov chain. *IEEE Trans. Softw. Eng.* **2017**, *45*, 489–506. [CrossRef]
13. Yue, T.; Tang, Y.; Yu, B.; Wang, P.; Wang, E. Learnafl: Greybox fuzzing with knowledge enhancement. *IEEE Access* **2019**, *7*, 117029–117043. [CrossRef]
14. Pham, V.T.; Böhme, M.; Roychoudhury, A. AFLNet: A greybox fuzzer for network protocols. In Proceedings of the 2020 IEEE 13th International Conference on Software Testing, Validation and Verification (ICST), Porto, Portugal, 24–28 October 2020; pp. 460–465.

15. Chen, H.; Xue, Y.; Li, Y.; Chen, B.; Xie, X.; Wu, X.; Liu, Y. Hawkeye: Towards a desired directed grey-box fuzzer. In Proceedings of the 2018 ACM SIGSAC Conference on Computer and Communications Security, New York, NY, USA, 15–19 October 2018; pp. 2095–2108.
16. Addresssanitizer. Available online: https://github.com/google/sanitizers/wiki/AddressSanitizer (accessed on 1 March 2021).
17. afl-Cmin. Available online: https://github.com/mirrorer/afl/blob/master/afl-cmin (accessed on 5 December 2020).
18. Stephens, N.; Grosen, J.; Salls, C.; Dutcher, A.; Vigna, G. Driller: Augmenting Fuzzing Through Selective Symbolic Execution. In Proceedings of the Network and Distributed System Security Symposium, San Diego, California, USA, 21-24 February 2016.
19. Peng, H.; Shoshitaishvili, Y.; Payer, M. T-Fuzz: Fuzzing by Program Transformation. In Proceedings of the 2018 IEEE Symposium on Security and Privacy (SP), San Francisco, CA, USA, 20–24 May 2018; pp. 697–710. [CrossRef]
20. Rawat, S.; Jain, V.; Kumar, A.; Cojocar, L.; Giuffrida, C.; Bos, H. VUzzer: Application-aware Evolutionary Fuzzing. *NDSS* **2017**, *17*, 1–14.
21. Lemieux, C.; Sen, K. Fairfuzz: A targeted mutation strategy for increasing greybox fuzz testing coverage. In Proceedings of the 33rd ACM/IEEE International Conference on Automated Software Engineering, Montpellier, France, 3–7 September 2018; pp. 475–485.
22. SanitizerCoverage. Available online: https://clang.llvm.org/docs/SanitizerCoverage.html (accessed on 2 February 2021).
23. Sun, B.; Wang, B.; Cui, B.; Fu, Y. Greybox Fuzzing Based on Ant Colony Algorithm. In Proceedings of the International Conference on Advanced Information Networking and Applications, Caserta, Italy, 15–17 April 2020; Springer: Berlin/Heidelberg, Germany, 2020; pp. 1319–1329.
24. Lemieux, C.; Padhye, R.; Sen, K.; Song, D. Perffuzz: Automatically generating pathological inputs. In Proceedings of the 27th ACM SIGSOFT International Symposium on Software Testing and Analysis, New York, NY, USA, 16–21 June 2018; pp. 254–265.
25. Song, C.; Yu, B.; Zhou, X.; Yang, Q. SPFuzz: A Hierarchical Scheduling Framework for Stateful Network Protocol Fuzzing. *IEEE Access* **2019**, *7*, 18490–18499. [CrossRef]
26. Dam, H.K.; Tran, T.; Pham, T.T.M.; Ng, S.W.; Grundy, J.; Ghose, A. Automatic feature learning for predicting vulnerable software components. *IEEE Trans. Softw. Eng.* **2018**. [CrossRef]
27. Reddy, S.; Lemieux, C.; Padhye, R.; Sen, K. Quickly generating diverse valid test inputs with reinforcement learning. In Proceedings of the 2020 IEEE/ACM 42nd International Conference on Software Engineering (ICSE), Seoul, Korea, 5–11 October 2020; pp. 1410–1421.
28. Godefroid, P.; Peleg, H.; Singh, R. Learn&fuzz: Machine learning for input fuzzing. In Proceedings of the 2017 32nd IEEE/ACM International Conference on Automated Software Engineering (ASE), Urbana-Champaign, IL, USA, 30 October–3 November 2017; pp. 50–59.
29. Fioraldi, A.; Maier, D.; Eißfeldt, H.; Heuse, M. AFL++: Combining incremental steps of fuzzing research. In Proceedings of the 14th USENIX Workshop on Offensive Technologies (WOOT), Online, 11 August 2020.
30. Fioraldi, A.; D'Elia, D.C.; Coppa, E. WEIZZ: Automatic grey-box fuzzing for structured binary formats. In Proceedings of the 29th ACM SIGSOFT International Symposium on Software Testing and Analysis, New York, NY, USA, 18–22 July 2020; pp. 1–13.
31. You, W.; Liu, X.; Ma, S.; Perry, D.; Zhang, X.; Liang, B. SLF: fuzzing without valid seed inputs. In Proceedings of the 2019 IEEE/ACM 41st International Conference on Software Engineering (ICSE), Montreal, QC, Canada, 25–31 May 2019; pp. 712–723.
32. Liu, S.; Mahar, S.; Ray, B.; Khan, S. PMFuzz: Test case generation for persistent memory programs. In Proceedings of the 26th ACM International Conference on Architectural Support for Programming Languages and Operating Systems, New York, NY, USA, 19–23 April 2021; pp. 487–502.
33. Cadar, C.; Dunbar, D.; Engler, D.R. Klee: Unassisted and automatic generation of high-coverage tests for complex systems programs. In Proceedings of the OSDI, Berkeley, CA, USA, 8–10 December 2008; Volume 8, pp. 209–224.
34. Ge, X.; Taneja, K.; Xie, T.; Tillmann, N. DyTa: Dynamic symbolic execution guided with static Verification results. In Proceedings of the 33rd International Conference on Software Engineering, Honolulu, HI, USA, 21–28 May 2011; pp. 992–994.
35. Chen, J.; Hu, W.; Zhang, L.; Hao, D.; Khurshid, S. Learning to Accelerate Symbolic Execution via Code Transformation. In Proceedings of the ECOOP, Amsterdam, The Netherlands, 16–21 July 2018.
36. Wang, M.; Liang, J.; Chen, Y.; Jiang, Y.; Jiao, X.; Liu, H.; Zhao, X.; Sun, J. SAFL: Increasing and Accelerating Testing Coverage with Symbolic Execution and Guided Fuzzing. In Proceedings of the 40th International Conference on Software Engineering: Companion Proceeedings, Gothenburg, Sweden, 27 May–3 June 2018; Association for Computing Machinery: New York, NY, USA, 2018; pp. 61–64. [CrossRef]

Article

A Situation-Aware Scheme for Efficient Device Authentication in Smart Grid-Enabled Home Area Networks

Anhao Xiang and **Jun Zheng** *

Department of Computer Science and Engineering, New Mexico Institute of Mining and Technology, Socorro, NM 87801, USA; anhao.xiang@student.nmt.edu
* Correspondence: jun.zheng@nmt.edu

Received: 27 May 2020; Accepted: 10 June 2020; Published: 13 June 2020

Abstract: Home area networks (HANs) are the most vulnerable part of smart grids since they are not directly controlled by utilities. Device authentication is one of most important mechanisms to protect the security of smart grid-enabled HANs (SG-HANs). In this paper, we propose a situation-aware scheme for efficient device authentication in SG-HANs. The proposed scheme utilizes the security risk information assessed by the smart home system with a situational awareness feature. A suitable authentication protocol with adequate security protection and computational and communication complexity is then selected based on the assessed security risk level. A protocol design of the proposed scheme considering two security risk levels is presented in the paper. The security of the design is verified by using both formal verification and informal security analysis. Our performance analysis demonstrates that the proposed scheme is efficient in terms of computational and communication costs.

Keywords: smart grids; device authentication; situational awareness; home area networks

---

1. Introduction

Smart grids offer many valuable benefits compared with traditional power grids. By enabling distributed power generation, distributed power storage, and microgrids in smart grids, more efficient and reliable power supply can be achieved [1]. The power generation of smart grids uses a mix of traditional fuel based power sources and renewable power sources such as wind farm and solar plant, which can significantly reduce the carbon footprint. The study in [2] shows that by 2030, CO_2 emissions can be reduced by 5% when adopting conservative approach to smart grids. The reduction can be nearly 16% if aggressive approach is adopted. The connection of home area networks (HANs) to smart grids enables the automation of home energy use. Smart grids also provide important infrastructure support for increased using of electric vehicles (EVs) through vehicle-to-grid (V2G) networks [3].

On the other hand, the implementation of smart grids faces major challenges in both physical and cyber domains. Since smart grids contain millions of nodes along with a complex control system, how to achieve the collaboration between components and the large-scale deployment of new devices and technologies becomes a crucial challenge [1]. Connecting power grids to cyber networks for advanced monitoring and control exposes the grids to cyber-attacks which can result in catastrophic damages as demonstrated by the 2015 Ukrine Blackout [4].

In this work, we concentrate on the security of smart grid-enabled HANs (SG-HANs), which connects many smart devices (SDs) of a smart home such as smart appliances, renewable energy sources and storage, EVs, etc. to smart grids. HANs are the most vulnerable part of smart grids since utilities have no direct control of this part [5]. Device authentication is one of the most important mechanisms to protect the security of SG-HANs against various attacks. In addition to the security consideration, the device authentication protocol must be lightweight since many of the SDs have limited computation power and memory storage. To this end, we propose a situation-aware scheme for efficient device authentication in SG-HANs. Unlike existing work, the proposed scheme selects a suitable authentication protocol based on the security risk information assessed by the smart home system. The aim of the scheme is to provide adequate security protection with reduced computational complexity, communication cost and power consumption. To the best of our knowledge, the proposed scheme is the first work that utilizes the situational awareness feature of smart home system for efficient device authentication in HANs.

The rest of this paper is organized as follows. Related work on device authentication in SG-HANs, situational awareness of smart home and situation-aware security schemes is described in Section 2. The system architecture of SG-HANs and the adopted attack model are introduced in Section 3. Section 4 presents the proposed situation-aware device authentication scheme for SG-HANs. The security analysis and performance analysis of the proposed scheme are provided in Sections 5 and 6, respectively. Finally, conclusions are drawn in Section 7.

2. Related Work

2.1. Device Authentication in SG-HANs

There are a number of works in the literature on device authentication in SG-HANs. Li proposed a ECC (Elliptic Curve Cryptography) based authenticated key establishment (EAKE) protocol for smart home energy management system in [6]. The EAKE protocol has two phases: a device or a security manager receives private/public key pair from the Certificate Agent (CA) through an out-of-band channel in the first phase; the initial session key is then established between the device and the security manager using the EAKE protocol in the second phase. In Ref. [7], Vaidya et al. also proposed a device authentication protocol for smart energy home area networks based on ECC. Both protocols of [6,7] are expensive for resource-limited devices due to the use of public key cryptography.

In Ref. [8], a secure key agreement protocol was proposed for radio frequency for consumer electronics (RF4CE) ubiquitous smart home systems based on symmetric key cryptography. In the proposed protocol, the initial unique secure information is pre-shared between the devices and manufacturers. The RF4CE-based controller receives the secret information from the manufacturer to authenticate a new device.

Ayday and Rajagopal [5] proposed three different device authentication mechanisms for the SG-HANs that provide (1) authentication between the gateway and the smart meter, (2) authentication between the smart appliances and the HAN, and (3) authentication between the transient devices and the HAN. The design of the three authentication mechanisms is based on symmetric key cryptography with the help of the trust center through the Internet.

Kumar et al. [9] proposed a lightweight and secure scheme for establishing session-key in smart home environments based on symmetric key cryptography. The smart home devices register with the security service provider offline to obtain security parameters including identity, a secret key with key identifier and a short authentication token. They also proposed a secure authentication and key agreement framework for smart home environments in [10] which realizes anonymity and unlinkability. The protocol is lightweight in comparison to other schemes because the design uses less encryption and decryption operations, and the number of exchanged messages is small.

Gaba et al. [11] proposed a robust and lightweight mutual authentication scheme called RLMA for distributed smart environments such as smart homes and smart buildings. The scheme utilizes implicit certificates to achieve simple and efficient mutual authentication and key agreement between smart devices in a smart environment.

2.2. Situational Awareness of Smart Home

Situational awareness is one of the essential features for smart homes [12]. The majority of the existing works for the situational awareness of smart homes are on activity recognition. For example, Wan et al. [13] proposed a dynamic sensor stream segmentation technology which helps the smart home system to categorize multiple sensor streams that belong to the same activity. Sensor correlation calculation and time correlation calculation are applied for the task. In Ref. [14], a data-driven approach based on neural network ensembles was developed for human activity recognition in smart home environments. Various approaches were explored to resolve conflicts between base models used in ensembles. Cicirelli et al. [15] proposed a framework for activity recognition under the cloud-assisted agent-based smart home environment (CASE). By using cloud computing technology, a smart home system can have greater analytic power. The work introduces an innovate approach, which embed activity recognition tasks including data acquisition, feature extraction, activity discovery, and activity recognition into different layers of CASE.

There are only a few works on the situational awareness of the smart home in cyberspace. A framework to measure the security risk of information leakage in IoT-based smart homes was proposed by Park et al. in [16]. The risk assessment is performed using the factor analysis of information risk (FAIR) method. The risk level for cyber situational awareness is obtained through risk grade clustering based on security scenarios.

2.3. Situation-Aware Security Schemes

There are a few recent works on developing situation-aware security schemes. Kim et al. [17] proposed DAoT, a dynamic and energy-aware authentication scheme for IoT devices. The scheme selects different key establishment (KE), message authentication code (MAC) and handshake operations to achieve energy efficient device authentication. The work evaluated the energy costs of different KE, MAC and handshake operations.

In Ref. [18], Hjelm and Truedsson investigated situation-aware adaptive cryptography for an IP camera. Situation parameters from WiFi and Bluetooth connections of the IP camera are used to determine the protection level. The cryptographic algorithms for encryption, hash and message authentication are then selected that are most suitable for the protection level. The power consumption, computational time and communication throughput were examined for different cryptographic algorithms.

Gebrie and Abie [19] proposed a risk-based authentication scheme for health care-related IoT authentication in smart homes. The channel characteristics in wireless body area network (WBAN) including Received signal strength indicator (RSSI), channel gain, temporal link signature, and Doppler measurement are used to determine risk level by using a naive Bayes algorithm. The authentication decision is then performed based on the risk level. For example, timeout and re-authentication will be performed if the risk level is determined as abnormal. It should be noted that there are no actual protocols designed in [17–19].

3. System Architecture and Attack Model

In this section, we introduce the system architecture of SG-HANs and the adopted attack model.

3.1. System Architecture of SG-HANs

The system architecture of SG-HANs considered in our work is shown in Figure 1, which consists of the infrastructure part and the HAN part. The infrastructure part controlled by utilities consists of smart meters (SMs), neighborhood area network (NAN) gateways, and control center. The HAN part in each house is controlled by the home owner, which consists of a number of SDs and one HAN gateway (HGW). A SD communicates with the HGW using a wireless protocol such as ZigBee or MQTT. In this work, we are interested in the authentication between SDs and HGW in the HAN part, which is helped by the control center. We assume that the smart home system is installed in the HAN with a situational awareness feature. Although the design of situational awareness feature is out of the scope of this work, we envision that the security risk assessment of the smart home system should combine activity recognition in physical domain [13–15] and risk analysis in cyber domain [16].

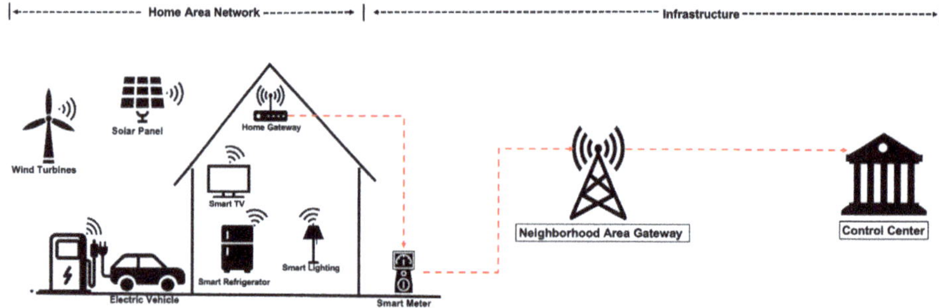

Figure 1. System architecture of SG-HANs.

3.2. Attack Model

The attack model considered in this work is the Dolev–Yao model [20]. In the model, the attacker can eavesdrop, intercept, inject, replay and modify messages exchanged on the open channel. Accordingly the attacker can launch various types of attacks including man-in-the-middle (MITM) attacks, replay attacks and impersonation attacks. Under this attack model, the proposed scheme will achieve security goals of message integrity, mutual authentication and session key establishment, and resistance against various attacks.

4. Proposed Scheme

In this section, we present a protocol design of the proposed situation-aware device authentication scheme for SG-HANs. Without loss of generality, we assume that the security risk assessed by the smart home system has two levels, low and high. The design can be easily extended to more than two security risk levels. The proposed scheme consists of two phases: device registration phase and device authentication and key agreement phase. Table 1 lists the notations and their descriptions that are used in the paper.

Table 1. Notations and their descriptions used in this paper.

Notation	Description
ID_A	Identity of SD A
ID_G	Identity of HGW
RC_A	Random number
R_A	Random number
R_G	Random number
S_i	Secret
A_i	Secret
SK_A	Session key
$H()$	one-way hash function
$E_K(M)$	Encrypt message M using key K
$D_K(M)$	Decrypt message M using key K
$\oplus$	XOR operation
$\|\|$	Message concatenation
T	Timestamp
ΔT	Maximum transmission delay

We have made the following assumptions for the proposed scheme: (1) SD has a clock which runs on its own battery and its assumed to be syAyday2013nchronized with the HGW's clock. (2) HGW is assumed to be authenticated before SD-HGW authentication takes place.

4.1. Device Registration Phase

Before installed in a SG-HAN, each SD needs to be registered offline at the control center. During the registration, the control center assigns an identification number ID_A to the registered SD A along with a random number RC_A. Furthermore, the control center computes secret $S_i = H(ID_A||RC_A)$. Finally, the control center sends ID_A and S_i to the SD A, and ID_A and RC_A to the HGW. The device registration phase is illustrated in Figure 2.

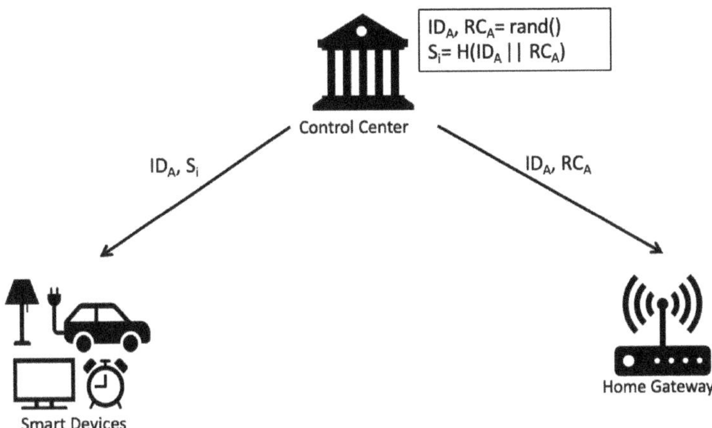

Figure 2. Illustration of device registration phase.

4.2. Device Authentication and Key Agreement Phase

After the registration, the SD A starts the authentication and key agreement process by sending the message MSG_1 to the HGW. MSG_1 includes an message header $HE_1 = \text{'}SD - AUTH\text{'}$ and ID_A as shown below:

$$MSG_1 = [HE_1||ID_A]$$

Upon receiving MSG_1, the HGW obtains the current security risk level from the smart home system. The following messages between the SD A and the HGW are generated based on the security risk level.

(a) ***Low security risk***

When the security risk is low, the HGW computes $S_i^* = H(ID_A^*||RC_A)$ and extracts current time stamp T_1. Then the HGW computes $C_{1,L} = (ID_G||T_1) \oplus S_i^*$ and $C_{2,L} = H(HE_{2,L}||ID_G||T_1||S_i^*)$. $HE_{2,L} = \text{'}HGW - LOW\text{'}$ is the header of the message $MSG_{2,L}$ that the HGW sends to the SD A.

$$MSG_{2,L} = [HE_{2,L}||C_{1,L}||C_{2,L}]$$

Upon receiving the message $MSG_{2,L}$ at time stamp T_1', the device A knows from the message header that the current security risk level is low. The ID of the HGW $ID_G{}^*$ and $T_1{}^*$ can be obtained by computing $ID_G{}^*||T_1{}^* = C_{1,L} \oplus S_i$. The device A also computes $C_{2,L}{}^* = H(HE_{2,L}{}^*||ID_G{}^*||T_1{}^*||S_i)$.

Then the SD A will check if $T_1' - T_1^* \leq \Delta T$ and $C_{2,L}^* == C_{2,L}$, where ΔT is the transmission delay. If not, the authentication process will be aborted. Otherwise, the SD A generates the secret $A_i = H(ID_G^*||H(ID_A||S_i))$ and extracts the current time stamp T_2. Then the SD A computes $C_{3,L} = (ID_A||T_2) \oplus A_i$ and $C_{4,L} = H(HE_{3,L}||ID_A||T_2||A_i)$, where $HE_{3,L} =$ '$SD - LOW$' is the header of the message $MSG_{3,L}$. Finally, the SD A sends $MSG_{3,L}$ to the HGW:

$$MSG_{3,L} = [HE_{3,L}||C_{3,L}||C_{4,L}]$$

The SD A computes the key $SK_A = H(T_1^*||T_2||S_i||A_i)$ which will be used as the shared session key between the device and the HGW.

When the HGW receives $MSG_{3,L}$ at time stamp T_2', it first computes $A_i^* = H(ID_G||H(ID_A||S_i^*))$ and then extracts ID_A^* and T_2^* by computing $C_{3,L} \oplus A_i^*$. The HGW checks if $T_2' - T_2^* \leq \Delta T$ and $C_{4,L}^* == C_{4,L}$, where $C_{4,L}^* = H(HE_{3,L}^*||ID_A^*||T_2^*||A_i^*)$. Assume all checks pass, the HGW adds ID_A to the trusted list of devices and computes the key $SK_A = H(T_1||T_2^*||S_i^*||A_i^*)$. After this step, both the SD A and the HGW have generated the symmetric session key which will be used for future data communication.

(b) ***High security risk***

When the security risk level obtained by the HGW is high, the message exchange between the SD A and the HGW needs higher security strength.

Upon receiving MSG_1 under high security risk, the HGW computes $S_i^* = H(ID_A^*||RC_A)$ and generates a random number R_G. Then the HGW extracts current time stamp T_1 and forms $MSG_{2,H}$ as following:

$$MSG_{2,H} = [HE_{2,H}||C_{1,H}||C_{2,H}]$$

where $HE_{2,H} =$ '$HGW - HIGH$' is the message header of $MSG_{2,H}$, $C_{1,H} = E_{S_i^*}(ID_G||T_1||R_G)$ and $C_{2,H} = H(HE_{2,H}||ID_G||T_1||R_G)$. Finally, the HGW sends $MSG_{2,H}$ to the SD A.

Upon receiving the message $MSG_{2,H}$ at time stamp T_1', the SD A learns from the message header that the security risk level is high. The SD A then uses S_i to decrypt $C_{1,H}^*$ to obtain ID_G^*, T_1^* and R_G^*. Then it checks if $T_1' - T_1^* \leq \Delta T$ and $C_{2,H}^* == C_{2,H}$, where $C_{2,H}^* = H(HE_{2,H}^*||ID_G^*||T_1^*||R_G^*)$. The authentication process will be terminated if the check is failed. Otherwise, the SD A generates the secret $A_i = H(ID_G^*||H(ID_A||S_i))$ and a random number R_A. Then the device extracts the current time stamp T_2 and computes $C_{3,H} = E_{A_i}(ID_A||T_2||R_A)$ and $C_{4,H} = H(HE_{3,H}||ID_A||T_2||R_A)$, where $HE_{3,H}$ = 'SD-HIGH' is the message header of $MSG_{3,H}$. The message $MSG_{3,H}$ is then formed and sent to the HGW:

$$MSG_{3,H} = [HE_{3,H}||C_{3,H}||C_{4,H}]$$

Finally, the SD A computes the shared key SK_A as $H(T_1^*||T_2||S_i||A_i||R_A||R_G^*)$.

After receiving $MSG_{3,H}$ at time stamp T_2', the HGW computes the secret $A_i^* = H(ID_G||H(ID_A||S_i^*))$ and extract ID_A^*, T_2^* and R_A^* by performing $D_{A_i^*}(C_{3,H})$. The HGW then computes $C_{4,H}^* = H(HE_{3,H}^*||ID_A^*||T_2^*||R_A^*)$ and checks if $T_2' - T_2^* \leq \Delta T$ and $C_{4,H}^* == C_{4,H}$. If all checks pass, the HGW adds ID_A to the trusted list of devices and computes the session key $SK_A = H(T_1||T_2^*||S_i^*||A_i^*||R_A^*||R_G)$.

Figures 3 and 4 show the message flows of the proposed scheme under low security risk and high security risk, which are denoted as two protocols P_L and P_H, respectively.

Authentication Process: Low Security Risk

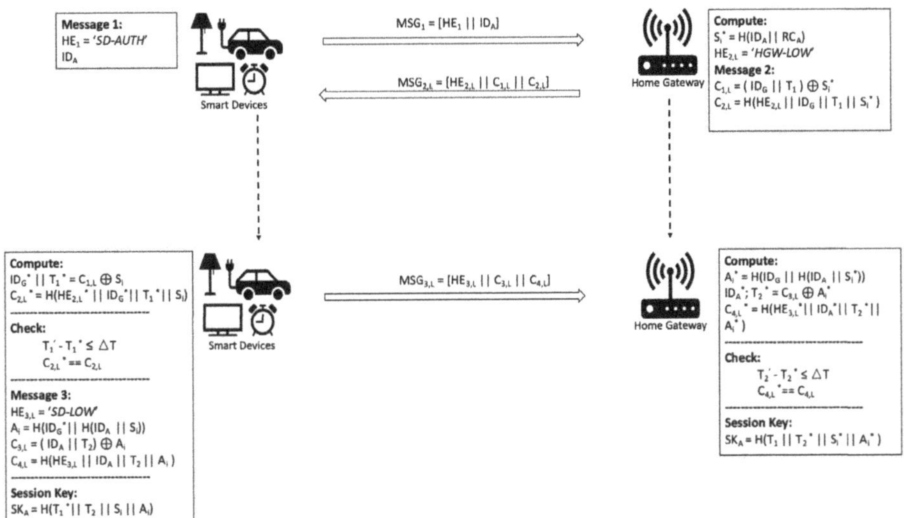

Figure 3. The message flow of the proposed scheme at low security risk (P_L).

Authentication Process: High Security Risk

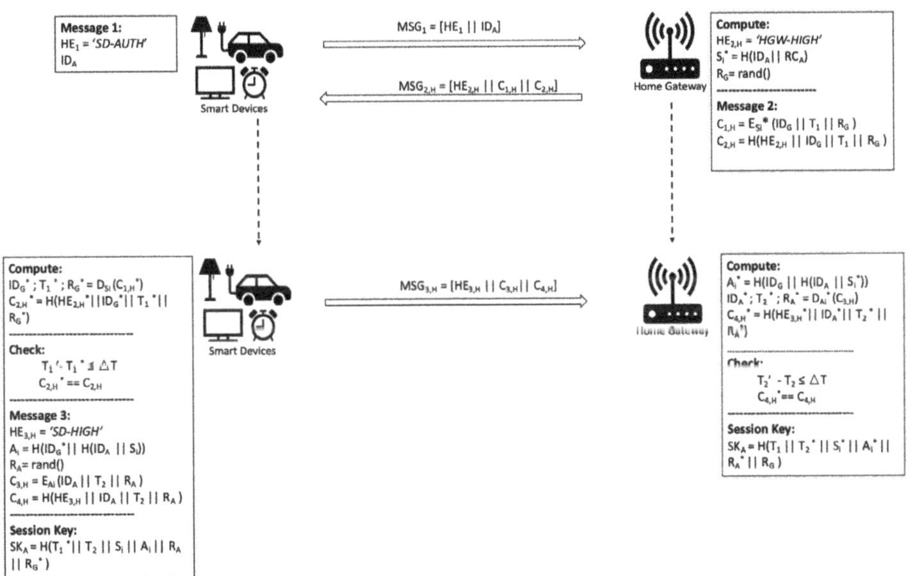

Figure 4. The message flow of the proposed scheme at high security risk (P_H).

5. Security Analysis

In this section, we verify the security of the proposed scheme using formal verification and informal security analysis.

5.1. Formal Security Verification

The formal security verification of the proposed scheme was done by using the automated validation feature of the Internet Security Protocols and Applications (AVISPA) tool [21], which is a push-button security analyzer tool designed for large scale internet security-sensitive protocols. AVISPA tool has been widely applied for formal security analysis of authentication protocols [9,10,22–24].

The architecture of AVISPA tool is illustrated in Figure 5. High Level Protocol Specification Language (HLPSL) is used to describe protocol design and specify security goals. AVISPA tool takes a HLPSL file as input and translates the file into intermediate format (IF) by using HLPSL2IF translator. The IF code becomes the input to the backend, where protocol security goals will be verified. Finally, the backend outputs the security report. As shown in Figure 5, the backend of AVISPA tool consists of four components: on-the-fly Model-Checker (OFMC), CL-based Attack Sercher (CL-AtSe), SAT-based Model-Check (SATMC), and Tree Automata-based Protocol Analyzer (TA4SP). Users can choose the backend components according to security requirements of their design. Notice that HLPSL is a role based language. The basic role states initial variables, constants, and transition steps. The composed role instantiate one or more basic roles. Finally, a top level role called environment role, states global constants and a composition of multiple sessions.

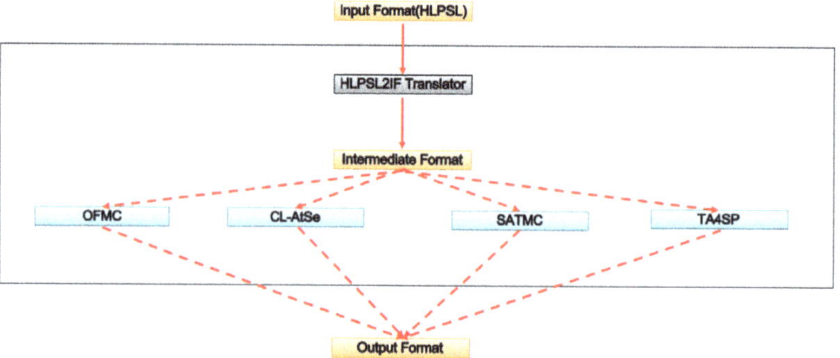

Figure 5. Architecture of the AVISPA tool [21].

The security goals of the proposed scheme are specified in Figure 6 as: (1) *secrecy_of sessionkey* means that the session key generated in the proposed scheme is kept secret between the SD and the HGW; (2) *authentication_on gateway_Si* means that secret S_i will be verified at the SD; (3) *authentication_on_device_Ai* means that secret A_i will be verified at the HGW; (4) *authentication_on_device_t2* means that the timestamp T_2 generated by the SD will be agreed between the SD and the HGW; (5) Similarly, *authentication_on_gateway_t1* verifies the agreement on timestamp T_1 between the HGW and the SD. The first security goal tests the strength and secrecy of the session key against various attacks such as MITM attack. The second and third security goals together confirm the establishment of mutual authentication, and the last two security goals test the protocol design against replay attacks. By running the HLPSL file through the backend, we test not only the protocol design against various attacks, but also whether the protocol satisfies specific requirements.

Figures 7 and 8 specify the roles of the SD and the HGW for low security risk, respectively. In the SD role, State 0 indicates the beginning of the authentication process. At State 0, the SD starts the authentication process by sending identity ID_A to the HGW through the $SND()$ function. On the other side, the HGW receives the device identity ID_A at State 0 by using the $RCV()$ function. Upon receiving ID_A, the HGW will move to State 1, where secret S_i is generated by using the built-in hash function $H()$, T_1 will be generated as random number by calling $new()$ function. Then the HGW uses built-in xor function to generate the response message. Similarly, after sending ID_A to the HGW, the SD will move to State 1 and wait for the response message from the HGW. Both SD and HGW generates the session key at State 2. Similar to low security risk, Figures 9 and 10 specify the SD and HGW roles for high security risk, respectively.

```
goal

 secrecy_of sessionkey
 authentication_on gateway_t1
 authentication_on device_t2
 authentication_on gateway_si
 authentication_on device_ai

end goal
```

Figure 6. Specification of security goals of the proposed scheme.

```
role device (A, B:      agent,
            H:          hash_func,
            SND,RCV:    channel(dy))
played_by A def=

  local State: nat,
    Device_id, Gateway_id, Rc, Si   :text,
    C0,C1,C2,C3,C4,C5               :message,
    T1, T2, Ai, Ks                  :text,

  init State := 0

  transition
    1.  State = 0   /\ SND(Device_id') =|>
        State' := 1

    2.  State = 1   /\ RCV(C0'.C1'.C2') =|>
        State' := 2 /\ Gateway_id' := xor(C0', Si)
              /\ T1' := xor(C1', Si)
              /\ T2' := new()
              /\ Verify' := H(Gateway_id'.T1'.Si)
              /\ Ai' := H(Gateway_id'.H(Device_id.Si))
              /\ C3' := xor(Ai',Device_id)
              /\ C4' := xor(Ai',T2)
              /\ C5' := H(Device_id.T2',Ai')
              /\ SND(C3'.C4'.C5')
              /\ Ks' := H(T1'.T2'.Si.Ai')
              /\ secret(Ks', sessionkey, {A,B})
              /\ witness(A,B,device_t2,T2')
              /\ witness(A,B,device_ai,Ai')

end role
```

Figure 7. Specification of the SD role for low security risk.

```
role gateway (A,B:      agent,
            H:          hash_func,
            SND,RCV:    channel(dy))
played_by B def=

  local State: nat,
    Device_id, Gateway_id, Rc, Si  :text,
    C0,C1,C2,C3,C4,C5              :message,
    T1, T2, Ai, Ks                 :text,

  init State := 0

  transition
   1.   State = 0 /\ RCV(Device_id') =|>
        State' := 1  /\ Si' := H(Device_id'.Rc)
              /\ T1' := new()
              /\ C0' := xor(Si',Gateway_id)
              /\ C1' := xor(Si',T1')
              /\ C2' := H(Gateway_id.T1'.Si')
              /\ SND(C0'.C1'.C2')
              /\ witness(B,A,gateway_t1,T1')
              /\ witness(B,A,gateway_si,Si')

   2.  State = 1 /\ RCV(C3', C4',C5') =|>
       State' := 2/\ Ai' := H(Gateway_id.H(Device_id.Si))
              /\ T2' := xor(Ai',C4)
              /\ Ks' := H(T1.T2'.Si.Ai')
              /\ secret(Ks', sessionkey, {A,B})

end role
```

Figure 8. Specification of the HGW role for low security risk.

```
role device (A,B        :   agent,
             H          :   hash_func,
             SND,RCV    :   channel(dy))
played_by A def=

  local State: nat,
     Device_id,Gateway_id,Rc,Si       : text,
    C0,C1,C2,C3,C4,C5                 : message,
    T1,T2,Ai,Ks,RG,RA                 : text,

  init State := 0

  transition
   1.    State = 0  /\ SND(Device_id') =|>
         State' := 1

   2.    State = 1   /\
    RCV({Gateway_id'.T1'.RG'}_Si.C2') =|>
         State' := 2 /\ T2' := new()
            /\ Verify' := H(Gateway_id'.T1'.Si)
            /\ Ai' := H(Gateway_id'.H(Device_id.Si))
            /\ RA' := new()
            /\ C4' := ({Device_id.T2'.RA'}_Ai')
            /\ C5' := H(Device_id.T2',Ai')
            /\ SND({Device_id.T2'.RA'}_Ai'.C5')
            /\ Ks' := H(T1'.T2'.Si.Ai'.RA'.RG')
            /\ secret(Ks', sessionkey, {A,B})
            /\ witness(A,B,device_t2,T2')
            /\ witness(A,B,device_ai,Ai')
end role
```

Figure 9. Specification of the SD role for high security risk.

```
role gateway (A,B       :  agent,
              H         :  hash_func,
              SND,RCV   :  channel(dy))
played_by B def=

  local State: nat,
        Device_id,Gateway_id,Rc,Si  : text,
        C0,C1,C2,C3,C4,C5           : message,
        T1,T2,Ai,Ks,RG,RA           : text,

  init State := 0

  transition
   1.    State = 0 /\ RCV(Device_id') =|>
         State' := 1 /\ RG' := new()
             /\ Si' := H(Device_id'.Rc)
             /\ T1' := new()
             /\ C1' := ({Gateway_id.T1'.RG'}_Si')
             /\ C2' := H(Gateway_id.T1'.Si')
             /\ SND({Gateway_id.T1'.RG'}_Si.C2')
             /\ witness(B,A,gateway_t1,T1')
             /\ witness(B,A,gateway_si,Si')

   2.    State = 1 /\ RCV({Device_id.T2'.RA'}_Ai'.C5') =|>
         State' := 2 /\ Ai' := H(Gateway_id.H(Device_id.Si))
             /\ Ks' := H(T1.T2'.Si.Ai'.RA'.RG)
             /\ secret(Ks', sessionkey, {A,B})
end role
```

Figure 10. Specification of the HGW role for high security risk.

Figure 11 specifies the protocol session role. In this role, we instantiate one instance of each basic role and compose them together to construct the whole protocol session. *Channel(dy)* declaration means that the intruder has full control over the channel, where *dy* stands for the Dolev–Yao attack model. Finally, the top-level environment role is defined in Figure 12. This role defines device ID, gateway ID, *rc* and *si* as global constants, and a composition of three sessions. Note that the intruder represented as constant *i*, will have names of all agents as initial knowledge.

```
role session(A,B  : agent,
             H    : hash_func)
def=
  local SC, CR, SD, DR, SG, GR: channel (dy)

  composition
       device(A,B,H,SD, DR)
    /\ gateway(B, A, H, SG, GR)
end role
```

Figure 11. Specification of the session role.

```
role environment()

def=

  const  a,b                                 : agent,
         h                                   : hash_func,
         device_id, gateway_id,rc, si        : text,

    sessionkey    : protocol_id,
    gateway_t1    : protocol_id,
    device_t2     : protocol_id,
    gateway_si    : protocol_id,
    device_ai     : protocol_id

  intruder_knowledge = {a,b}

  composition
     session(a,b,h)
     /\ session(i,b,h)
     /\ session(a,i,h)

end role
```

Figure 12. Specification of the environment role.

The outputs of the OFMC and CL-AtSe backends for P_L and P_H of the proposed scheme are shown in Figures 13–16. The results show that the proposed scheme is safe in the OFMC and CL-AtSe backends. This means that the proposed scheme successfully meets specified security goals.

```
SUMMARY
  SAFE
DETAILS
  BOUNDED_NUMBER_OF_SESSIONS
PROTOCOL
  /home/span/span/testsuite/results/protocol1.if
GOAL
  as_specified
BACKEND
  OFMC
```

Figure 13. Output of OFMC backend for low security risk.

```
SUMMARY
  SAFE
DETAILS
  BOUNDED_NUMBER_OF_SESSIONS
PROTOCOL
  /home/span/span/testsuite/results/protocol2.if
GOAL
  as_specified
BACKEND
  OFMC
```

Figure 14. Output of OFMC backend forhigh security risk.

```
SUMMARY
  SAFE
DETAILS
  BOUNDED_NUMBER_OF_SESSIONS
  TYPED_MODEL
PROTOCOL
  /home/span/span/testsuite/results/protocol1.if
GOAL
  As Specified
BACKEND
  CL-AtSe
```

Figure 15. Output of CL-AtSe backend for low security risk.

```
SUMMARY
  SAFE
DETAILS
  BOUNDED_NUMBER_OF_SESSIONS
  TYPED_MODEL
PROTOCOL
  /home/span/span/testsuite/results/protocol2.if
GOAL
  As Specified
BACKEND
  CL-AtSe
```

Figure 16. Output of CL-AtSe backend for high security risk.

5.2. Informal Security Analysis

In this section, we perform an informal security analysis to show how the proposed scheme achieves different security objectives.

5.2.1. Message Integrity

Both P_L and P_H of the proposed scheme use one-way hash functions to achieve the message integrity. To tamper the transmitted messages, the attacker needs to learn the secrets S_i and A_i which can not be obtained through the eavesdropped messages. Thus, the attacker cannot compute a valid hash value for a message, which means that the proposed scheme achieves the message integrity properly.

5.2.2. Mutual Authentication

Mutual authentication is an important property to verify the legitimacy of the SD and HGW to each other. In the proposed scheme, the SD authenticates the HGW by verifying the validity of the value $C_{2,*}$ using the secret S_i. The HGW then authenticates the SD by verifying the validity of the value $C_{4,*}$ using the secret A_i. As the secrets S_i and A_i cannot be obtained from the eavesdropped messages, the proposed scheme support the mutual authentication between the SD and HGW.

5.2.3. Resistance against MITM Attack

An attacker can launch the MITM attack by relaying and manipulating the messages exchanged between the SD and HGW. In the proposed scheme, the attacker needs to learn the secret S_i to manipulate the messages successfully. Since the secret S_i cannot be obtained from the previously eavesdropped messages, the propose scheme can resist the MITM attack.

5.2.4. Resistance against Replay Attack

In the replay attack, the attacker can replay previously eavesdropped messages to establish an authenticated session with the targeted entity. The proposed scheme uses the timestamp to verify if a received message is valid or not. Since the replayed message has the old timestamp, it cannot pass the verification. Thus, the proposed scheme can resist the replay attack.

5.2.5. Resistance against Impersonation Attack

An attacker may impersonate a SD by forging the request message MSG_1 with a fake/stolen ID as MSG_1 is in plain text. However, the response message $MSG_{2,*}$ from the HGW cannot be interpreted by the attacker since the secret S_i is unknown to the attacker. Therefore, the attacker cannot continue the authentication process. There is also no way for the attacker to impersonate the HGW by forging the response message since the HGW identity ID_G is protected with the secret S_i during the transmission. Thus, the proposed scheme can resist the impersonation attack.

6. Performance Analysis

Since a SD is usually resource limited, the design of authentication scheme should not overwhelm the SD's computational and communication resources. In this section, we perform an analysis of the computational and communication costs of the proposed scheme.

6.1. Communication Cost

The communication cost of the proposed scheme is evaluated using the total number of bits sent and received by the SD and the communication energy cost. In the analysis, we assume that message header is 3 bits in length, device ID and HGW ID are 8 bits, timestamp and random number are 32 bits, and outputs of hash and encryption operations are 128 bits.

Table 2 compare the proposed scheme with [6,8,9] in terms of total number of exchanged messages. Both P_L and P_H of the proposed scheme require three messages exchanged between the SD and the HGW, which is comparable to that of [9] and less than those of [6,8].

Table 2. Comparison of total number of exchanged messages.

Scheme	Total Number of Messages
Li [6]	4
Han et al. [8]	6
Kumar et al. [9]	3
P_L	3
P_H	3

The communication overheads of P_L and P_H of the proposed scheme in terms of total number of bits are shown in Table 3, which are calculated using aforementioned parameters. Figure 17 shows the communication overhead of the proposed scheme with different percentages of P_L and P_H being used. Generally, the higher chance that P_L is used, the lower the communication overhead of the proposed scheme. The communication overheads of three existing works [6,8,9] are also plotted in Figure 17. It is obvious that the proposed scheme achieves the lowest communication overhead even only P_H is used.

Besides communication overhead, communication energy cost is another important factor when evaluating communication cost. In order to simulate a resource limited SD, we used the TelosB platform which embeds a 16-bit processor running at 8 MHz clock frequency. TelosB also has limited amount of memory: 48 KB of ROM and 10 KB of RAM [25]. To measure the communication energy cost, we obtained the energy costs of sending and receiving one bit of data on TelosB platform as 0.72 μJ and 0.81 μJ from [26]. Then the communication energy costs of P_L and P_H are obtained as 269.55 μJ and 403.47 μJ (Table 4). Table 5 compares the communication energy cost of the proposed scheme with those of [6,8,9]. We assume that P_L and P_H have equal chance to be used for the proposed scheme. The results indicate that the proposed scheme is more efficient than other schemes in terms of communication energy cost.

Table 3. Communication overhead (in bits).

Message	P_L	P_H
MSG_1	11	11
MSG_2	171	259
MSG_3	171	259
Total	353	529

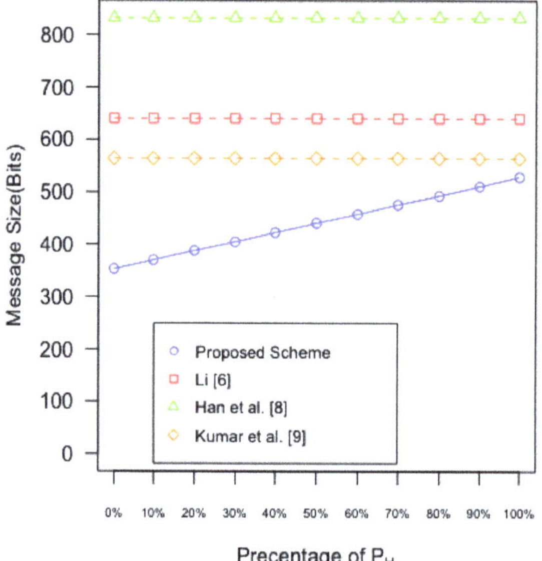

Figure 17. Communication overhead of the proposed scheme compared with those of three existing works [6,8,9].

Table 4. Communication energy cost.

P_L	Energy Cost (μJ)	P_H	Energy Cost (μJ)
MSG_1	7.92	MSG_1	7.92
$MSG_{2,L}$	138.51	$MSG_{2,H}$	209.79
$MSG_{3,L}$	123.12	$MSG_{3,H}$	185.76
Total:	269.55	Total:	403.47

Table 5. Comparison of communication energy cost.

Scheme	Communication Energy Cost (μJ)
Li [6]	483.84
Han et al. [8]	656.64
Kumar et al. [9]	430.22
Proposed Scheme (50% P_L + 50% P_H)	336.51

6.2. Computational Cost

Table 6 compares the computational cost of the proposed scheme with those of [6,8,9]. In the table, 'H' represents the time to execute one hash function. 'XOR' represents the time to perform an exclusive-or operation. 'E' and 'D' represent the times to perform encryption and decryption, respectively. 'MAC' and 'HMAC' represent the times used to compute the message authentication code and the hashed message authentication code, respectively. 't' is the time to perform a point multiplication operation. As shown in Table 6, P_L of the proposed scheme requires five hash operations and two XOR operations while P_H requires five hash operations, one encryption operation and one decryption operation. Since both P_L and P_H use five hash operations, a time and memory efficient hash algorithm such as BLAKE2 [27] is recommended for the proposed scheme. In comparison, the scheme proposed in [6] requires two point multiplication operations, one MAC operation, one encryption operation, one decryption operation, and one hash operation. Note that the point multiplication operation has high computational complexity compared with other operations. The scheme proposed in [8] requires seven MAC operations, four encryption operations, four decryption operations, and five hash operations. Finally, two hash operations, one MAC operation, one HMAC operation, one encryption operation and one decryption operation are required for the scheme of [9]. Overall, the proposed scheme is computational efficient and easy to implement compared with other schemes.

Table 6. Comparison of computational costs.

Operation	Li [6]	Han et al. [8]	Kumar et al. [9]	P_L	P_H
Hash	1H	5H	2H	5H	5H
XOR	–	–	–	2XOR	–
Cryptosystem	1E + 1D	4E + 4D	1E + 1D	–	1E + 1D
MAC	1MAC	7MAC	1MAC	–	–
HMAC	–	–	1HMAC	–	–
Point Multiplication	2t	–	–	–	–

We also analyzed the computational energy cost of the proposed scheme using a similar method of [9]. The energy consumption of a SD (E) is calculated by using the formula $E = V \times I$, where V is the voltage of the new batteries and I is the current of the circuit. Both V and I were retrieved from the TelosB datasheet [25]. The energy costs of executing hash function and encryption algorithm on TelosB platform can be computed based on the work of [28]. To compare with other schemes, we also obtained the energy costs of MAC and HMAC operations and point multiplication operation from [9,26], respectively. Since the time of executing XOR operation is negligible compared with other operations, it was excluded from the evaluation. The computational energy costs of different operations are shown in Table 7. Table 8 compares the total computational energy cost of the proposed scheme (50% P_L and 50% P_H) with those of [6,8,9]. The results indicate that the proposed scheme is more efficient than other schemes in terms of computational energy cost.

Table 7. Computational energy costs of different operations.

Operation	Energy Cost (μJ)
Hash	8.1
Encryption	14.9
MAC	45.36
HMAC	210.6
Point Multiplication	17,000

Table 8. Comparison of computational energy costs.

Scheme	Computational Energy Cost (μJ)
Li [6]	34,068.36
Han et al. [8]	417.62
Kumar et al. [9]	287.06
Proposed Scheme (50% P_L + 50% P_H)	55.4

7. Conclusions

Situation awareness is the essential feature of a smart home system which can be used to develop various smart applications. In this paper, we propose an efficient device authentication scheme for SG-HANs that can adapt to the security risk information assessed by the smart home system. The scheme selects a suitable authentication protocol based on the assessed security risk level that provides adequate security protection with reduced computational and communication costs. We presents a protocol design of the proposed scheme by considering two security risk levels. A formal security verification using AVISPA tool and an informal security analysis are performed to prove the security of the design. The performance analysis demonstrates that the proposed scheme is efficient for device authentication in SG-HANs in terms of both computational and communication costs. In future, we will research how to use the information collected by the smart home system in both physical and cyber domains to assess the security risk level, which is the key to enable the proposed scheme.

Author Contributions: Conceptualization, J.Z.; methodology, A.X. and J.Z.; formal analysis, A.X. and J.Z.; software, A.X.; writing–original draft preparation, A.X. and J.Z.; writing–review and editing, A.X. and J.Z.; supervision, J.Z.; funding acquisition, J.Z. All authors have read and agreed to the published version of the manuscript.

Funding: This material is based upon work funded by the National Science Foundation EPSCoR Cooperative Agreement OIA-1757207.

Conflicts of Interest: The authors declare no conflict of interest.

References

1. Fang, X.; Misra, J.; Xue, G.; Yang, D. Smart grid—The new and improved power grid: A survey. *IEEE Commun. Surv. Tutor.* **2012**, *14*, 944–980. [CrossRef]
2. Hledik, R. How green is the smart grid? *Electr. J.* **2009**, *22*, 29–41. [CrossRef]
3. Shaukat, N.; Khan, B.; Ali, S.M.; Mehmood, C.A.; Khan, J.; Farid, U.; Majid, M.; Anwar, S.M.; Jawad, M.; Ullah, Z. A survey on electric vehicle transportation within smart grid system. *Renew. Sustain. Energy Rev.* **2018**, *81*, 1329–1349. [CrossRef]
4. Liang, G.; Weller, S.; Zhao, J.; Luo, F.; Dong, Z. The 2015 Ukraine blackout: Implications for false data injection attacks. *IEEE Trans. Power Syst.* **2017**, *32*, 3317–3318. [CrossRef]
5. Ayday, E.; Rajagopal, S. *Secure Device Authentication Mechanisms for the Smart Grid-Enabled Home Area Networks;* Technical Report; 2013; pp. 1–18. Available online: https://infoscience.epfl.ch/record/188373/files/smart_grid_tech_report.pdf (accessed on 20 May 2020)
6. Li, Y. Design of a key establishment protocol for smart home energy management system. In Proceedings of the 2013 Fifth International Conference on Computational Intelligence, Communication Systems and Networks, Madrid, Spain, 5–7 June 2013; pp. 88–93.
7. Vaidya, B.; Makrakis, D.; Mouftah, H.T. Device authentication mechanism for smart energy home area networks. In Proceedings of the 2011 IEEE International Conference on Consumer Electronics (ICCE), Las Vegas, NV, USA, 9–12 January 2011; pp. 787–788.
8. Han, K.; Kim, J.; Shon, T.; Ko, D. A novel secure key pairing protocol for RF4CE ubiquitous smart home systems. *Pers. Ubiquit. Comput.* **2013**, textit17, 945–949. [CrossRef]
9. Kumar, P.; Gurtov, A.; Iinatti, J.; Ylianttila, M.; Sain, M. Lightweight and secure session-key establishment scheme in smart home environments. *IEEE Sens. J.* **2016**, *16*, 254–264. [CrossRef]
10. Kumar, P.; Braeken, A.; Gurtov, A.; Iinatti, J.; Ha, P.H. Anonymous secure framework in connected smart home environments. *IEEE Trans. Inf. Forensics Secur.* **2017**, *12*, 968–979. [CrossRef]
11. Gaba, G.S.; Kumar, G.; Monga, H.; Kim, T.-H.; Kumar, P. Robust and lightweight mutual authentication scheme in distributed smart environments. *IEEE Access* **2020**, *8*, 69722–69733. [CrossRef]
12. Lee, S.-Y.; Lin, F.J. Situation awareness in a smart home environment. In Proceedings of the 2016 3rd World Forum on Internet of Things (WF-IoT), Reston, VA, USA, 12–14 December 2016; pp. 678–683.
13. Wan, J.; O'grady, M.J.; O'hare, G.M. Dynamic sensor event segmentation for real-time activity recognition in a smart home context. *Pers. Ubiquit. Comput.* **2015**, *19*, 287–301. [CrossRef]

14. Irvine, N.; Nugent, C.; Zhang, S.; Wang, H.; Ng, W.W.Y. Neural network ensembles for sensor-based human activity recognition within smart environments. *Sensors* **2020**, *20*, 216. [CrossRef] [PubMed]
15. Cicirelli, F.; Fortino, G.; Giordano, A.; Guerrieri, A.; Spezzano, G.; Vinci, A. On the design of smart homes: A framework for activity recognition in home environment. *J. Med. Syst.* **2016**, *40*, 200. [CrossRef] [PubMed]
16. Park, M.; Oh, H.; Lee, K. Security risk measurement for information leakage in IoT-Based smart homes from a situational awareness perspective. *Sensors* **2019**, *19*, 2148. [CrossRef] [PubMed]
17. Kim, Y.; Yoo, S.; Yoo, C. DAoT: Dynamic and energy-aware authentication for smart home appliances in internet of things. In Proceedings of the 2015 IEEE International Conference on Consumer Electronics (ICCE), Las Vegas, NV, USA, 9–12 January 2015; pp. 196–197.
18. Hjelm, V.; Truedsson, M. Situation-Aware Adaptive Cryptography. Master's Thesis, Lund University, Lund, Sweden, 2018.
19. Gebrie, M.T.; Abie, H. Risk-based adaptive authentication for internet of things in smart home ehealth. In Proceedings of the 11th European Conference on Software Architecture (ECSA), Canterbury, UK, 11–15 September 2017; pp. 102–108.
20. Dolev, D.; Yao, A. On the security of public key protocols. *IEEE Trans. Inf. Theory* **1983**, *29*, 198–208. [CrossRef]
21. Viganò, L. Automated security protocol analysis with the AVISPA tool. *Electron. Notes Theor. Comput. Sci.* **2006**, *155*, 61–86. [CrossRef]
22. Chen, C.; He, D.; Chan, S.; Bu, J.; Gao, Y.; Fan, R. Lightweight and provably secure user authentication with anonymity for the global mobility network. *Int. J. Commun. Syst.* **2011**, *24*, 347–362. [CrossRef]
23. Nicanfar, H.; Jokar, P.; Beznosov, K.; Leung, V. Efficient authentication and key management mechanisms for smart grid communications. *IEEE Syst. J.* **2014**, *8*, 629–640. [CrossRef]
24. Mohammadali, A.; Haghighi, M.S.; Tadayon, M.H.; Nodooshan, A.M. A novel identity-based key establishment method for advanced metering infrastructure in smart grid. *IEEE Trans. Smart Grid* **2018**, *9*, 2834–2842. [CrossRef]
25. *TelosB Datasheet*. Available online: http://www.memsic.com/userfiles/files/Datasheets/WSN/telosb_datasheet.pdf (accessed on 20 May 2020).
26. de Meulenaer, G.; Gosset, F.; Standaert, F.-X.; Pereira, O. On the energy cost of communication and cryptography in wireless sensor networks. In Proceedings of the 2008 IEEE International Conference on Wireless and Mobile Computing, Networking and Communications, Avignon, France, 12–14 October 2008; pp. 580–585.
27. Fast Secure Hasing. Available online: htps://blake2.net (accessed on 20 May 2020).
28. Pereira, G.; Alves, R.; de Silva, F.; Azevedo, R.; Albertini, B.; Margi, C. Performance evaluation of cryptographic algorithms over IoT platforms and operating systems. *Secur. Commun. Netw.* **2017**, *2017*. [CrossRef]

MDPI
St. Alban-Anlage 66
4052 Basel
Switzerland
Tel. +41 61 683 77 34
Fax +41 61 302 89 18
www.mdpi.com

Electronics Editorial Office
E-mail: electronics@mdpi.com
www.mdpi.com/journal/electronics

www.ingramcontent.com/pod-product-compliance
Lightning Source LLC
LaVergne TN
LVHW071304170726
843515LV00006BA/1487

* 9 7 8 3 0 3 6 5 4 1 4 6 4 *